ENZYMES

ENZYMES

NOVEL BIOTECHNOLOGICAL APPROACHES FOR THE FOOD INDUSTRY

Edited by

SELIM KERMASHA

Professor, McGill University, Montreal, Quebec, Canada

MICHAEL N.A. ESKIN

Professor, University of Manitoba, Winnipeg, Manitoba, Canada

ELSEVIER

ACADEMIC PRESS

An imprint of Elsevier

Academic Press is an imprint of Elsevier
125 London Wall, London EC2Y 5AS, United Kingdom
525 B Street, Suite 1650, San Diego, CA 92101, United States
50 Hampshire Street, 5th Floor, Cambridge, MA 02139, United States
The Boulevard, Langford Lane, Kidlington, Oxford OX5 1GB, United Kingdom

Library of Congress Cataloging-in-Publication Data
A catalog record for this book is available from the Library of Congress

British Library Cataloguing-in-Publication Data
A catalogue record for this book is available from the British Library

ISBN: 978-0-12-800217-9

For information on all Academic Press publications
visit our website at https://www.elsevier.com/books-and-journals

Publisher: Charlotte Cockle
Acquisitions Editor: Patricia Osborn
Editorial Project Manager: Devlin Person
Production Project Manager: Kumar Anbazhagan
Designer: Miles Hitchen

Typeset by Thomson Digital

Contents

Contributors

Yusuf Chisti
School of Engineering, Massey University, Palmerston North, New Zealand

Michael N.A. Eskin
University of Manitoba, Winnipeg, Manitoba, Canada

Liuping Fan
Jiangnan University, Wuxi, China

Jagpreet K. Gill
McGill University, Montreal, Quebec, Canada

Selim Kermasha
McGill University, Montreal, Quebec, Canada

Sunil K. Khare
Indian Institute of Technology, Delhi, India

Sumit Kumar
Indian Institute of Technology, Delhi, India

Gregory S. Ladics
DuPont Nutrition & Bioscience, Wilmington, DE, United States

Vincent J. Sewalt
DuPont Nutrition & Bioscience, Palo Alto, CA, United States

Armin Spök
Graz University of Technology, Graz, Austria

Henry-Eric Spinnler
AgroParis Tech, Paris, France

Preface

Enzymes, biological catalysts, are indispensible to the metabolism of all living species, whether animals, plants, or microorganisms. They are responsible for the myriad of biochemical reactions required for all organisms to survive and function optimally. A number of these enzymes, primarily from microbial sources, are used extensively in the production of food products and natural food ingredients as well as pharmaceuticals, detergents, biofuels, and paper and pulp industry. This book provides an in-depth and comprehensive up-to-date coverage of the biotechnological applications of microbial enzymes in the food industry. After an initial review of their discovery, classification, and kinetics in Chapters 1 and 2, the book covers the production of enzymes by fermentation and genetic engineering and their recovery and purification for industrial use in Chapters 3 and 4. Since enzymes are mainly composed of protein they are very susceptible to denaturation. Chapters 5 and 6 discuss efforts made to enhance their stability by chemical treatment and immobilization. While traditionally, enzyme reactions are conducted in an aqueous environment, Chapter 7 outlines additional benefits found when such reactions are carried out in nonaqueous or nonconventional media. The utilization of enzymes in food production, like additives, must meet important safety standards, and these are discussed in detail in Chapter 8. The final chapter details the current applications of enzymes in the production of a variety of food products.

This book includes novel approaches for the use of enzymes in food production that we hope will be useful to both academic and industrial research. We are particularly grateful to the important chapter contributions made by leading scientists in the area of enzymes. We also acknowledge the outstanding editorial assistance by the staff of Elsevier/Academic Press in producing this book.

Selim Kermasha
Michael N.A. Eskin

CHAPTER ONE

Introduction

Selim Kermasha[a], Michael N.A. Eskin[b]
[a]McGill University, Montreal, Quebec, Canada
[b]University of Manitoba, Winnipeg, Manitoba, Canada

1.1 Introduction

Enzymes are three-dimensional protein biomacromolecules composed of many peptide chains. Of importance are the supramolecular interactions involved inside these biomacromolecules that play a role in both the recognition of the substrate and the catalytic process itself (Wang et al., 2018). Such enzymes are highly valued as very efficient biological catalysts with distinct advantages over chemical catalysts by operating under mild conditions, exhibiting high selectivity, and having lower environmental and physiological toxicity (Bommarius and Paye, 2013; Choi et al., 2015; Madhavan et al., 2017; Roy and Prasad, 2017). The commercial applications of enzymes are broad, covering pharmaceutical, food and beverage, detergent, and biofuel industries as listed in Table 1.1.

There is a great demand for such enzymes as they provide "greener" alternatives to chemical synthesis. In 2017, the industrial enzyme market was valued at around $7.0 billion and is expected to rise to $8.5 billion by 2022 (Ferrer et al., 2016). Of all these applications, however, it is their use in the production of food and beverages that is the primary focus of this book. This segment continues to be the largest market for enzymes globally as shown in Fig. 1.1 and is projected to continue in this position for the foreseeable future.

1.2 Historical development of enzymology

While the production of bread, cheese, and alcohol have been recorded for thousands of years, recognition of the role that enzymes play in these processes is less than 200 years old. Payen and Persoz (1833) were the first to report a substance in an alcohol precipitate from a malt extract that broke down starch to sugar which they called diastase. They subsequently proposed the use of the last three letters *ase* as the suffix to indicate what

Enzymes: Novel Biotechnological Approaches for the Food Industry.
http://dx.doi.org/10.1016/B978-0-12-800217-9.00001-0

Table 1.1 Industrial applications of enzyme catalysis.

Sector	Enzymes	Applications
Pharmaceuticals	Nitrile hydratase, transaminase monoamine oxidase, lipase penicillin acylase	Synthesis of intermediates for production of active pharmaceutical ingredients
Food processing	Trypsin, amylase, glucose isomerase, papain, pectinase	Conversion of starch to glucose, production of high fructose corn syrup, production of prebiotics, debittering of fruit juice
Detergent	Protease, lipase, amylase, cellulase	Stain removal, removal of fats and oils, color retention
Biofuels	Lipase, cellulase, xylanase	Production of fatty acid methyl esters, decomposition of lignocellulotic material for bioethanol production
Paper and pulp	Lipase, cellulase, xylanase	Removal of lignin for improved bleaching, improvement in fiber properties

Source: From Chapman et al. (2018).

substance was acted on. Earlier work by Kirchhoff (1812) reported the hydrolysis of starch to sugar by dilute acids.

The connection between these two discoveries, plus related observations, led Jon Jacob Berzelius to coin the word catalysis in 1834 to describe these actions. The word *enzyme* derived from the Greek meaning "*in yeast*" was later introduced by Wilhelm Kuhn in 1876. This followed exciting developments in the field of fermentation involving Theodor Schwann,

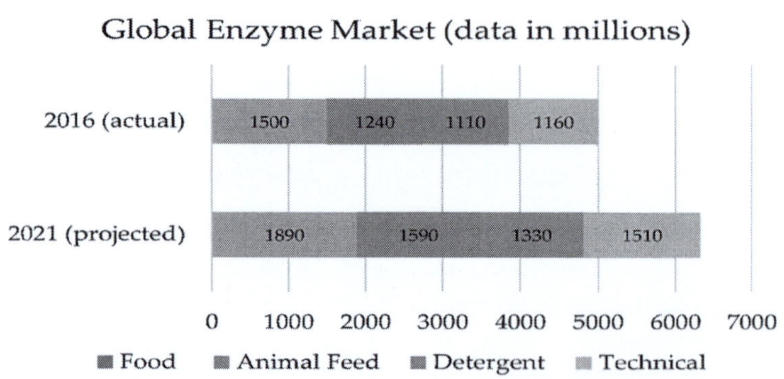

Figure 1.1 Global enzyme market in 2016 with projected global enzyme market in 2021 (Chapman et al., 2018).

Julius Leibig, Louis Pasteur, and Wilhelm Kuhn. Leibig attributed the fermentation properties of yeast to the presence of chemical substances while Pasteur believed fermentation could not be separated from the living yeast cell. Both hypotheses were rejected by Kuhn who suggested there was something in yeast responsible for this activity.

In their review of the history of enzyme nomenclature, Tipton and Boyce (2000) pointed out that a controversy persisted with considerable squabbles over who deserved the credit for this discovery. For example, Marie von Manassein, a Russian pioneer in physiological chemistry, contradicted Louis Pasteur's theory of fermentation by demonstrating that it could take place with substances isolated from yeast and did not require the living yeast cell itself. This claim, presented in an 1871 presentation, was dismissed by Eduard Buchner in 1897 on the grounds that her sterile solutions were contaminated with micrococci and as a result it was not a cell-free system. Manassein was known to be extremely diligent in her research but, as a result of these unsubstantiated allegations, was never recognized for her pioneering work. Buchner, on the other hand, received the Nobel Prize in 1907 four years after Manassein's death, for discovering the chemical nature of fermentation.

With the discovery of biological catalysts or enzymes capable of functioning outside the living cell, subsequent research focused on their composition. The association of enzymatic activity and proteins suggested to Richard Willstater that they were the carriers of enzymes crucial for functioning as catalysts. This idea was eliminated in 1926 when James B. Sumner isolated and crystallized the enzyme urease and showed it was a pure protein (Sumner, 1926). He repeated this again for catalase in 1937. The fact that enzymes were proteins was also confirmed by John Howard Northrop and Wendell Meredith Stanley for pepsin, trypsin, and chymotrypsin. These scientists, together with Sumner, all received the Nobel Prize in Chemistry in 1946 for this important work.

Since then, many thousands of enzymes have been identified and isolated, with a great deal of work determining their structure and kinetic properties. Until the 1980s the general belief was that all enzymes were globular proteins responsible for the myriad of metabolic processes essential for sustaining life. This proved incorrect, however, when Sidney Altman and Thomas Cech both independently examined how the genetic code of DNA was transcribed into RNA for the synthesis of proteins. The DNA strands had regions (introns) not needed for protein synthesis that were transcribed to the corresponding RNA molecules. Their removal from the

RNA molecules was required for them to catalyse shearing and splicing needed to eliminate those extra and unnecessary pieces of nucleic acids. Cech and Altman both made the surprising discovery that the enzyme responsible for this was a complex between one protein and one RNA molecule, later referred to by biochemists as RNAsP. Altman studied the RNAsP enzyme in *Escherichia coli* and found that the removal of protein eliminated any catalytic activity from the RNA molecule. Cech investigated the splicing of RNA in *Tetrahydra thermophilia*, but to his surprise found, that even in the absence of protein, the RNA molecule still functioned as a catalyst. Subsequent work by Altman, in the early 1980s, also confirmed the catalytic activity of the protein-free RNA molecule by its ability to shear and splice RNA. Close to one hundred RNA enzymes or ribozymes, as they are now called, have since been identified (Fedor and Williamson, 2006). For this major discovery both Cech (1990) and Altman (1990) were awarded the Nobel Prize in Chemistry in 1987. To-date, however, no other nonprotein enzymes have since been reported.

1.3 Biotechnology

Biotechnology is a term that encompasses the application of scientific and engineering principles to the processing of materials by biological agents, living cells, and their constituents, to provide products that are economically viable. It embraces a wide range of scientific disciplines, mainly biochemistry, microbiology, genetics, and bioengineering (Amarakoon et al., 2017).

The use of enzymes as tools in the development of food products and their preservation can be traced back to the ancient times. A list of historical examples includes barley malting, vinegar, bread, fruit fermentation, and cheese ripening (Vasic-Racki, 2006). However, the discovery of genetic engineering via recombinant deoxyribonucleic acid (DNA) technology in the second half of 20th century is responsible for the current biotechnology boom (Smith and Shaw, 2008). These biotechnological innovations have resulted in food products with higher nutritional and organoleptic qualities at minimum cost (McNeil et al., 2013). The biotechnological processes also improved the production of many secondary metabolites, used as natural food ingredients and additives, such as flavors (Aziz et al., 2014, 2016) and colorants (Mustafa et al., 2005). Bioprocesses, involving the use of enzymes, could also be used to produce novel added-value biomolecules of high nutritional quality as nutraceuticals (Aziz and Kermasha, 2014) and structured products (Khodadadi et al., 2013; Khodadadi and Kermasha, 2014).

DNA technology through protein engineering (Perkel, 2019) will enhance enzyme production and the development of novel enzymes with unique specificities to meet the need of the industry (Smith and Shaw, 2008). The overall objective of fermentation is not only to optimize the biomass yield, but also as a source of enzymes with high activities and selected specificities.

1.4 Food biotechnology

Food biotechnology is the application of modern biotechnological techniques to the manufacture and processing of food products as well as food ingredients and food additives. Although agricultural biotechnology, such as crop breeding and animal production, are generally regarded as food biotechnology, it is expected to have a great impact on the agri-food industry (Lee, 2015).

The food industry has gone through major changes in recent years. These changes are in response to consumers' demand for food products low in calories and made with natural ingredients and additives, as opposed to synthetic ones. The overall efforts are directed for the development of novel approaches to produce food products that will meet the consumer's demand for healthy foods (Sanders, 1991).

There is a spectacular increase in the use of enzymes as useful tools in the food industry (Singh et al., 2019). These applications cover a wide range of biotechnological applications as well as food production, food processing, and food preservation (Raveendran et al., 2018).

There are currently amazing new developments in fermentation, enzyme technology, genetic engineering, bioengineering, and food processing. Although biotechnology may involve the use of one or many living forms, microorganisms have an important and significant role in its development, due to their rapid production of large amounts of biomasses (Vasic-Racki et al., 2011). In addition, such production could be obtained with an added-value approach by using inexpensive culture media of agri-food waste and residual materials (Bull, 1983).

Using microorganisms, the innovations in genetic engineering (Perkel, 2019) contributed greatly to the development of many desirable enzymes for the food industry (Smith and Shaw, 2008). Different bioconversion processes are carried out by enzymes, including de novo synthesis throughout precursor fermentation as well as single and multistep bioconversions. However, most of enzymes of interest in food industry are often extracellular rather than intracellular.

Overall, research in food biotechnology is mainly focused on the improvement of enzymes, currently used by the industry, in terms of their stabilities and specificities at different stages of the processes (Raveendran et al., 2018). In addition, the development of novel and unique enzymes as well as their industrial applications is a race for knowledge. With environment and economic concerns regarding agri-food wastes and residues, there is increasing interest in the development of novel approaches using a single enzyme or multienzyme system in bioprocessing for the bioconversion of desirable added-value products.

1.5 Use of enzymes in bioconversion as biotechnological tools

In addition to their higher specificities, the main objective of using enzymes as tools in biotechnological processes in the food industry is to rapidly produce high volumes of products at lower costs (Hughes and Lewis, 2018). To achieve such an objective, it is essential to determine the rate of bioconversion and the maximum yields of end-products. Investigating the process conditions in terms of pH, ionic strength, temperature, and other kinetic parameters for all enzymes is essential (Segel, 1993; Straathof and Kasche, 2000).

The use of enzymes in biotechnological applications is based on their role as biocatalysts (Schmid et al., 2001). A biocatalyst or an enzyme that is cell-free or in the whole-cell lowers the activation energy for a given chemical reaction thereby accelerating the bioconversion process (Aehle, 2007). Although a chemical process requires a shorter time of reaction and uses less expensive raw materials, it involves harsher conditions, is difficult to control, unspecific, and produces unacceptable synthetic by-products. In comparison, an enzymatic process is characterized being milder, easy to control, highly specific and more importantly is considered natural. Regardless of the different factors that could be considered when evaluating an enzymatic process versus a chemical one, the most significant one is the specificity of enzyme reactions.

In order to develop a biotechnological process using enzymes, it is important to select the most appropriate biocatalyst (Wubbolts et al., 2000). Consequently, investigating enzyme structure, properties and its mechanism of action, as well as the environment needed for its action, is fundamental for developing a well-defined industrial bioconversion process (Fersht, 1984).

Although the number of enzymes in nature is estimated in the thousands, a limited number of them are used in biotechnological applications (Bhatia, 2018). It is also important to note, however, that there are multiple applications for each single enzyme (Robinson, 2015). In addition, the selection of an enzyme for a given industrial biotechnological application is related to the nature of the substrate and ultimately the recovery of the product at the end of the process. The reaction media (Illanes et al., 2012) and all kinetic parameters (Segel, 1993; Illanes et al., 2008) are important factors in determining the viability of industrial enzymes.

High productivities are essential for sustained industrial success. The success of biocatalysts is demonstrated by the rapidly expanding sales of industrial enzymes. All sectors of the industrial enzyme market are growing rapidly.

1.6 Industrial enzymes

In order to use a given enzyme in an industrial application, both scientific and business backgrounds are required to achieve such an objective. In addition to the overall cost of an industrial enzyme, there are major steps that should be taken into consideration when developing an industrial enzyme. These are listed as follow:

1.6.1 Enzyme specificity

One of the major benefits of using enzyme technology in industrial applications is specificity. However, the potential of using a given enzyme in multiple applications may require further research efforts and resources for achieving this goal (Robinson, 2015).

1.6.2 Fermentation and biomass production

Fermentation processes generally lead to the biomass production of enzymes that could be used in biotechnological applications (Vasic-Racki et al., 2011; Stanbury et al., 2016). The development in the use of fermentation, through innovations in molecular genetics, cell fusion, and enzyme technology, provided very powerful biotechnological techniques that laid out the ground for a novel approach in the use of enzymes in the food industry.

Enzyme production through fermentation is the most biotechnological approach for the food industry. This generates bioconversion processes *in situ* and the production of end-products, such as alcohol and flavors, but also to the recovery of exo- and endoenzymes as well as food ingredients, such as gums, vitamins, and food preservatives.

1.6.3 Recovery of enzymes and their enrichments

The recovery of enzymes from microbial biomass cultures is a major step in the industrial production of industrial biocatalysts (Kroner et al., 1987; Vijayaraghavan et al., 2016). The main challenges in this process are to preserve the enzyme's integrity, stability, and biocatalytic activity (Echavarria et al., 2012). Many enzymatic extracts are subjected to further enrichment, partial purification and purification. The degree of purity of an industrial enzyme depends on the objective of its use. In addition, the cost of all these procedures is a major factor that will impact the potential use of a given industrial enzyme in a well-defined bioprocess. It is obvious that exoenzymes are easier to be recovered from the fermentation media, with the least technical and cost problems, as compared to the endoenzymes.

1.6.4 Stabilization of enzymes

The stabilization of a native enzyme extract in terms of its integrity as a protein molecule and its biocatalytic activity during storage and its subsequent use in an industrial bioprocess is a crucial issue (Chang and Yeung, 2010; Stepankova et al. 2013; Silvaa et al., 2018). In addition, the industrial bioprocess environment could alter the enzyme structure and consequently its active site (Wang et al., 2018). The enzyme's stability could be retained using a wide range of chemical formulations in order to maintain and to enhance its biocatalytic activity (Gianfreda and Scarfi, 1991). The types of these formulations depend on the nature of the enzymes due to the formation of chemical crosslinks between these chemicals and part(s) of the enzymatic protein molecule; these crosslinks maintain the three-dimensional structure of the enzyme or its unfolded structure (Papinutti et al., 2008).

1.6.5 Biocatalysis media

Although there is a wide range of potential biotechnological applications for enzymes in industrial processes, there is a limitation to their uses. This is due to the surrounding environment of the bioprocess and its effect on the biocatalytic activity of the enzyme. The enzymes are mostly protected by using certain chemical stabilizers or immobilized either located on a surface or confined, allowing them to be used in various conventional aqueous as well as nonconventional reaction systems (Illanes et al., 2012). A more recent approach was developed to proceed with the biocatalysis in neat organic solvent media (Kuldamrong et al., 2013).

Over the past few decades, nonconventional, nonaqueous systems, have been used as the reaction media for enzyme biocatalysis (Wang et al., 2016). The use of organic media (Sheldon and Pereira, 2017; Priyadarshini and Pandey, 2018) significantly increased substrate specificity, relatively reduced reaction time, and waste generation compared to the conventional methods used for producing chemical catalysts (Sheldon, 2016).

1.6.6 Immobilization of enzymes

The immobilization of an enzyme or a cell refers to its confinement during the bioprocess. Among the advantages of immobilization are the multiple use of a given batch of enzyme, better process control, enhanced stability and enzyme-free products. There is a wide range of approaches and methods of immobilization that are defined by the nature of the enzyme and substrate as well as by the bioprocess environment and cost (Datta et al., 2017).

The use of free enzymes in industrial process is generally subjected to the surrounding conditions, which may limit their stability. Moreover, the recovery and reusability of an active enzyme is extremely difficult, and as a result, immobilization may reduce the cost of the industrial bioprocess (Brena et al., 2013). Hence, enzyme immobilization is a procedure that addresses these concerns in any given industrial application (Ammann et al., 2014; Mohamad et al., 2015). Although the immobilization of enzymes involves the confinement of the enzyme in a well-defined region, the biocatalyst should retain its catalytic activity (Khan and Alzohairy, 2010; Mohamad et al., 2015). In addition, the microencapsulation of an enzyme may be an interesting approach for its confinement (Kermasha et al., 2017), and may provide better protection (Gill et al., 2018). However, despite the advantages of immobilization, the technique may alter the inherent catalytic properties of the enzyme. Such alteration, however, will depend on the process design (Mateo et al., 2007).

1.6.7 Safety regulations of food enzymes

With world-wide economic globalization and international trade agreements, safety regulations are essential (Ladics and Sewalt, 2018). Since many industrial microbial enzymes are used in the food industry as part of industrial food bioprocesses, the regulations are increasingly rigorous (Agarwal and Sahu, 2014). Nevertheless, the different governmental legislations and roles that are specific to individual nations and regional organizations present increasing challenges to ensure the safe use of industrial microbial enzymes. There are increasing efforts world-wide to harmonize these legislations and guidelines to ensure the nontoxicity and safety of industrial microbial enzymes.

1.7 Selected industrial of food enzymes

In addition to certain enzymes that are currently used in the food industry, there are many others with potential applications. The major groups of industrial enzymes include hydrolases (carbohydrases, proteases, esterases, and lipases), oxidoreductases (glucose oxidase, lipoxygenase, and laccase), isomerases (glucose isomerase), and lyases (pectate lyase, hydroperoxide lyase, and acetolactate decarboxylase).

1.8 Conclusion

There is an increasing interest in biotechnology, particularly in its development of enzymes for food products, nutraceuticals, natural ingredients, and additives. Such developments must meet the consumer demands for natural, safe, nutritional, and acceptable food products. This will depend, not only on the improvement of the current industrial enzymes, but also on the development of new ones. The incredible progress made in the scientific knowledge of enzymology and associated disciplines will hopefully lead to better quality food products, in which enzymes will continue to play a major role.

References

Aehle, W. 2007. Enzymes in Industry: Production and Applications, third ed. Wiley-VCH Verlag, Darmstadt, Germany.

Agarwal, S., Sahu, S., 2014. Safety and regulatory aspects of food enzymes: An industrial perspective. Inter. J. Interduce. Multidiscip. Studies (IJIMS) 1, 253–267.

Altman, S., 1990. Nobel lecture. Enzymatic cleavage of RNA by RNA. Biosci. Rep. 10, 317–337.

Amarakoon, I.I., Hamilton, C.L., Mitchell, S.A., Tennant, P.F., Roye, M.E., 2017. Biotechnology. In: Badal, S., Delgoda, R. (Eds.), Pharmacognosy: Fundamentals, Application and Strategy. Academic Press, New York, N.Y., pp. 549–563.

Ammann, E.M., Gasser, C.A., Hommes, G., Corvini, P.F.X., 2014. Immobilization of defined laccase combinations for enhanced oxidation of phenolic contaminants. Appl. Microbiol. Biotechnol. 98, 1397–1406.

Aziz, M., Ben Akacha, N., Husson, F., Kermasha, S., 2014. Synthesis of linoleic acid hydroperoxides as flavor precursors, using selected substrate sources. J. Am. Oil Chem. Soc. 91, 1867–1876.

Aziz, M., St-Louis, R., Husson, F., Kermasha, S., 2016. Selected dehydrogenases in *Yarrowia lipolytica* JMY 861: their role in the synthesis of flavor compounds. Biosci. Biotechnol. Biochem. 80, 2184–2191.

Aziz, S., Kermasha, S., 2014. Assessment of the antioxidant capacity and oxidative stability of esterified phenolic lipids in selected edible oils. J. Food Sci. 79, 730–737.

Bhatia, S., 2018. Introduction to enzymes and their applications. In: Introduction to Pharmaceutical Biotechnology, vol 2, Enzymes, Proteins and Bioinformatics. IOPscience, Bristol, U.K., pp. 1–29.

Bommarius, A.S., Paye, M.F., 2013. Stabilizing biocatalysts. Chem. Soc. Rev. 42, 6534–6565.

Brena, B., González-Pombo, P., Batista-Viera, F., 2013. Immobilization of enzymes: A literature survey. Immobilization of Enzymes and Cells. Humana Press, Totowa, N.J., pp. 15–31.

Bull, D.N., 1983. Fermentation and genetic engineering: Problems and prospects. Biotechnol. 1, 847–856.

Cech, T.R., 1990. Nobel Lecture. Self-slicing and enzymatic activity of an intervening sequence RNA from Tetrahymena. Biosci. Rep. 10, 239–261.

Chang B.S., Yeung B., 2010. Physical Stability of Protein Pharmaceuticals. Formulation and Process Development Strategies for Manufacturing Biopharmaceuticals. California: John Wiley & Sons, Inc., pp. 69–104.

Chapman, J., Ismail, A.E., Dinu, C.Z., 2018. Industrial applications of enzymes: Recent advances, techniques, and outlooks. Catalysts 8, 238–271.

Choi, J.-M., Han, S.S., Kim, H.-S., 2015. Industrial applications of enzyme biocatalysts. Current status and future. Biotechnol. Adv. 33, 1443–1454.

Datta, S., Christena, L.R., Rajaram, Y.R.S., 2017. Enzyme immobilization: An overview on techniques and support materials. Biotech. 3, 1–9.

Echavarria, A.P., Albaz, A., Conde, J., Pegan, J., 2012. Enzyme recovery and effluents generated in the enzymatic elimination of clogging of pectin cake in filtration process. J. Food Eng. 111, 52–56.

Fedor, M.J., Williamson, J.R., 2006. The catalytic diversity of RNAs'. Nature Reviews. Mol. Biol. 6, 399–412.

Ferrer, M., Marlinez-Martinez, M., Bargleta, R., Streit, W.R., Golyshina, O.V., Golyshin, P.N., 2016. Estimating the success of enzyme bioprocessing through metagenomics: Current status and future trends. Microb. Biotechnol. 91, 22–34.

Fersht, A., 1984. In Enzyme Structure and Mechanism, second ed. W. H. Freeman & Co, New York, N.Y.

Gianfreda, L., Scarfi, M.R., 1991. Enzyme stabilization: State of the art. Mol. Cell. Biochem. 100, 97–128.

Gill, J.K., Orsat, V., Kermasha, S., 2018. Screening trials for the encapsulation of laccase enzymatic extract in silica sol-gel. J. Sol-Gel Sci. Technol. 85, 653–657.

Hughes, G., Lewis, J.C., 2018. Introduction: Biocatalysis in industry. Chem. Rev. 118, 1–3.

Illanes, A., Cauerhff, A., Wilson, I., Castro, G.R., 2012. Recent trends in biocatalysis engineering. Bioresource. Tecnol. 115, 48–57.

Illanes, A., Fernandez-Lafuente, R., Guisan, J.M., Wilson, L., 2008. Heterogeneous enzyme kinetics. In: Illanes, A. (Ed.), Enzyme Biocatalysis: Principles and Applications. Springer, New York, N.Y, pp. 155–197.

Kermasha, S., Aziz, S., Gill, J., Neufeld, R., 2017. Microencapsulation of esterified krill oil, using complex coacervation. J. Microencapsul. 35, 36–48.

Khan, A.A., Alzohairy, M.A., 2010. Recent advances and applications of immobilized enzyme technologies: A review. Res. J. Biol. Sci. 5, 565–575.

Khodadadi, M., Kermasha, S., 2014. Modeling lipase-catalyzed interesterification of flaxseed oil and tricaprylin for the synthesis of structured lipids. J. Mol. Catal. B: Enz. 102, 33–40.

Khodadadi, M., Aziz, S., St-Louis, R., Kermasha, S., 2013. Lipase-catalyzed synthesis and characterization of flaxseed oil-based structured lipids. J. Functional Foods 5, 424–433.
Kirchhoff, C.G., 1812. On the preparation of sugar from starch. (Teckhnologichesky Zhurnal) Technol. Magazine 9, 3–26.
Kroner, K.H., Hustedt, H., Cordes, A., Schutte, H., Rechtewald, A., Papamichad, N., Kula, M.-R., 1987. New trends in enzyme recovery. Annal. N.Y. Acad. Sci. 501, 403–442.
Kuldamrong, W., Husson, F., Kermasha, S., 2013. Biocatalysis with hydroperoxide lyase in extracts from *Penicillium camemberti* in neat organic solvent media. Biocatal. Biotransform. 31, 94–99.
Ladics, G.S., Sewalt, V., 2018. Industrial microbial enzyme safety: What does the weight-of-evidence indicate? Regulatory Toxicol. Pharmacol. 98, 151–154.
Lee, B.H., 2015. What is Food Biotechnology. Fundamentals of Food Biotechnology. second ed. Wiley-Blackwell, Hoboken, N.J, pp. 1–2.
Madhavan, A., Sindhu, R., Binod, P., Sukumaran, R.K., Pandey, A., 2017. Strategies for design of improved biocatalysts for industrial applications. Bioresour. Technol. 245, 1304–1313.
McNeil, B., Archer, D. and, Giavasis, I., Harvey, L. 2013. Microbial Production of Food Ingredients, Enzymes and Nutraceuticals. In: Woodhead Publishing Series in Food Science, Technology and Nutrition, Elsevier, New York, N.Y.
Mohamad, N.R., Marzuki, N.H.C., Buang, N.A., Huyop, F., Wahab, R.A., 2015. An overview of technologies for immobilization of enzymes and surface analysis techniques for immobilized enzymes. Biotechnol. Biotechnol. Equip. 29, 205–220.
Mateo, C., Palomo, J.M., Fernandez-Lorente, G., Guisan, J.M., Fernandez-Lafuente, R., 2007. Improvement of enzyme activity, stability and selectivity via immobilization techniques. Enzyme Microb. Technol. 40, 1451–1463.
Mustafa, R., Muniglia, M., Rovel, B., Girardin, M., 2005. Phenolic colorants obtained by enzymatic synthesis using a fungal laccase in a hydro-organic biphasic system. Food Res. Int. 38, 995–1000.
Papinutti, L., Dimitriu, P., Forchiassin, F., 2008. Stabilization studies of fomes sclerodermeus laccases. Bioresources Technol. 99, 419–424.
Payen, A., Persoz, J.F., 1833. Memoire sur la diastase, les principaux produits de ses reactions, et leurs applications aux arts industriels. Ann. Chim. Phys. 53, 73–92.
Perkel, J.M., 2019. The race for enzymatic DNA synthesis heats up: An alternative to chemical oligonucleotide synthesis inches closer to reality. Nature 566, 565.
Priyadarshini, A., Pandey, P., 2018. Enzymes non-conventional media. Biocatalysis and Agricultural Biotechnology: Fundamentals, Advances, and Practices for a Greener Future. Apple Academic Press, CRC Press, Boca Raton, FL.
Raveendran, S., Parameswaran, B., Beevi Ummalayama, S., Abraham, A., Kuruvilla Mathew, A., Madhavan, A., Rebello, S., Pandey, A., 2018. Applications of microbial enzymes in food industry. Food Technol. Biotechnol. 56, 16–30.
Robinson, P.K., 2015. Enzymes: Principles and biotechnological applications. Essays Biochem. 59, 1–41.
Roy, I., Prasad, S., 2017. Converting enzymes into tools of industrial importance. Recent Pat. Biotechnol. 12, 35–56.
Sanders, J.P.M., 1991. Introduction. Biotechnological Innovation in Food Processing. Biotolol, Butterworth-Heinemann, London, U.K., pp. 1–5.
Schmid, A., Dordick, J.S., Hauer, B., Wubbolts, M., Witholt, B., 2001. Industrial biocatalysis today and tomorrow. Nature 409, 258–268.
Segel, I.H. 1993. Enzyme Kinetics: Behavior and Analysis of Rapid Equilibrium and Steady-State Enzyme Systems, third ed. John Wiley, New York, N.Y.
Sheldon, R.A., 2016. Biocatalysis and biomass conversion in alternative reaction media. Chem. 22, 12984–12999.

Sheldon, R.A., Pereira, P.C., 2017. Biocatalysis engineering: The big picture. Chem. Soc. Rev. 46, 2678–2691.
Silvaa, C., Martinsa, M., Jingb, S., Fuc, J., Cavaco-Paulo, A., 2018. Practical insights on enzyme stabilization. Crit. Rev. Biotechnol. 38, 335–350.
Singh, R., Singh, A., Sachan, S. 2019. Enzymes used in the food industry: friends or foes? In:Enzymes in Food Biotechnology: Production, Applications and Future Prospects, Academic Press, New York, N.Y., pp. 827-843.
Smith, A., Shaw, K., 2008. Discovering the relationship between DNA and protein production. Nat. Edu. 1, 112.
Stanbury, P., Whitaker, A., Hall, S. 2016. Principles of Fermentation Technology, third ed. Elsevier, New York, N.Y.
Stepankova, V., Bidmanova, S., Koudelakova, T., Prokop, Z., Chaloupkova, R., Damborsky, J., 2013. Strategies for stabilization of enzymes in organic solvents. ACS Catal. 3, 2823–2836.
Straathof, A.J.J., Kasche, V., 2000. Process concepts for bioconversion. In: Straathof, A.J.J., Adlercreutz, P. (Eds.), Applied Biocatalysis. second ed. Harwood Academic Publishers, Amsterdam, The Netherlands, pp. 317–338.
Sumner, J.B., 1926. The isolation and crystallization of the enzyme urease. J. Biol. Chem. 69, 435–441.
Tipton, K., Boyce, S., 2000. History of the enzyme nomenclature system. Bioinformatics 16, 34–40.
Vasic-Racki, D., 2006. History of industrial transformation: Dreams and realities. Industrial Biotransformation. Wiley-VCH, New York N.Y., pp. 1–36.
Vasic-Racki, D., Findrik, Z., Vrsalovic´ Presecˇki, A., 2011. Modelling as a tool of enzyme reaction engineering for enzyme reactor development. Appl Microb. Biotechnol. 91, 845–856.
Vijayaraghavan, P., Raj, S.R.F., Vincent, S.G.P., 2016. Industrial enzymes: Recovery and purification challenges. Agro-Industrial Waste as Feedstock for Enzyme Production. Academic Press, New York, N.Y., pp. 95–110.
Wang, T., Fan, X., Hou, C., Lui, S., 2018. Design of artificial enzymes by supramolecular strategies. Curr. Opin. Struct. Biol. 51, 19–27.
Wang, S., Meng, X., Zhou, H., Liu, Y., Secondo, F., Liu, Y., 2016. Enzyme stability and activity in non-aqueous systems: A mini review. Catalysts 6, 32. doi: 10.3390/catal6020032.
Wubbolts, M.G., Bucke, C., Bielecki, S., 2000. How to get the biocatalyst. In: Straathof, A.J.J., Adlercreutz, P. (Eds.), Applied Biocatalysis. second ed. Harwood Academic Publishers, Amsterdam, The Netherlands, pp. 153–211.

CHAPTER TWO

Enzymes

Selim Kermasha[a], Michael N.A. Eskin[b]
[a]McGill University, Montreal, Quebec, Canada
[b]University of Manitoba, Winnipeg, Manitoba, Canada

2.1 Introduction

The majority of enzymes are proteins, which determine their catalytic capacity. This capacity is due to the sequence of amino acid residues of a given protein molecule and its three-dimensional conformation. With exceptional cases, amino acid residues of a given enzyme molecule are L-isomers rather than the D-ones. The large number of amino acid residues and their sequence in the peptide chains determine the unique catalytic activity of each enzyme molecule. The constant discovery of new enzymes has placed considerable demands on their classification. What started as a relatively simple classification has evolved into an encyclopedic production that requires a specific website to accommodate all the details ascribed to each of over 5000 enzymes identified so far.

Investigating the nature and the structure of the three-dimensional macromolecule of a protein is a challenge. Add to that the catalytic nature of an enzymatic protein, with a well-defined active site makes it even more challenging. Nevertheless, exciting progress in enzymology has been achieved in elucidating the mechanisms for selected enzyme using sophisticated analytical techniques as well as computer modeling. These discoveries not only provide important scientific information but also expands the range of applications particularly, the biotechnological ones. These applications are determined not only by the specificity of the enzyme but also by its rate of reaction and other kinetic parameters.

One of the main drawbacks of enzymes is their limited heat stability. A detailed examination of enzymes operating at extreme temperatures revealed unique protein tertiary structures capable of tolerating such conditions. Incorporating such structural changes in selected enzymes substantially enhanced their heat stability. Alternatively, there has been considerable activity developing artificial enzymes. The catalytic properties of such enzymes operate quite differently from conventional enzymes as they have no

Enzymes: Novel Biotechnological Approaches for the Food Industry.
http://dx.doi.org/10.1016/B978-0-12-800217-9.00002-2

active site. Recent research reported that a single amino acid phenylalanine, in the presence of zinc ions, spontaneously and rapidly assembled into an amyloid-like structure that exhibited catalytic activity.

This chapter covers the basic structure of enzymes, their classification and kinetics. The remarkable progress in modifying the heat stability of enzymes as well as the development of artificial enzymes are unique developments that are changing our fundamental understanding of enzyme catalysis.

2.2 The nature of enzymes

2.2.1 Enzyme structure

Enzymes, predominantly protein in nature, have been extensively researched with the genetically determined sequence of amino acids covalently linked by peptide bonds to form a ribbon-like arrangement known as the primary structure. These ribbons can be straight or fold in sheets or helical arrangements to form secondary structures maintained by hydrogen bonds. The secondary structures then interweave with each other to form three-dimensional tertiary structures that are held together by hydrogen bonds, salt bridges, hydrophobic interactions, and covalent bonds such as disulfide linkages between cysteine residues in adjacent polypeptide chains (See Chapter 5, Section 4). The structures of enzymes have been extensively examined by spectroscopic methods (Busenlehner and Armstrong, 2005), X-ray crystallography (Ogata et al., 2005) and currently using multidimensional NMR (Chan et al., 2013). Monasterio (2014) reviews the effectiveness of studying enzyme structure, dynamics and function using NMR. In addition to structures and ligand-binding thermodynamics, NMR provides information on the dynamics of the structural regions of the protein. In particular, NMR relaxation results from the fluctuating magnetic fields generated by the molecular motions at the atomic level.

Enzymes are soluble globular proteins in which their tertiary structures are generally rounded and ball shaped. Each protein bends or folds into a three-dimensional shape that is different for each protein. The bending and twisting involved in forming the tertiary structure is specific for every protein and ensures its maximum stability and lowest energy state. Under normal physiological conditions, the hydrophobic side chains of the neutral or nonpolar amino acids, phenylalanine, and isoleucine, are generally found in the interior of the protein's tertiary structure to shield them from the aqueous environment. The enzyme's catalytic activity is dependent on its unique structure which is lost if the protein is denatured or dissociated into its individual units (Seibert and Raushel, 2005).

2.2.2 Catalytic activity at the active site

The most important part of the enzyme is a small region in its protein, the active site, where the catalytic reaction takes place. It is here where the substrate molecule binds and undergoes the chemical change to form the product, which is then released leaving the enzyme unchanged. The specificity of each enzyme is determined by the arrangement of the amino acids in the active site, which forms a groove or pocket in the protein (Pravda et al., 2014). The active site represents a very small part of the enzyme with the remaining protein supporting it like a scaffold. For example, the ability of chymostrypsin to cleave peptide bonds involves the serine 195 residue, which briefly covalently bonds with the substrate. The subsequent hydrolysis is catalyzed with histidine 57 and aspartic acid-102. These three amino acids at the active site are crucial for chymotrypsin's catalytic properties (Appel, 1986).

2.3 Classification of enzymes and nomenclature

With the discovery of enzymes, trivial names were initially assigned which were very confusing and ambiguous. The similarity in the names of some enzymes only further added to the confusion. The use of the suffix "*ase*" at the end of the substrate also added to the confusion. As Tipton and Boyce (2000) pointed out, urease, the first enzyme crystallized, had a descriptive name that was both helpful and represented its catalytic action. Other names, however, were distinctly misleading. For example, the enzyme rhodanese (thiosulfate sulfurtransferase) neither described the nature of the substrate or its mode of action. It was clear that a systematic method was needed to differentiate the new enzymes being identified. The first attempt by Otto Hoffmann-Ostenhof (1953) used a system based on the number of molecules involved in the enzymic-catalyzed reaction. He proposed the following three general classes:

1. Hydrolases, transferases and oxidoreductases (A + B = C + D)
2. Lyases and synthases (A = B + C)
3. Racemases (A = B)

At the same time Dixon and Webb compiled a list of all known enzymes for their classic book on Enzymes (Dixon and Webb, 1958). In it, they classified a total of 659 into three groups based on the type of reaction catalyzed hydrolyzing, transferring and others. They also included a list of cofactors and prosthetic groups needed by each enzyme which formed the

Class	Name	Reactions catalyzed
EC1	Oxidoreductases	AH + B = A + BH (reduced)
		A + O = AO (oxidized)
EC2	Transferases	AB + C = A + BC
EC3	Hydrolases	AB + H_2O = AOH + BH
EC4	Lyases	Cleaving C-C, C-N, C-O or C-S bond
EC5	Isomerases	AB = BA
EC6	Ligases	A + B + ATP = A-B + ADP + Pi

Figure 2.1 Six categories of enzymes.

basis for the current enzyme classification and nomenclature system. After being contacted by both Dixon and Hoffmann-Ostenhof, The President of the International Union of Biochemistry (now known as the International Union of Biochemistry and Molecular Biology (IUBMB) Marcel Florkin, set up the International Commission on Enzymes in 1956. The Nomenclature Commission was given the task of implementing the systematic naming of old enzymes so that they could be clearly classified. In it, the original classification of Dixon and Webb was extended to include the following six categories (Fig. 2.1):

The individual classes were then further subdivided, with each enzyme having its own four digits code, with a distinct Enzyme Commission (EC) Number. The second digit (subclass) provides information about the type of compound or group involved. The third digit or sub-subclass specifies the type of reaction involved. The fourth digit gives the serial number that identifies the individual enzyme within a subclass.

The Nomenclature Committee (NC) of IUBMB published the first enzyme classification list in 1961 (IUB, 1961) with five updated versions subsequently released between 1965 and 1992. A further expansion of this list was presented by Edwin Webb in 1993 (Webb, 1993) in his "Personal Perspective" of its development. A more recent review by McDonald and Tipton (2014) pointed out the dramatic increase in the number of enzymes classified. As a result, a paper edition of the enzyme list would be extremely cumbersome and impractical by taking up over 2000 pages. Fortunately, the availability of on-line versions allows for its continuous updating with the newly discovered enzymes through the ExplorEnz site. Such a large and ever-expanding data base allows for a very detailed presentation of each enzyme. Nevertheless, there remain many challenges with such an encyclopedic list of close to 5000 enzymes identified so far. McDonald and Tipton (2014) emphasized that the modest aim of the NC-IUBMB enzyme system was to provide correct and unambiguous information on enzymes according the reaction(s) catalyzed. Related information on protein structure, however, could be obtained from other linked databases as well as cofactor requirements. More information on reaction mechanisms is available in links such as MACIE (MACIE) database. Details on a particular enzyme's kinetic parameters, are not provided in the IUBMB list but are available from databases such as BRENDA (BRENDA), the world's enzyme information system (Schomburg et al., 2014), and SABIO-RK (SABIO-RK), the database for biochemical reactions (Wittig et al., 2012).

2.4 Mechanism of action

2.4.1 Fit of substrate and active site

The importance of enzymes is their ability to reduce the activation energy (E_a) so that the myriad of physiological reactions, from digestion to reproduction, can be catalyzed at a safe and low temperature. Enzyme reactions are extremely efficient and may be millions of times faster than the corresponding noncatalyzed reactions (Bugg, 2012). The mode of action, for over a century, was attributed to the structural relationship between and enzyme and substrate referred to as a "lock-and-key" model (Fischer, 1894). This is very rigid model which described the enzyme's active site as the lock in which the substrate, the key, fits perfectly (Daniel et al., 2003). This model proved far too limited and was amended by Haldane (1930) who suggested that the key does not necessarily fit the lock perfectly but might exercise a strain on it. Around the same time Eyring and Polanyi (1931) proposed the *Transition State theory* (TST) for chemical reactions in which a chemical equilibrium existed between reactants and activated transition state complexes. Based on this theory Pauling (1948) suggested that the catalytic power of enzymes also comes from the highly specific binding of transition states during the reaction. The rigid and very limited *lock-and key theory* was eventually replaced by the *induced-fit model* which described the formation of an enzyme-substrate (E-S) complex in a flexible active site (Koshland, 1958). In other words, a conformation change occurred in the active site of the enzyme which facilitated its interaction with the different functional groups on the substrate.

2.4.2 Balancing bond breaking and bond making

The breakthrough in enzymology came with the analysis of the structure of the first enzyme isolated from hen egg-white, lysozyme (Blake et al., 1965). The 2A° resolution crystal structure of this enzyme enabled atomic resolution of its structure and any changes responsible for its catalytic mechanism. As a result, integration of protein structure, dynamics, and function are all considered responsible for the dynamic activity of enzymes with internal protein motions closely linked to the functioning of these catalysts. The movement of protein parts and their role in enzyme catalysis has since been extensively studied using a variety of different techniques including NMR, hydrogen-deuterium exchange, neutron scattering, and biochemical and mutational analysis. An increase in computational power methods enabled scientists to simulate enzymatic reactions and correlate with their functional properties (Lifson and Warshel, 1968; Warshel and Lifson, 1970; Warshel and Levitt, 1976). Ariel Warshel (Warshel and Levitt, 1976; Warshel, 2002), a pioneer in this field, was one of the first to conduct molecular dynamic simulation of a biological process. For this work, he was awarded, with Martin Karplus and Michael Levitt, the Nobel prize in Chemistry in 2013. This new field of computational enzymology is crucial for understanding the phenomenal rate enhancement by enzymes as well as elucidating their mechanisms of action (Carvalho et al., 2014).

2.5 Kinetics

2.5.1 Generality

The kinetics of a given enzyme, used as biocatalyst, determine the rates of the biochemical reactions (Conn et al., 1987; Lu et al., 1998; Bugg, 2012; Schnell and Maini, 2004; Berg et al., 2007; Purich, 2010; Cornish-Bowden, 2012).

The kinetics of an enzyme would provide significant information not only on the catalytic mechanism, but also on the degree of its purity and its efficiency (Danson and Eisenthal, 2002; Cleland and Cook, 2007). The kinetics of an enzyme would also provide valuable information on its validity for a given industrial application not only by its specificity but also economically.

The enzyme molecule carries out biocatalytic activity by binding to the substrate molecule and proceeding with the formation of a series of enzyme-substrate intermediate complexes which results in the formation of the product and a regenerated enzyme molecule. This formation of the enzyme complex takes place at the active site of the enzyme molecule. It is therefore important to elucidate the mechanistic action of the enzyme. Indeed, understanding the enzyme mechanism is crucial for defining the step that determines the rate of the reaction (Kraut et al., 2003). The enzyme may bind with one substrate or many substrates at the same time, which result in a single substrate or multisubstrates kinetics (Cleland, 1963; Fromm and Hargrove, 2012).

Indeed, measuring enzyme kinetics is important to elucidate the chemical mechanism of an enzyme reaction and to determine the sequence of chemical steps that are involved in the bioconversion of a given substrate into a product as well the rate of formation of intermediates, even if they cannot be identified (Price and Stevens, 2000).

At the present time, there is an amazing progress in elucidating the mechanism of a given enzyme not only using selected substrates and inhibitors as well as the physicochemical analyses but also the development of software modeling and statistical analyses (Bar-Even et al., 2011).

In contrast to a chemical reaction, the rate of an enzymatic reaction increases in the presence of a higher substrate concentration and hence increases the enzyme-substrate concentrations, until all active sites are saturated for the rate to reach its maximum.

The two most important kinetic properties of an enzyme are the velocity of the saturation of the active site and the formation of the enzyme-substrate complex, as well as the maximum rate that can be achieved. Nevertheless, the kinetics of a given enzyme depend on the environment of the enzyme reaction, which are dramatically different whether it is in vivo or in vitro. Other parameters important, include pH, temperature, activators, and inhibitors.

Studying enzyme kinetics is important for many reasons, but basically it helps to explain the mechanistic action of the enzyme and to predict its behavior. The different kinetic constants are critical for a better understanding of mechanistic enzymatic reaction for controlling the bioconversion process.

2.5.2 Reaction order

As in the laws of thermodynamics, enzyme catalytic reactions follow different mechanisms pending on the substrate [S] concentration.

2.5.2.1 Zero-order kinetics

In a zero-order mechanism, the reactant has no effect on the initial rate of reaction, whereas in a first-order one, the initial rate of the reaction doubles when the reactant is doubled. For the second-order mechanism,

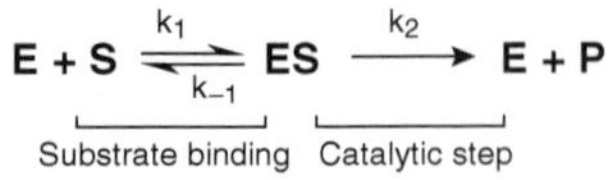

Figure 2.2 Single-substrate mechanism for an enzyme reaction (Wikipedia).

the initial rate of the reaction does increase by four times when the reactant is doubled.

2.5.2.2 First-order kinetics

A general enzyme-catalyzed reaction (Fig. 2.2) could be stated as follows:

The substrate [S] interacts and binds with the active of the enzyme [E] to form the enzyme-substrate [ES] complex, with a reaction rate constant (k_1) and with reaction rate (v). That reaction is followed by the dissociation of ES, with a reaction rate constant (k_2), to produce the product [P] and regenerate the free enzyme [E]. The rate of the reaction depends on the concentration of [ES]. In a first-order mechanism, the reaction rate depends on the concentration of one of the reactants (Engel, 2018). However, as [S] gets higher, the enzyme becomes saturated with substrate and the rate reaching its V_{max}, or the enzyme's maximum rate (Fig. 2.3).

Although most enzymes with a single = substrate mechanism follow this scheme reaction, some of them do not fall into this class of mechanism. A good example of an exceptions is catalase, where the enzyme would bind first with the substrate molecule of hydrogen peroxide, which is oxidized and then reduced by a second molecule of substrate. Although there is only a single substrate in this reaction the modification of enzyme intermediate resulted from a type of ping-pong mechanism which can be found in multisubstrate reactions (Cleland, 1963).

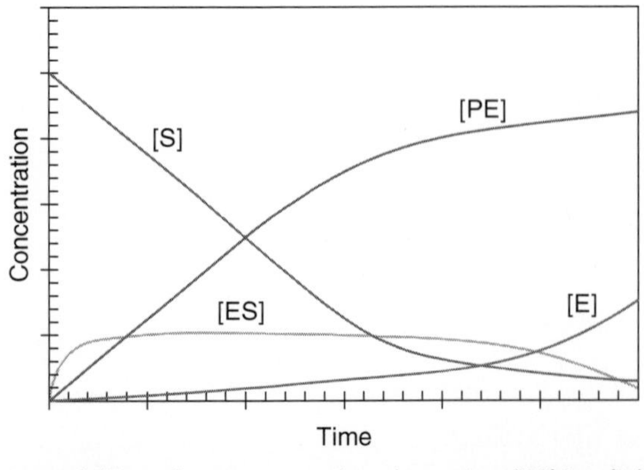

Figure 2.3 Kinetic modeling of enzyme-catalyzed reaction (Wikipedia).

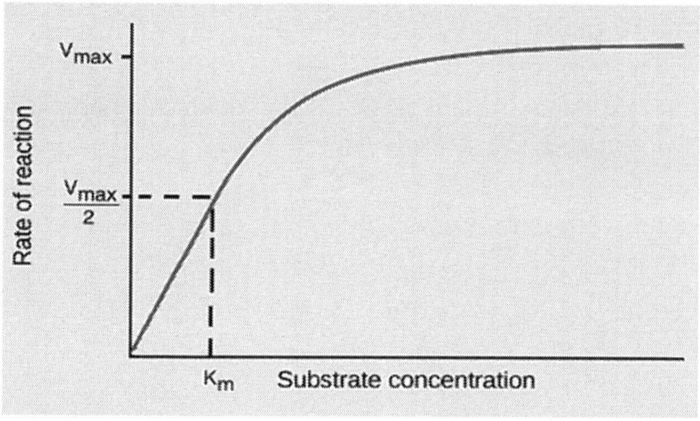

Figure 2.4 Michael-Menten kinetics model of enzymatic reaction (Wikipedia).

The Michaelis-Menten kinetic model of a single-substrate reaction is shown in Fig. 2.4 (Michaelis and Menten, 1913; Xue et al., 2006), where there is an initial bimolecular reaction between the enzyme [E] and the substrate [S] to form the enzyme-substrate [ES] complex, where it is approximately equal to the total enzyme $[E]_{tot}$.

The Michaelis-Menten equation indicates that the reaction rate (v) depends on the position of the substrate-binding equilibrium and the rate constant k_2. Michaelis and Menten showed that when k_2 is much less than k_{-1}, they could derive a mathematical calculation (Segel, 1976; Ricard and Cornish-Bowden, 1987; Berberan-Santos, 2010).

Although the mechanism of a single-substrate enzyme reaction is simplified by the general equation of E + S → E + P, the real rate of the production of [P] is determined by the value of rate constant k_2:

$$v = k_2\,[ES] \tag{2.1}$$

where k_2 is also defined as k_{cat} or the turnover number, which is the maximum number of enzymatic reactions catalyzed per second.

$[E]_{tot}$ is the total enzyme concentration, or $[E]_{tot}$ = [E] + [ES], which is approximately equal to [ES] under saturating conditions.

$$[E]_{tot} = [E] + [ES] \tag{2.2}$$

At low substrate concentrations [S], [E] is in an equilibrium between both free [E] and [ES] complex. However, increasing [S] would result in a simultaneous increase in [ES] and a decrease in [E], shifting the binding equilibrium toward the right.

Since the rate of reaction depends on the concentration [ES], a small change in [S] results in a change in the rate of reaction. Nevertheless, at very high [S], [E] is entirely saturated with [S] and the only form of it is as [ES]. Under these conditions, any increase in [S] has no effect on the rate of reaction and v becomes equal to V_{max} and can be defined as $v \approx k_2\,[E]_{tot} = V_{max}$.

The Michaelis-Menten equation describes not only the reaction rate v, but also the affinity of the enzyme to the substrate as well the rate constant k_2. At equilibrium, the Michael-Menten equation shows that k_2 is much less than k_{-1}, This equation is the basis for most single-substrate enzyme kinetics, and can be written as follow:

$$v = V_{max}\,[S] / K_m + [S] \tag{2.3}$$

where, K_m or the Michaelis constant is defined as [S] at which the rate of the enzyme reaction is ½ V_{max}. The K_m is approximately, but relatively rare, the dissociation constant (k_2) of the [ES] complex.

The most normal mechanism, where $k_2 > k_{-1,}$ is also called Briggs-Haldane kinetics (Briggs and Haldane, 1925). Nevertheless, the Michaelis–Menten equation is valid under the more general conditions, as may be derived from the steady-state approximation (Segel, 1993; Walsh et al., 2010; Engel, 2018).

The specificity constant k_{cat}/K_m is used as quantitative measure for the determination of the efficiency of a given enzyme in the bioconversion of a substrate [S] into a product [P]. Using the term K_m, the Michaelis-Menten equation may be written as follow:

$$v = k_2[ES] = k_2 / k_3([E]\,[S]) \tag{2.4}$$

The specificity constant is an effective bimolecular rate constant for [E] to react with [S] to form [P]. The specificity constants could be used for a quantitative comparison of the efficiency of a given enzyme between two substrates. The slope of the Michaelis-Menten equation at low [S], when [S] << K_m, also yields the specificity constant. It is primordial to optimize kinetically the enzyme reaction to obtain its highest catalytic efficiency (Stroppolo et al., 2001).

2.5.2.3 Second-order kinetics

A second-order reaction is one, which depends on the concentration [S] of two first-order reactants reaction mechanism or one second order reactant mechanism (Taylor, 2002). In other words, the overall catalysis occurs by a sequence of step-by-step reaction. In a second-order reaction, the rate is proportional to the square of the concentration of a reactant [S] or the product [P] or the concentrations of two reactants.

2.5.3 Linear plots of the Michaelis-Menten equation

It is difficult to estimate accurately the K_m and V_{max} values from the nonlinear curve of interactive Michaelis-Menten kinetics (Bravo et al., 2001). However, this nonlinearity could be explored and developed in double-reciprocal linear plots (Bevc et al., 2011). One of these common approaches is the Lineweaver-Burk plot (Lineweaver and Burk, 1934). that is often used to illustrate the kinetic data (Fig. 2.5). It is the linear form of the Michaelis–Menten equation by plotting 1/v versus 1/[S], with y intercept $(1/V_{max})$, and x intercept $(1/K_m)$:

$$1/v = (K_m / V_{max} + [S]) + 1/V_{max} \tag{2.5}$$

It is obvious there is no experimental data that can be gathered at a negative 1/[S]. Indeed, the lower limiting value 1/[S] = 0 corresponds to an infinite [S], where $1/v = 1/V_{max}$. The *x*-intercept is an extrapolation of the experimental data taken at positive concentrations. In general, the Lineweaver-Burk plot skews the importance of measurements taken at low substrate concentrations and, hence yielding inaccurate estimates of V_{max} and K_m. Although a more accurate linear plotting method is the Eadie-Hofstee plot (Fig. 2.6). Nevertheless, current modern research uses more reliable methods of linearization such as nonlinear regression methods.

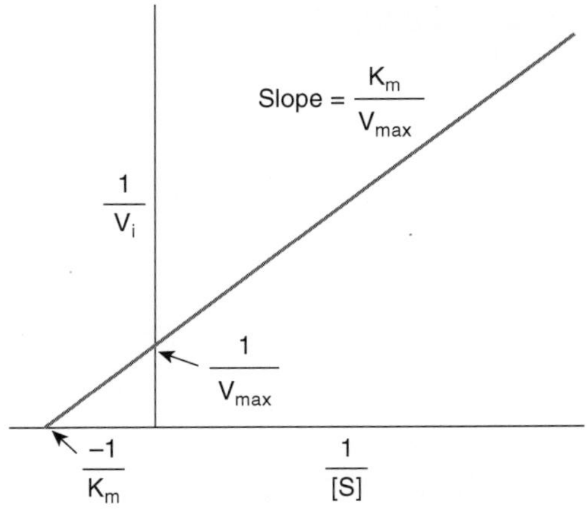

Figure 2.5 Lineweaver-Burk plot of enzyme kinetic data (Wikipedia).

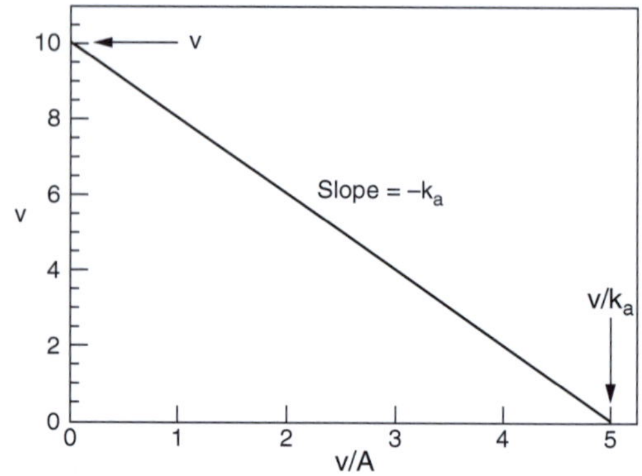

Figure 2.6 Eadie-Hofstee plot (Wikipedia).

2.5.4 Multisubstrate kinetics

2.5.4.1 Multisubstrate/single product kinetics

In the multisubstrate enzyme mechanism, both substrates must bind to the biocatalyst before proceeding with reaction and obtaining the product (Cleland, 1963). The kinetics of multi-substrate reactions follow complex rate equations that indicate the order of substrates binding.

With a constant concentration of substrate [A] and a varied one of substrate [B], the kinetics of the reaction will follow a single-substrate enzyme mechanism (Kraut et al., 2003) and v can be plotted versus [S] to obtain the apparent K_m and V_{max} constants for substrate B. When the enzymatic reaction initiated with variant [S] of substrate A and constant one of substrate B, kinetic analyses can also be performed. In both cases, Lineweaver-Burk plots will provide a series of intersecting lines and derivative curves that can be plotted to obtain the different kinetic constants.

2.5.4.2 Multisubstrate/product kinetics

The mechanism of multisubstrate reactions follows complex rate equations that indicate the nature of the substrate binding and its sequence. For an enzyme that takes two substrates [A] and [B] and turns them into two products [P] and [Q], there are two types of mechanism: ternary complex and ping–pong (Cleland, 1963).

2.5.4.2.1 Ternary-complex mechanism

In this mechanism, both substrates (A and B) bind to the enzyme molecule (E) at the same time resulted in (EAB) ternary intermediate complex. This binding can either be a random or ordered mechanism (Fig. 2.7).

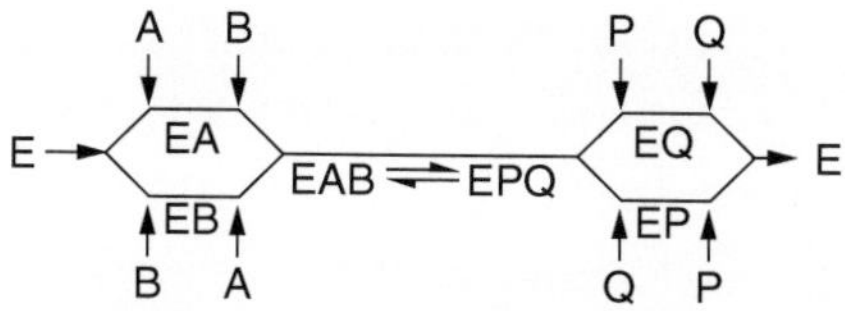

Figure 2.7 Random-order ternary-complex mechanism for an enzyme reaction. The reaction path is shown as a line and enzyme intermediates containing substrates [A] and [B] or products[P] and [Q] are written below the line (Wikipedia).

The mechanism of ternary complex kinetics is due to the formation of a complex between two substrate molecules [A] and [B] and an enzyme [E]. This type of mechanism can be found in multisubstrate enzyme-catalyzed reactions, where two substrates and two products can be formed. The ternary complex is an intermediate between the product formation in this type of enzyme-catalyzed reactions.

2.5.4.2.2 Ping-pong mechanisms

Ping-pong kinetics involve a nonsequential mechanism, where product [A] is released after the first substrate is bound (Fig. 2.8). Once the product [P] is released, the second substrate [B] is bound. This indicates that the product [P] is obtained while the enzyme changes its conformation into intermediate form to bind with the second substrate [B] to produce the second product [Q].

The enzymes with a ping-pong mechanism (Walsh, 1979) can be found in either the free form (E) or as a chemically modified intermediate one [E★]. In such mechanisms, the binding of substrate [A] to [E] will change the form of the enzyme to [E★], by interaction of a selected chemical with the active site, before regenerating [E] form again to interact with the substrate [B].

Enzymes with ping-pong mechanism are a very common such as the diverse family of enzymes, the serine proteases. In these enzymes, the [E★] intermediate is an acyl-enzyme species formed by the binding of the active site serine residue with the peptide bond of the protein substrate.

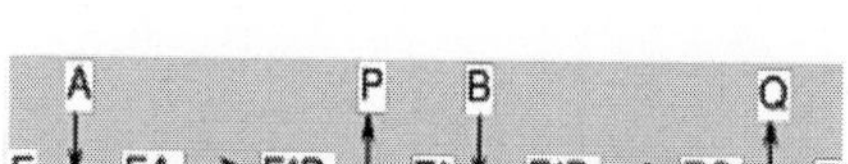

Figure 2.8 Ping-pong mechanism for an enzyme reaction. Enzyme intermediates contain substrates [A] and [B] or products [P] and [Q] (Wikipedia).

2.5.5 Multisite and allosteric enzymes

In certain enzyme reactions, the analysis of kinetic data, when v is plotted against [S], resulted by a sigmoid plot of [S] versus v (Srinivasan et al., 2017). This type of plot (Fig. 2.9) indicates a highly cooperative binding of the substrate at the active site of the enzyme (Duggleby, 1995). This suggests that the binding of one substrate molecule influences the binding of the subsequent substrate molecules. Enzymes providing such data analyses are referred to as allosteric and have multiple active sites (Ricard and Cornish-Bowden, 1987).

These multiple active sites exhibit the property of cooperative mechanism, where the binding of the substrate at one active site affects the affinity of other active sites on the enzyme. This type of mechanism is mostly found in multimeric enzymes, with several interacting active sites. However, this cooperative mechanism could be a positive one where the binding of the first substrate molecule *increases* the affinity of the other active sites for substrate, whereas the negative one occurs when the binding of the first substrate *decreases* the affinity of the enzyme for other substrate molecules (Fersht, 1999).

Allosteric enzymes are important biocatalysts since they are involved in regulating the response of enzymes to changes in the concentrations of their substrates (Hammes, 2002). Indeed, a positive cooperative mechanism makes enzymes much more sensitive to [S] as a limited increase in substrate concentration greatly affects their activities. In contrast, a negative cooperative mechanism makes enzymes insensitive to small changes in [S].

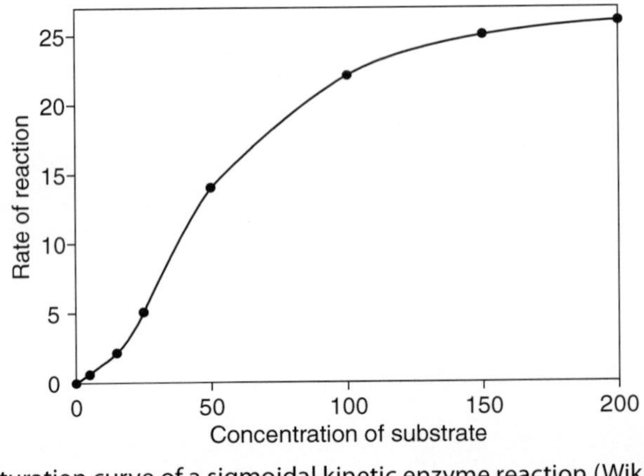

Figure 2.9 Saturation curve of a sigmoidal kinetic enzyme reaction (Wikipedia).

In order to obtain linear correlation between v and [S] for allosteric enzymes, the Hill equation (Hill, 1886) is most likely to determine quantitatively the degree of cooperation, where the binding of substrate to one active site could affect its binding of substrate to the other active sites. The Hill equation corresponds to two closely related equations that would reflect the nature of the binding of substrates to the enzyme protein, as a function of [S] concentration. It can also be derivatized in a linear correlation (Segel, 1976).

2.5.6 Selected enzyme kinetic terms

2.5.6.1 Enzyme specificity

Enzyme specificity indicates the affinity of the chemical group at the active site of a given enzyme to interact with a selected substrate of well-defined chemical structure (Fersht, 1999). This affinity depends not only on the chemical nature of the active site but also on the structural conformation of the enzyme molecule's substrate. It could be defined as the number of [moles] of formed product [P] or converted substrate [S] per mg protein (E) per min.

2.5.6.2 Enzyme unit

The enzyme unit (U) or international unit (IU) for an enzyme is a unit of enzyme's catalytic activity. 1 U (μmol/min) is defined as the amount of the enzyme that catalyzes the conversion of one micromole of substrate per min under the specified conditions of the assay method.

2.5.6.3 Enzyme turnover

The turnover number (k_{cat}) or catalytic rate constant of a given enzyme is defined as the maximum number of substrate molecules (moles) that are bioconverted into product molecules per active site per time unit (min), when the enzyme is the rate-limiting factor.

2.5.7 Factors governing catalytic activity

Enzyme catalytic activity in vivo as well as in vitro could be affected by many parameters (Bar-Even et al., 2011). These parameters may have positive or negative effects in proceeding with the enzymatic catalysis reaction.

2.5.7.1 Enzyme activation

Enzyme activators are molecules that bind to enzymes and increase their activity. These activators may include metal ions, organic molecules, and cofactors. They work opposite to enzyme inhibitors.

2.5.7.1.1 Metal ions

Selected metal ions activate certain enzymes by affecting their three-dimensional conformation but are not involved in the catalytic reaction. Examples of common metal ions, involved in enzyme activation, are Cu, Zn, and Mn.

2.5.7.1.2 Organic molecules

Organic molecules are used as enzyme activators to increase the rate of enzymatic reactions. These types of molecules cover a range of peptides, proteins, and lipids.

2.5.7.1.3 Enzyme cofactors

Certain enzymes (protein) need a nonprotein molecule to carry out their biocatalysis processes. These nonprotein molecules are known as cofactors that fit in the three-dimensional macromolecule of the inactive protein (apoenzyme). These cofactors cover a wide range of ions, such as zinc and iron, as well as organic molecules such as vitamins or their derivatives.

2.5.7.2 Enzyme inhibition

Although the activity of a given enzyme could be inhibited by physicochemical conditions, such as temperature and pH, the term *enzyme inhibitor* refers to a chemical molecule that will bind either to the active site of the enzyme or with the substrate to decrease or even to stop the enzymatic reaction (Walsh et al., 2011).

Enzyme inhibitors are classified according to the nature of their binding and consequently on their mechanism of action. They can be classified generally as either reversible inhibitors where the enzyme activity can be restored by removing the inhibitor molecule or irreversible inhibitors where the enzyme is denatured and inactivated permanently.

2.5.7.2.1 Reversible inhibitors

Reversible enzyme inhibitors can be classified as competitive, noncompetitive, uncompetitive, or mixed, according to their mechanism of action and consequently on their effects on K_m and V_{max}. These different effects result from the nature of the binding of the inhibitor [I] with the free enzyme [E], to the enzyme-substrate [ES] complex, or to both, as shown in Figs. 2.10 and 2.11. Inhibition can be elucidated by investigating the enzyme kinetics as a function of the inhibitor concentration. The different types of inhibition are clearly demonstrated by the Lineweaver–Burke and Eadie-Hofstee plots.

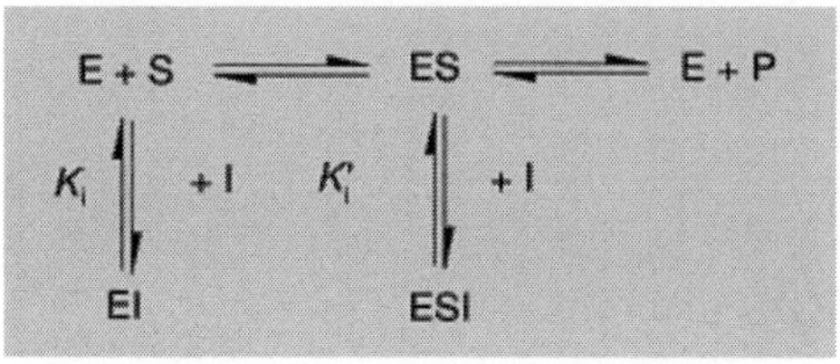

Figure 2.10 Kinetic scheme for reversible enzyme inhibitors (Wikipedia).

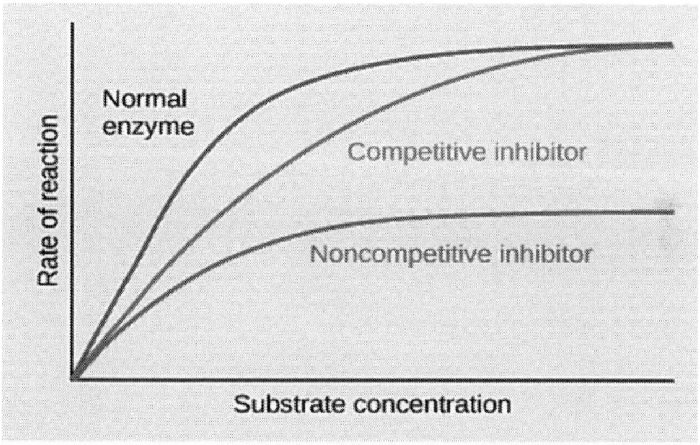

Figure 2.11 Michael-Menten plots of reversible enzymatic inhibitors (Wikipedia).

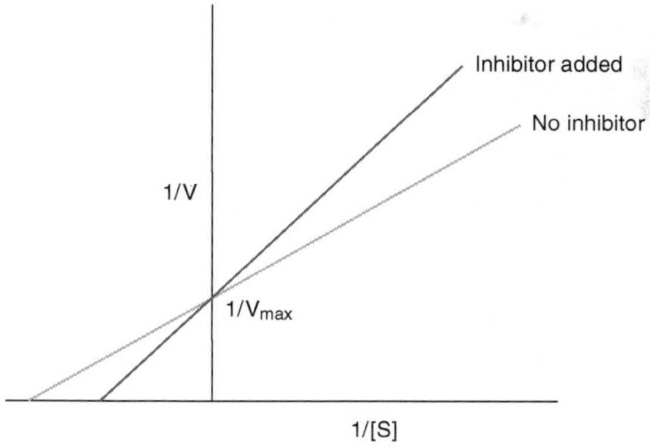

Figure 2.12 Lineweaver-Burk plot of enzyme competitive inhibitor (Wikipedia).

2.5.7.2.1.1 Competitive inhibition A competitive inhibitor of an enzyme is more likely to be a chemical compound that has a structure and certainly a functional group similar to that of the substrate (Fig. 2.12). Therefore, both substrate and inhibitor are competing for binding with the enzyme at its active site. The enzyme allows only one of them to bind at a given moment, but never lets both bind at the same time. The preference of the enzyme for one of them depends on their concentration. In other words, enzyme inhibition [I] could be prevented or decreased by increasing the substrate concentration, while inhibition could be increased by increasing the inhibitor concentration. Most competitive inhibitions are reversible by increasing the substrate concentrations [S].

Kinetically, a competitive inhibitor [I] would alter only the affinity of the enzyme for a substrate[S] and since K_m is a measurement of enzyme affinity, it is only K_m that would be affected by this type of inhibition, as V_{max} would not be affected.

2.5.7.2.1.2 Noncompetitive inhibition In the non-competitive inhibition of an enzyme the inhibitor decreases activity by binding to the enzyme, whether it is already bound to the substrate or not. In other words, noncompetitive inhibition indicates that both inhibitor and substrate may be bound to the enzyme at any given time. This suggests that the inhibitor[I] has an equal affinity for the enzyme [E] and the enzyme-substrate [ES] complex.

In this case, there is an inhibition of the enzyme reaction that prevents the production of [P]. Indeed, there would be only a dissociation of the enzyme-substrate-inhibitor [E-S-I] back to the enzyme substrate [ES] complex and inhibitor [I] as demonstrated in Fig. 2.13,

Kinetically, non-competitive inhibition [I] decreases the maximum rate of the enzymatic reaction (V_{max}), without affecting the affinity of the enzyme for a given substrate (K_m).

The main difference between competitive and noncompetitive inhibition is that the competitive inhibitor would affect the enzyme's ability to

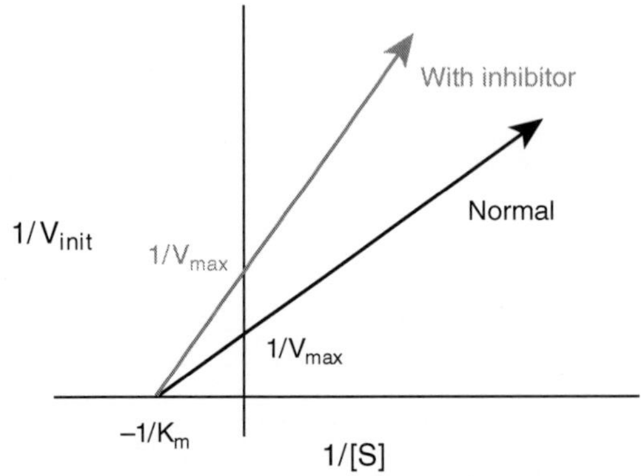

Figure 2.13 Lineweaver-Burk plot of enzyme noncompetitive inhibitor (Wikipedia).

bind to the substrate [S] by competing for the active site of the enzyme [E], whereas a non-competitive inhibitor binds to an allosteric active site preventing the formation of an enzyme-substrate [ES] complex that is required for the enzymatic reaction to proceed.

On the other hand, a noncompetitive inhibition differs from uncompetitive inhibition by allowing the substrate to bind to the enzyme-inhibitor complex and form an enzyme-substrate-inhibitor complex, whereas an uncompetitive inhibitor prevents the substrate from binding to the enzyme inhibitor complex.

2.5.7.2.1.3 Uncompetitive enzyme inhibition An uncompetitive inhibitor is characterized by only binding to the enzyme-substrate (ES) complex. Uncompetitive inhibition is typically associated with enzymatic reactions that have two or more substrates or products.

Uncompetitive inhibition differs from competitive inhibition as it cannot be reversed by increasing the substrate concentration [S] so that the Lineweaver-Burk plot results in parallel lines rather than intersecting ones. Compared to non-competitive inhibition, uncompetitive inhibition is conditional to the formation of enzyme-substrate [ES] complex. Noncompetitive inhibitions, however, can occur in the presence or absence of the substrate [S].

The binding of the uncompetitive inhibitor to the enzyme-substrate [ES] complex indicates (Fig. 2.14) that the binding of the functional group of the inhibitor[I] is available only **to** the enzyme-substrate complex [ES] and not to the enzyme [E].

Kinetically, the binding of the uncompetitive inhibitor [I] lowers the concentration of [ES] complex by forming the enzyme-substrate-inhibitor [ESI] complex.The decrease in [ES] would result in a decrease in the maximum enzyme activity (V_{max}). Since the decrease in the Michael-Menten rate constant (K_m) value, which is mathematically equal to the substrate concentration [S]

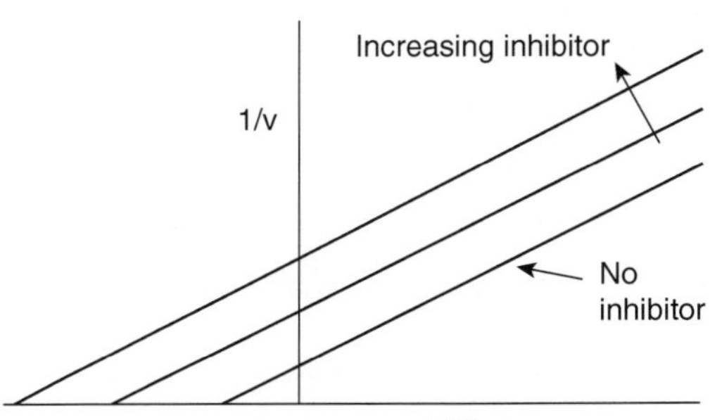

Figure 2.14 Lineweaver-Burk plot of enzyme uncompetitive inhibitor (Wikipedia).

at half of the reaction maximal velocity (V^{max}), there would be a decrease in the [ES] complex. The thermodynamics of the enzymatic reaction would proceed to generate free enzyme [E], which in turn would lead to the further formation of the [ES] complex (English et al., 2006). In other words, it is most likely that the presence of high concentrations of the [ES] complex as well as a lower K_m value are useful indicators of a higher affinity of the enzyme for its substrate. Although the mechanism of uncompetitive inhibition does not require the substrate molecule to be relatively similar to the inhibitor, a high concentration of [S] is more appropriate for this type of inhibition.

2.5.7.2.1.4 Enzyme linear mixed-type inhibition The mixed inhibition term indicates that the inhibitor [I] may bind either to the enzyme in its free form [E] or in its complex with the substrate [ES]. The preference of the inhibitor [I] depends on its affinity for one or the other, [E] or [ES]. The term mixed is a result of the status of the binding of the inhibitor [I] as with the enzyme it is competitive inhibition when it binds with the free enzyme [E] form and uncompetitive inhibition when it binds with the enzyme-substrate [ES] complex. When the affinity of the inhibitor [I] is to bind equally to [E] or [ES], the inhibition becomes a noncompetitive one. It is most likely that in the case of a mixed inhibition, the inhibitor would bind to an allosteric site other than that designed for the substrate (Ricard and Cornish-Bowden, 1987; Srinivasan et al., 2017).

Kinetically, an increase in K_m value or a decrease in the affinity of the enzyme [E] for the substrate [S] may indicate that the inhibitor [I] has more affinity to bind with the free enzyme [E] than with the enzyme-substrate [ES] complex, suggesting the precharacteristic of a competitive binding of the inhibitor.

On the other hand, a decrease in K_m value or an increase in the affinity of the enzyme [E] for the substrate [S] may indicate that the inhibitor [I] has more affinity to bind with the enzyme-substrate [ES] complex rather than with the free enzyme [E], suggesting the characteristic of an uncompetitive binding for the inhibitor.

In both cases, the mixed-type inhibition would result in a decrease in the maximum enzyme reaction rate (V_{max}).

2.5.7.2.2 Irreversible enzyme inhibition

An irreversible inhibitor would covalently bind the enzyme at its active site preventing the formation of the enzyme substrate complex and result in enzyme denaturation.

2.5.7.2.3 Feedback inhibition

Biological systems containing multiple biochemical processes, pathways, and enzyme systems, need a cellular control mechanism. The feedback inhibition is therefore the mechanism that enables the cell to regulate its metabolic activity by inhibiting the enzyme's end-product. Typically, the feedback inhibition would act on the first enzyme in each pathway and hence prevent the production of an excess of any given end-product.

2.6 Latest developments

Natural enzymes are macromolecular biocatalysts that exhibit both high catalytic activity and specificity. While the use of such enzymes by industry and medicine is extensive, their high cost and poor stability have limited their applications. In an effort to overcome these limitations considerable research has been conducted over the past two decades for improving their thermostability as well as developing artificial enzymes (Rigoldi et al., 2018; Shi et al., 2019).

2.6.1 Enhancing enzyme thermostability

Identification of thermophilic organisms has led to the development of more thermostable enzymes. The ability of these organisms to survive under extreme conditions of temperature, pressure, pH, and salt concentration, is attributed to their cellular components including enzymes (Adams et al., 1995). Such enzymes, referred to as extremozymes, tolerate high temperatures and are resistant to denaturation and proteolysis (Kumar and Nussinov, 2001; Vieille and Zeikus, 2001). Enzymes currently used in industry, however, are mostly from fungi or mesophilic bacteria. The only industrial extremozyme is the thermostable DNA polymerase obtained from thermophilic organisms, *Thermus aquaticus*, *Pyrococcus furiosus*, and *Thermococcus litoralis* used in the polymerase chain reaction (PCR) (Rigoldi et al., 2018). The key to developing thermostable enzymes is to understand the sequence and structural characteristics of their proteins. Unlike mesophilic proteins, thermophilic proteins are composed mostly of hydrophobic and charged residues with a much smaller proportion of uncharged polar residues and somewhat higher in valine and isoleucine residues (Han et al. 2013; Kim et al., 2001). The tendency of the hydrophobic residues to cluster and shield themselves from the aqueous environment is considered a key factor in the folding and thermal stability of these proteins. A higher structural rigidity in the protein is general associated with enhanced thermostability (Arcus et al., 2016). This contrasted with earlier work by Fitter et al. (2001) who found that α-amylase from thermophilic organisms had a more flexible protein structure than the corresponding enzyme from mesophilic organisms. The higher thermostability in the latter case, however, was attributed to entropic stabilization. A review of protein engineering strategies for producing industrial thermostable enzymes by Rigoldi et al. (2018) included both directed evolution and rational design.

2.6.2 Directed evolution

This method, a fundamental strategy in protein engineering, involves selecting novel enzymes with improved features and then subjecting the parent protein to mutagenesis followed by screening. By identifying the desired mutant, and eliminating unproductive mutations, the resulting variants can be tested for the desired enzymatic function. Examples of thermal stabilization obtained using directed evolution include a more thermostable amylase from *Thermus* sp. strain [IM650] (Arnold and Volkov, 1999; Shoichet et al. 1995) and a more thermostable endo-β-1,4-xylanase from *Thermomyces lanuginosus* (Elizabeth et al., 2006). A Multiplex Automated Genome Engineering (MAGE) approach was subsequently developed by Wang et al. (2009) as a rapid, specific and efficient tool for genome modifications. Applications include changing the genetic code (Isaacs et al., 2011), incorporation of non-standard amino acids (Lajoie et al., 2013; Rovner et al., 2015) and insertion of His-tags (Wang et al., 2012). This is a very precise and portable genome engineering method for comparing mutational effects across bacterial species (Nyerges et al., 2016). Using this technique, it is anticipated that novel enzymes with greater thermostability will be produced for industry.

2.6.3 Rational design

Directed evolution has produced some mutant enzymes with improved thermostability. However, this is a time-consuming and labor-intensive random process that has limited success. An alternative approach is the application of computational or rational design for synthetic enzymes (Linder, 2012; Welborn and Head-Gordon, 2019). Designing a new enzyme involves selecting a catalytic mechanism and then creating an active site. Rothlisberger et al. (2008) used a Rosetta Match hashing algorithm that searched for suitable protein backbone positions to support idealized active sites in a large set of stable protein scaffolds. The positioning of protein functional groups is important for maximizing transition state stabilization. Procedures developed for designing thermostable enzymes include the Framework for Rapid Enzyme Stabilization to Computational Libraries (FRESCO) (Wijma et al., 2014) and Protein Repair One Stop Shop (PROSS) (Goldenzweig et al., 2016).

A review of rational and semi rational protein design by Korendovych (2018) pointed out that the explosive growth in structural information from the advances in computation provides robust tools for constructing protein structures. The generation of well-designed thermostable protein catalysts would require greater structural rigidity and hydrophobic packing at the protein core (Borgo and Havranek, 2012) plus a favorable network of positive and negative charges on the protein surface (Gribenko et al., 2009).

2.6.4 Artificial enzymes

The term, artificial enzymes, was first coined by Ronald Brewslow (Breslow and Overman, 1970) to describe enzyme mimics. Matsuo and Hirota (2014) outlined the construction strategies for artificial enzymes as genetic mutation and chemical modification or a combination of both. Such modifications of natural and readily available enzymes improved their functions, as well as converted nonenzymatic proteins into enzymes. The creation of a variety of these artificial biocatalysts, or artificial enzymes, was achieved using protein scaffolds such as hemoprotein, heme b (protoheme IX 1)-containing proteins. The successful replacement of the noncovalently attached heme cofactor to the protein matrix with a synthetic metallaporphyrin-derivative cofactor led to creation of a variety of artificial oxidases (Hayashi and Hisaedi, 2002; Matsuo and Hayashui, 2009). Digianantonio and Hecht (2016) recently encoded an artificial enzyme from protein using synthetic genes that functioned in the cells of *E. coli*.

Wang and coworkers (2018) discussed the different approaches for designing artificial enzymes using supramolecular strategies as outlined in Fig. 2.15. In addition to proteins, other scaffolds involved macrocyclic molecules, container molecules, and DNA. The ability of macrocyclic molecules, cyclodextrins, cucurbituils, calixaremes to protect the enzymatic reaction from the surrounding environment, influenced both their catalytic ability and selectivity. For example, simple artificial metalloenzymes were synthesized from cyclodextrin diacids with copper iron, and zinc. Of these, only the iron complexes exhibited a Fenton-like oxidation of benzylic alcohol with a catalytic rate acceleration of up to 2700.

In the last few years container molecules have also been constructed to mimic the enzymic pockets and simulate natural enzymes. Kapthan et al.

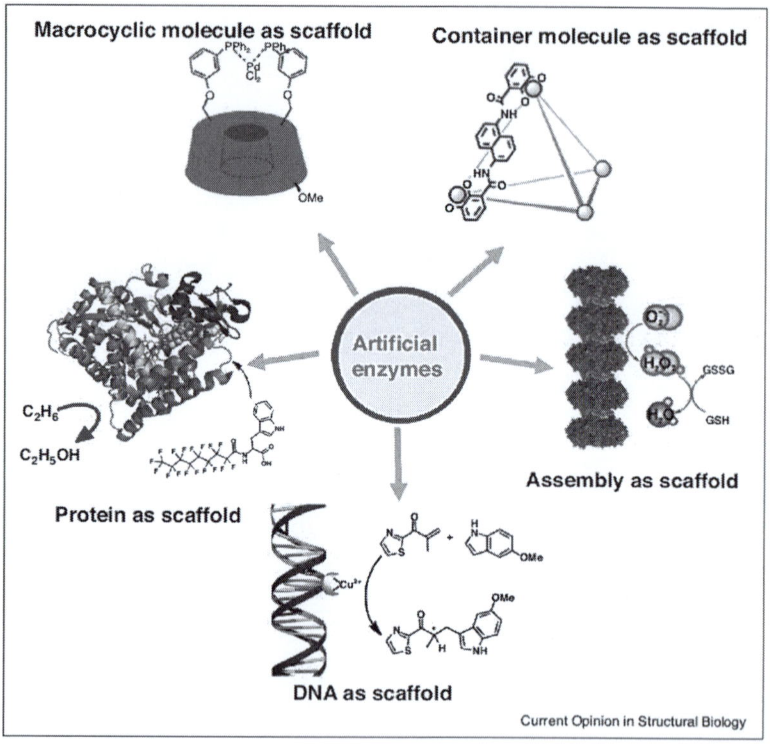

Figure 2.15 Different approaches for designing artificial enzymes using supramolecular strategies. (Wang et al., 2018).

(2015) constructed a tetrahedral cage that catalyzed the bimolecular Aza-Prins cyclization, an important reaction in the synthesis of natural products such as alkaloids. The ability of nucleic acid complexes to serve as quality platforms provides another strategy for engineering artificial enzymes. For example, the DNA-based metalloenzymes complex of DNA G quadruplex and Cu (II) was first shown by Cheng et al. (2016) to catalyze enantioselective sulfoxidation, an important reaction in synthetic chemistry. Finally, the three-dimensional peptide polymers of enzymes can be used to construct a variety of nanometer scale assemblies by providing excellent platforms for designing artificial enzymes. A recent paper by Komiyama and Ariga (2019) reviewed the application of nano architectonics for the preparation of practical artificial enzymes. Nanoparticles with very high catalytic activity were discovered that mimicked oxidase, catalase, and superoxide dismutase enzymes (Wei and Wang, 2013; Wang et al., 2016; Zhou et al., 2017; Salarizadeh et al., 2018;

Wang et al., 2018). They are referred to as nanozymes and bind the substrate by noncovalent adsorption as they have no specific substrate binding site. These nanoenzymes exhibited higher stability and lower costs than the corresponding natural enzymes with the potential for wide applications. Stasyuk et al. (2019) recently described the construction of oxidase-based bioelectrodes for use in food analysis using nanoenzymes as peroxidase mimetics.

Compared to industrial catalysts, enzymes are quite remarkable in terms of their catalytic efficiency (Korendovych and DeGrado, 2014). Nevertheless, these enzymes are very large polypeptides than can be expensive and difficult to produce and utilize (Zozulia et al., 2018). The assumption that these complex molecules might have evolved from simpler molecules led an examination of shorter peptides. As a result, even very short peptides were found to exhibit a wide range of chemical transformations. For example, Kolundzic et al. (2011) reported the effectiveness of aspartyl peptides to catalyze the enantioselective oxidation of indoles, a key step in the biosynthesis of natural alkaloids. An obvious conclusion from this discovery is that catalytic activity may not be dependent on a supramolecular structure. I remember as an undergraduate student in biochemistry, over half a century ago, that the enzyme papain was shown to be a very large protein. The question raised was whether such a huge protein was really needed for its catalytic activity? Clearly, recent research answered this question by showing that peptides with as few as 7 amino acids exhibited catalytic activity greater than those of enzymes by weight (Zozulia et al., 2018). Makam et al. (2019) recently designed, de novo, a simple nonproteinaceous hydrolase. Phenylalanine molecules spontaneously coordinated with zinc ions forming a robust, layered, supramolecular amyloid type of architecture. This artificial supramolecule exhibited considerable carbonic hydrase catalytic activity and had the lowest molecular weight. In the same publication, Makhlynets and Korendovych (2019) reported the production of an enzyme from a single amino acid, phenylalanine. In the presence zinc ions, phenylalanine spontaneously and rapidly self-assembled into needle-like crystals. The crystals showed a β-sheet-like arrangement held by hydrogen bonds between the amino acids with the hydrophobic side chains tightly packed against each other. This amyloid-like structure exhibited catalytic activity that hydrolyzed *p*-nitrophenylacetate.

These remarkable new developments in artificial enzymes will undoubtedly play an increasing role not only in the food and beverage industry but also in the manufacture of pharmaceuticals and in medicine.

2.7 Conclusion

This chapter has attempted to provide a limited but comprehensive overview of the nature of enzymes .The development of novel approaches and technologies not only in enzymology but also in other areas, such as bioengineering, crystallography, software modeling, and statistical analyses, has led to great progress in elucidating the catalytic mechanisms of enzyme action.This progress has resulted in widening the horizon of novel potential biotechnological applications of enzymes. Hence, providing this summarized review on enzymes was aimed at presenting basic knowledge in terms of their nature, mechanistic actions and kinetics. It was also important to report briefly on the latest developments in the area of enzymology, including enhancement of enzyme thermostability, and artificial enzymes.

References

Adams, M.W., Perler, F.B., Kelley, R.M., 1995. Extremozymes: Expanding the limits of biocatalysis. Nat. Biotechnol. 13, 662–668.

Appel, W., 1986. Chymotrypsin: Molecular and catalytic properties. Clin. Biochem. 19, 317–322.

Arcus, V.L., Prentice, E.J., Hobbs, J.K., Mulholland, A.J., Van der Kamp, M.W., Pudney, C.R., et al., 2016. On the temperature dependence of enzyme-catalyzed rates. Biochem. 55, 1681–1688.

Arnold, F.H., Volkov, A.A., 1999. Directed evolution of catalysts. Curr. Opin. Chem. Biol. 31, 54–59.

Bar-Even, A., Noor, E., Savir, Y., Liebermeister, W., Davidi, D., Tawfik, D.S., et al., 2011. The moderately efficient enzyme: Evolutionary and physicochemical trends shaping enzyme parameters. Biochem. 50, 4402–4410.

Berberan-Santos, M.N., 2010. A general treatment of Henri Michaelis Menten enzyme kinetics: Exact series solution and approximate analytical solutions. Communic. Mathemat. Computer Chem. 63, 283.

Berg, J.M., Tymoczko, J. and Stryer, L., 2007. Biochemistry, seventh ed. W.H. Freeman & Co., New York, N.Y.

Bevc, S., Konc, J., Stojan, J., Hodoščeк, M., Penca, M., Praprotnik, M., et al., 2011. ENZO: A web tool for derivation and evaluation of kinetic models of enzyme catalyzed reactions. PLOS ONE 6, e22265.

Blake, C.C., Koenig, D.F., Mair, G.A., North, A.C., Phillips, D.C., Sarnia, V.R., 1965. Structure of hen egg-white lysozyme. A three-dimensional Fourier synthesis at 2 angstrom resolution. Nature 206, 757–761.

Borgo, B., Havranek, J.J., 2012. Automated selection of stabilizing membrane in designed and natural proteins. Proc. Natl. Acad. Sci. U. S. A. 109, 1494–1499.

Bravo, I.G., Busto, F., De Arriaga, D., Ferrero, M.A., Rodríguez-Aparicio, L.B., Martínez-Blanco, H., et al., 2001. A normalized plot as a novel and time-saving tool in complex enzyme kinetic analysis. Biochem. J. 358, 573–583.

Breslow, R., Overman, L.C., 1970. Artificial enzyme combining a metal catalytic group and a hydrophobic binding capacity. J. Am. Chem. Soc. 92, 1975–1977.

Briggs, G.E., Haldane, J.B., 1925. A note on the kinetics of enzyme action. Biochem. J. 19, 338–339.

Bugg, T.D.H., 2012, Introduction to Enzyme and Coenzyme Chemistry, third ed. John Wiley & Sons, New York, N.Y.
Busenlehner, L.S., Armstrong, R.N., 2005. Insights into enzyme structure and dynamics elucidated by amide H/D exchange mass spectrometry. Arch. Biochem. Biophys. 435, 34–46.
Carvalho, A.T.P., Barrozo, A., Doron, D., Kilshtain, A.V., Major, D.T., Kamerlin, S.C.L., 2014. Challenges in computational studies of enzyme structure, function and dynamics. J. Mol. Graph. Model. 54, 62–79.
Chan, P.H., Cheung, A.H., Okon, M., Chen, H.-M., Withers, S.G., McIntosh, L.P., 2013. Investigating the structural dynamics of α-1,4-galactosyltransferase C from *Neisseria meningitides* by nuclear magnetic resonance spectroscopy. Biochem. 52, 320–332.
Chang, M.P., Li, V.H., Zhou, J., Jia, G.D., Lu, S.M., Young, L., et al., 2016. Enantoselective sulfoxidation reaction catalyzed by a G-quadruplex DNA metalloenzyme. Chem. Commun. 52, 9644–9647.
Cleland, W.W., Cook, P., 2007. Enzyme Kinetics and Mechanism, Garland Science, New York, N.Y.
Cleland, W.W., 1963. The kinetics of enzyme-catalyzed reactions with two or more substrates or products. III. Prediction of initial velocity and inhibition patterns by inspection. Biochim. Biophys. Acta 67, 188–196.
Conn, E.E., Stumpf, P.K., Bruening, G., Doi, R.H., 1987. Outlines of Biochemistry, fifth ed. John Wiley & Sons, New York, N.Y.
Cornish-Bowden, A., 2012. Fundamentals of Enzyme Kinetics, fourth ed. John Wiley-Blackwell, Hoboken, N.J.
Daniel, r.M., Dunn, B.V., Finney, J.L., Smith, J.C., 2003. The role of dynamics in enzyme activity. Ann. Rev. Biophys. Bimolecular Struct. 32, 69–92.
Danson, M. and Eisenthal, R., 2002. Enzyme Assays: A Practical Approach. Oxford University Press, Oxford, U.K.
Digianantonio, K.M., Hecht, M.H., 2016. A protein constructed de novo enables cell growth by altering gene regulation. PNAS 113, 2400–2405.
Dixon, M. and Webb, E.C., 1958. Enzymes, Longmanns Green, London, U.K., pp. 183-277.
Duggleby, R.G., 1995. Analysis of enzyme progress curves by nonlinear regression. Methods Enzymol. 249, 61–90.
Elizabeth, D., Rumbold, K., Permaul, K., Alexander, B., Singh, S., 2006. J. Biotechnol. 27, 348–354.
Engel, P.C., 2018. Enzyme Kinetics: The Steady-State Approach, Springer, New York, N.Y.
English, B.P., Min, W., van Oijen, A.M., Lee, K.T., Luo, G., Sun, H., et al., 2006. Ever-fluctuating single enzyme molecules: Michaelis-Menten equation revisited. Nature Chem. Biol. 2, 87–94.
Eyring, H., Polanyi, M., 1931. Uber einfache Gasreaktionen. Z. Phys. Chem. B12, 270–311.
Fersht, A., 1999. Structure and Mechanism in Protein Science: A Guide to Enzyme Catalysis and Protein Folding, W.H. Freeman & Co., San Francisco, CA.
Fischer, E., 1894. Einfluss der configuration aif der wirkung der enzyme. Ber. Dtsch. Chem. Ges. 27, 2984–2993.
Fitter, J., Hermann, R., Hauβ, T., Lechner, R.E., Dencher, N.A., 2001. Dynamic properties of α-amylase in the folded and unfolded state: The role of thermal equilibrium fluctuations for conformational entropy and protein stabilization. Physica B. 301, 1–7.
Fromm, H.J., Hargrove, M.S., 2012. Enzyme kinetics. Essentials of Biochemistry. Springer-Verlag, Berlin, Germany, pp. 81–122.
Goldenzweig, A., Goldsmith, M., Hill, S.E., Sussman, J.L., Tawfik, D.S., Fleishman, S.J., et al., 2016. Automated structure and sequence-based design of proteins for high bacterial expression and stability. Mol. Cell. 63, 337–346.
Gribenko, A.V., Patel, M.M., Liu, J., McCallum, S.A., Wang, C., Makhatadze, G.I., 2009. Rational stabilization of enzymes by computational redesign of surface charge-charge interactions. Proc. Natl. Acad. Sci. U.S.A. 106, 2601–2606.

Haldane, J.B.S., 1930. Enzymes, Longmans, Green, London, U.K.
Hammes, G.G., 2002. Multiple conformational changes in enzyme catalysis. Biochem. 41, 8221–8228.
Han, T., Zeng, F.M., Li, Z., Wei, M., Guan, Q., Liang, X., et al., 2013. Biochemical characterization of a recombinant pullulanase from *Thermococcus kodakarensis* KOD1. Lett. Appl. Microbiol. 57, 336–343.
Hayashi, T., Hisaedi, Y., 2002. New fuctionalization of myoglobin by chemical modification of heme proteins. Accounts Chem. Res. 30, 35–43.
Hill, G.W., 1886. On the part of the motion of the lunar perigee which is a function of the mean motions of the sun and the moon. Acta Math. 8, 1–36.
Hoffmann-Ostenhof, O., 1953. Suggestions for a more rational classification a nomenclature of enzymes. Adv. Enzymol. 14, 219–260.
Isaacs, F.J., Carr, P.A., Wang, H.H., Lajoie, M.J., Kraal, L., Tolonen, A.C., et al., 2011. Precise manipulation of chromosomes in vivo enables genome-wide colon replacement. Science 333, 348–353.
Kapthan, D.M., Toste, F.D., Bergmann, R.G., Raymond, K.N., 2015. Enabling new modes of reactivity via constructive binding in a supramolecular-assembly-catalyzed AZA-Prinscyclization. J. Am. Chem. Soc. 137, 9202–9205.
Kim, J.M., Flowers, L.D., Whitely, M., Peeples, T.L., 2001. Biochemical confirmation and characterization of the family-57-like α-amylase of *Methanococcus jannaschii*. Folia Microbiol. 46, 467–473.
Kolundzic, F., Noshi, M.N., Tjandra, M., Movassaghi, M., Miler, S.J., 2011. Chemoselective and entioselective oxidation of indoles employing aspartyl peptide catalysts. J. Am. Chem. Soc. 133, 9104–9111.
Komiyama, M., Ariga, K., 2019. Nanoarchitectonics to prepare practically useful artificial enzymes. Mol. Catal. 475, 110492.
Korendovych, I.V., DeGrado, W.F., 2014. Catalytic efficiency of designed catalytic proteins. Curr. Opin. Struct. Biol. 27, 113–121.
Korendovych, I.V., 2018. Rational and semirational protein design. Methods Mol. Biol. 1685, 15–23.
Koshland, D.E., 1958. Application of a theory of enzyme specificity protein synthesis. Proc. National Acad. Sci. USA 44, 98–104.
Kraut, D.A., Carroll, K.S., Herschlag, D., 2003. Challenges in enzyme mechanism and energetics. Ann. Rev. Biochem. 72, 517–571.
Kumar, S., Nussinov, R., 2001. How dothermophilic proteins deal with heat. Cell. Mol. Life Sci. 58, 1216–1233.
Lajoie, M.J., Rovner, A.J., Goodman, D.B., Aerni, H.R., Haimovich, A.D., Kuznetsov, G., et al., 2013. Genomically recoded organisms expand biological functions. Science 342, 357–360.
Lifson, S., Warshel, A., 1968. A consistent force field for calculation of conformations, vibrational, spectra and enthalpies of cycloalkanes and n-alkane. J. Chem. Phys. 49, 5116.
Linder, M., 2012. Computational enzyme design: Advances, hurdles and possible ways forward. Comp. Struct. Biotechnol. J. 2, e201209009.
Lineweaver, H., Burk, D., 1934. The determination of enzyme dissociation constants. J. Am. Chem. Soc. 56 (3), 658–666.
Lu, H.P., Xun, L., Xie, X.S., 1998. Single-molecule enzymatic dynamics. Science 282, 1877–1882.
Makam, P., Yamijala, S.S.R.K.C., Tao, K., Shimon, L.J.W., Eisenberg, D.S., Sawaya, M.R., et al., 2019. Non-prteinaceous hydrolase comprised of a phenylalanine metallo-supramolecular amyloid-like structure. Nat. Catal. 2, 977–985.
Makhlynets, O.V., Korendovych, I.V., 2019. A single amino acid enzyme. Nat. Catal. 2, 949–950.

Matsuo, T., Hayashui, T., 2009. Electron transfer and oxidase activities in reconstituted hemoproteins with chemically modified cofactors. J. Porphyrins Phthalocyanins 13, 1082–1089.
Matsuo, T., Horota, S., 2014. Artificial enzymes and protein scaffolds. Structural design and modification. Biorg. Medicinal Chem. 22, 5638–5656.
McDonald, A.G., Tipton, K.F., 2014. Fifty-five years of enzyme classification: advances and difficulties. FEBS J. 281, 583–592.
Michaelis, L., Menten, M.L., 1913. Kinetik der invertinwirkung (The kinetics of invertase action). Biochem. Z. 49, 333–369.
Monasterio, O., 2014. Nomenclature for the applications of nuclear magnetic resonance to the study of enzymes. Perspec. Sci. 1, 88–97.
Nyerges, A., Csorgo, B., Nagy, I., Balint, B., Bihari, P., Lazar, V., et al., 2016. A highly precise and portable genome engineering method allows comparison of mutational effects across bacterial species. PNAS 113, 2502–2507.
Ogata, H., Hirata, S., Nakahara, A., Komori, H., Shibata, N., Kato, T., et al., 2005. Activation process of [NiFe] hydrogenase elucidated by high-resolution X-ray analyses: Conversion of the ready to the unready state. Structure 13, 1635–2164.
Pauling, L., 1948. The nature of forces between large molecules of biological interest. Proc. Royal Inst. of Great Britain. 34, 181–187.
Pravda, L., Berka, K., Svoboova Varenkova, R., Banas, P., Laskowski, R.A., Olypeka, M., 2014. Anatomy of enzyme channels. BMC Bioinformatic 15, 379–387.
Price, N.C. and Stevens, L., 2000. Fundamentals of Enzymology: The Cell and Molecular Biology of Catalytic Proteins, third ed. Oxford University Press, Oxford, U.K.
Purich, D.L., 2010. Enzyme Kinetics: Catalysis and Control: A Reference ofTheory and Best Practice Methods, 3rd edition Academic Press, New York, N.Y.
Ricard, J., Cornish-Bowden, A., 1987. Co-operative and allosteric enzymes: 20 years on. Eur. J. Biochem. 166, 255–272.
Rigoldi, F., Donini, S., Redaelli, S., Parisini, E., Gautieri, A., 2018. Review: Engineering of thermostable enzymes for industrial applications. APL Engineer 2, 001501.
Rothlisberger, D., Khersonky, O., Wollacott, A.M., Jiang, L., DeChancle, J., Becker, J., et al., 2008. Kemp elimination catalysts by computational enzyme design. Nature 454, 190–195.
Rovner, A.J., Haimovich, A.D., Katz, S.R., Li, Z., Grome, M.W., Gassaway, B.M., et al., 2015. Recorded organisms engineered to depend on synthetic amino acids. Nature 518, 83–93.
Salarizadeh, N., Sadry, M., Sajedi, R.H., 2018. Synthesis and catalytic evaluation of Fe_3O_4/MNCNTs nanozyme as recyclable mimetics: Biochemical and biophysical characterization. Appl. Organometal. Chem. 32, e4018.
Schnell, S., Maini, P.K., 2004. A century of enzyme kinetics: Reliability of the K_M and V_{max} estimates. Comments Theoretical Biol. 8, 169–187.
Schomburg, I., Chang, A., Schomburg, D., 2014. Standardization in enzymology-data integration in the world's information system BRENDA. Perspect. Sci. 1, 15–23.
Segel, I.H., 1976. Biochemical Calculations, second ed. John Wiley & Sons, Inc., New York, N.Y.
Segel, I.H., 1993. Enzyme Kinetics: Behavior and Analysis of Rapid Equilibrium and Steady-State Enzyme Systems, John Wiley & Sons, Inc., New York, N.Y.
Seibert, C.M., Raushel, F.M., 2005. Structural and catalytic diversity within the amidohydrolase superfamily. Biochem. 44, 6383–6391.
Shi, H., Liu, Y., Qu, R., Li, Y., Ma, R., An, Y., et al., 2019. A facile one-pot method to prepare perioxidase-like nanogel artificial enzymes for highly efficient and controllable catalysis. Colloids Surfaces B: Biointerfaces 174, 352–359.
Shoichet, B.K., Baase, W.A., Kuroki, R., Matthews, B.W., 1995. A relationship between protein stability and protein function. Proc. Nat. Acad. Sci. U.S.A. 92, 452–456.

Srinivasan, B., Rodrigues, J.V., Tonddast-Navaei, S., Shakhnovich, E., Skolnick, J., 2017. Rational design of novel allosteric dihydrofolate reductase inhibitors showing antibacterial effects on drug-resistant *Escherichia coli* escape variants. ACS Chem. Biol. 12, 1848–1857.
Stasyuk, N., Goyda, G., Zakalsky, O., Serkiz, P., Gonchar, M., 2019. Ampometric biosensors based on oxidases and PtRu nanoparticles as artificial peroxidase. Food Chem. 285, 213–220.
Stroppolo, M.E., Falconi, M., Caccuri, A.M., Desideri, A., 2001. Superefficient enzymes. Cell. Mol. Life Sci. 58, 1451–1460.
Taylor, K.B., 2002. Enzyme Kinetics and Mechanisms, Springer, New York, N.Y.
Tipton, K., Boyce, S., 2000. History of the enzyme nomenclature system. Bioinformatics 16, 34–40.
Vieille, C., Zeikus, G.J., 2001. Hyperthermophilic enzymes: Sources, uses, and molecular mechanisms for thermostability. Microbiol. Mol. Biol. Rev. 65, 1–43.
Walsh, R., Martin, E., Darvesh, S., 2010. A method to describe enzyme-catalyzed reactions by combining steady state and time course enzyme kinetic parameters. Biochim. Biophys. Acta 1800, 1–5.
Walsh, R., Martin, E., Darvesh, S., 2011. Limitations of conventional inhibitor classifications. Integr. Biol. 3, 1197–1201.
Walsh, C., 1979. Enzymatic Reaction Mechanisms, W. H. Freeman, San Francisco, CA.
Wang, H.H., Huang, P.Y., Xu, G., Haas, W., Marblestone, A., Li, J., et al., 2012. Multiplexed in vivo His-tagging of enzyme pathways for in vitro single-pot multienzyme catalysis. ACC Synth. Biol. 1, 43–52.
Wang, H.H., Isaacs, F.J., Carr, P.A., Sun, Z.Z., Xu, G., Forest, C.R., et al., 2009. Programming cells by multiplex engineering and accelerated evolution. Nature 460, 894–989.
Wang, S., Meng, X., Zhou, H., Liu, Y., Secundo, F., Liu, Y., 2016. Enzyme stability and activity in non-aqueous reaction systems: A mini review. Catalysts 6, 32.
Wang, T., Fan, X., Hou, C., Lui, S., 2018. Design of artificial enzymes by supramolecular strategies. Curr. Opin. Struct. Biol. 51, 19–27.
Warshel, A., Levitt, M., 1976. Theoretical studies of enzymatic reactions: Dielectric, electrostatic and steric stabilization of the carbonium ion in the reaction of lysozyme. J. Mol. Biol. 103, 227–249.
Warshel, A., Lifson, S., 1970. Consistent force field calculations II. Crystal structure, sublimation energies, molecular lattice vibrations, molecular conformations and enthalpies of alkanes. J. Phys. Chem. 53, 582.
Warshel, A., 2002. Molecular dynamics simulations of biological reactions. Acta Chem. Res. 35, 385–395.
Webb, E.C., 1993. Enzyme nomenclature: A personal perspective. FASEB J. 12, 1192–1196.
Wei, H., Wang, C., 2013. Nanomaterials with enzyme-like characteristics (nanozymes): Next generation artificial enzymes. Chem. Soc. Rev. 42, 6060–6093.
Welborn, V.V., Head-Gordon, T., 2019. Computational design of synthetic enzymes. Chem. Rev. 119, 6613–6630.
Wijma, H.J., Floor, F.J., Jekel, P.A., Baker, D., Marrink, S.J., Janssen, D.B., 2014. Computationally designed libraries for rapid enzyme stabilization. Protein Engineer. Des. Sel. 27, 49–58.
Wittig, U., Kania, R., Golebiewski, M., Rey, M., Shi, L., Jong, L., et al., 2012. SABIO-RK-database for biochemical reaction kinetics. Nuclear Acid Res. 40, D790–D796, (DOI: 10.1093/nar/gkr1046).
Xue, X., Liu, F., Ou-Yang, Z.C., 2006. Single molecule Michaelis-Menten equation beyond quasistatic disorder. Physical Rev. E 74 (3 Pt 1).
Zhou, Y., Liu, B., Yang, R., Liu, J., 2017. Filling in the gaps between nanozymes and enzymes: Challenges and opportunities. Bioconjugate Chem. 28, 2903–2908.
Zozulia, O., Dolan, M.A., Korendovych, V.I., 2018. Catalytic peptide assembles. Chem. Soc. Rev. 47, 3621–3637.

CHAPTER THREE

Production of enzymes: Fermentation and genetic engineering

Henry-Eric Spinnler
AgroParis Tech, Paris, France

3.1 Introduction

Because enzymes are largely used as processing aids in food production, in washing powders and in chemical manufacturing, their production by fermentation has been seen as an opportunity to produce enzyme concentrates on a large scale. The progress in genetic engineering and DNA edition has been used to build strains with a modified genome particularly efficient for these production processes. One of the goals has been to maximize the specific enzymatic activity of certain strains. This has been done by incorporating a gene or increasing the number of copies of the gene coding the enzyme protein sequence, into the genome. Another step has involved the overexpression of these genes in the transcription and translation phases to obtain an increased enzymic activity. However, as these operations have been intracellular, a third goal has been to improve the efficiency of the enzymatic recovery. This has been done by selecting organisms that have the capacity to secrete the protein outside the cell, into a culture broth. Fig. 3.1 shows the key steps in a process using specific organisms. Bacteria such as *Bacillus sp.* (Cai et al., 2019) and fungi like *Trichoderma reesei*, *Aspergillus orizae,* or *Yarrowia lipolytica* (Archer and Peberdy, 1997) are species commonly used. By increasing the enzyme titer produced, it has reduced the risk of unexpected secondary activities in the culture broth and has facilitated an efficient process of enzyme extraction/ purification, which is a fourth step. The fifth step which is not the easiest has been to harvest and stabilize the enzyme produced.

Due to their ability to secrete the protein outside the cell, manufacturing companies have preferred using organisms mentioned above. In addition, companies have controlled and optimized the culture media used for

Enzymes: Novel Biotechnological Approaches for the Food Industry.
http://dx.doi.org/10.1016/B978-0-12-800217-9.00003-4

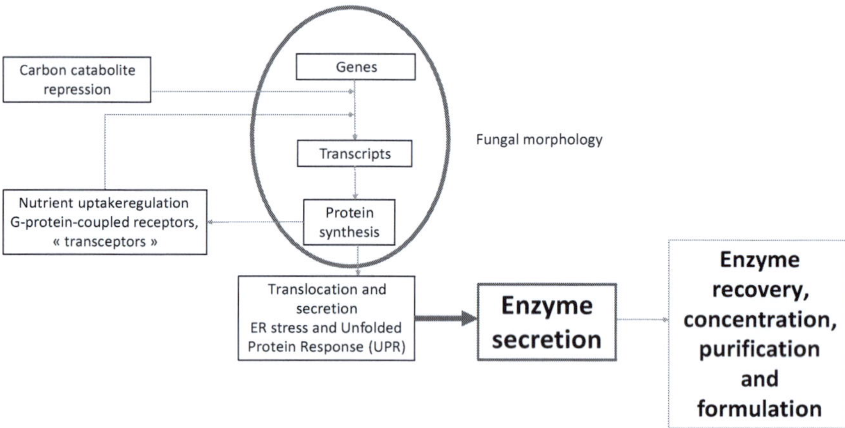

Figure 3.1 Key steps to overproduce a ready to use specific enzyme (Baker, 2018; Fiedler et al., 2018).

growing these organisms. It has been done for safety reasons (no toxins produced), and to optimize harvesting conditions and in general for easy implementation on an industrial scale. This chapter will attempt to cover the different aspects of this process so as to facilitate the commercialization of these enzymes (Fig. 3.2).

3.2 Enzyme screening

Several steps are necessary to achieve efficient enzyme production but the first is to develop an efficient screening method to easily select strains of microorganism producing the researched enzyme.

Traditional systems using screening on agar plates were shown to be efficient. This has now been surpassed when automatic microplate readers with a capacity to read over 96 wells became available. With both systems the main challenge was to obtain a semiquantitative indication of the enzyme activity, obtained via a signal such as a color reaction for example. If necessary, a quantitative measurement of the activity could be obtained subsequently, by more accurate and reliable systems. This could include labeled substrates such as fluorogenic and chromogenic substrates that generate a reporter molecule by elimination, fluorescence resonance energy transfer (FRET) substrates, and isotopic labels for screening of enantioselectivity.

Depending on the enzyme researched, several strategies to detect activity have been used (Goddard and Reymond, 2004). These include:

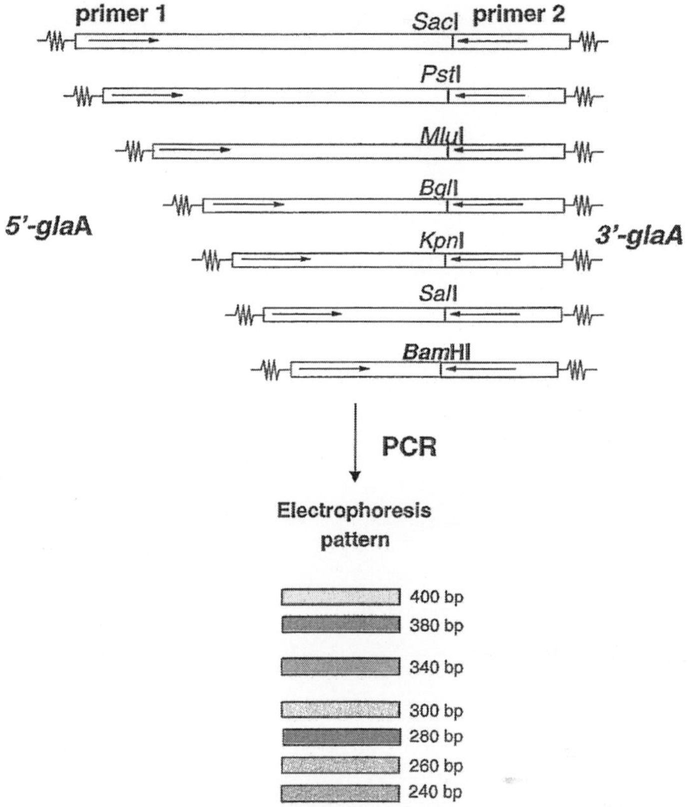

Figure 3.2 Visualization of all seven marked glaA loci of *A. niger* strains, in which the glaA genes were removed. All gene sequences of the seven glucoamylase loci were removed by using proprietary technology. Subsequently, they were partly filled using the same methodology with sequences from the 30 regions of the original locus. Due to the fact that these sequences varied in length and were decorated with a specific restriction site the seven marked loci can be detected individually by PCR using a set of specific primers and separation on a gel electrophoretic system (van Dijck et al, 2003).

- using a substrate releasing a chromophore or a fluorophore when the activity is present, as is the case for sugar hydrolase (with paranitrophenyl glycosides as substrates), proteases (with azocasein), peptidases (using para nitro-anilide derivated peptides or amino-acids releasing a colored nitro-aniline when a peptidase was present).
- using a specific reactivity of the reaction product (e.g., $CuCl_2$ because Cu^{++} reacts with fatty acids, resulting in a bright blue color, therefore enabling the detection of lipolytic activities)

- using pH indicators when a change in pH is expected and
- using fluorochromes such as Sybr Green or coumarin derivatives (Goddard and Reymond, 2004)

3.3 Microbial strain selection

When an efficient screening method is usable it becomes easier to screen culture collections. At first a diversity of wild type strains has to be screened to obtain high-yielding strains. However, the ability to modify genomes opened up possibilities in the second part of the 20th century. Several companies used this technology to create strains dedicated to the production of specific enzymes. In addition, fungi were preferred in the process because of their ability to grow in low pH environments. Fungi have been used to limit the growth of "short generation time" bacterial contaminants that usually grow at a neutral pH.

3.4 Strain improvement

To develop the first generation of hypersecreting strains three main steps are necessary. One needs to (1) prepare the receiving strain that is to be used industrially; (2) construct a vector in *E. coli* (the donor strain), and (3) transfer the researched gene(s) from the donor strain into the chromosomes of the receiving strain.

The strategy usually used is to insert the gene coding for the enzyme into cassettes constructed in the genome of the production strain. In *Trichoderma reesei* or in *Aspergillus niger*, genetic cassettes have been built by excision of known enzymes from the genome of the wild type strain (e.g., *Trichoderma* cellulases or *Aspergillus* glucanases) (van Dijck et al., 2003).

3.4.1 Preparation of the strain of production (Cassette construction)

On the sites of known genes of the wild type fungal strain, a zone, sensitive to a different restriction enzyme was inserted, while the gene coding for the known enzymes was removed. Up to seven cassettes were built on the fungal genome which means, on each cassette a sequence sensitive to a different restriction enzyme was positioned. This zone was flanked by small nucleotidic sequences that helped the insertion of the gene of interest at the right place, including a nucleotidic zone sensitive to a restriction enzyme other than the one initially present in the cassette (flag). The insertion was done from a vector and provided a high number of copies of the targeted gene in the production strain (up to 120 copies).

3.4.2 Preparation of the vector

The vector used to transfer the gene coding the enzyme is usually built in *E. coli*. If the original gene is coming from a species very different from the species used for the overproduction (e.g., initial gene coming from a prokaryote and produced by a yeast), a change of the sequence in order to avoid problems with codon bias that could happen between very different species should be applied using tables before to insert the gene of interest in *E. coli*. The plasmid usually bears a marker gene and is in targeted zones that can be opened with specific restriction enzymes, so that foreign genes could be inserted easily. Markers need to be installed when constructing a vector so that the transformed clones could be easily identified.

For a while, genes generating resistance to antibiotics were used as markers but this approach should be avoided for enzymes used for food applications. If this approach is unavoidable, these genes should be deleted in the final strain.

It should be stressed that this marker gene can enable the use of a specific substrate. For example, the addition of the amdS gene in the plasmid coding the acetamidase, gives the strain the capability to uptake acetamide as carbon source. In this case, the capability of the transformed strains to use acetamide indicates the efficiency of the plasmid transfer (Selten et al., 1995). Fig. 3.3 shows the construction of such a plasmid.

3.4.3 Construction of the overproducing strain

This construction should meet several objectives to ultimately obtain a stable overproducing strain. This means (1) successfully inserting the "gene coding the enzyme" on at least one of the chromosomes, (2) being able to have a hyperproduction of the enzyme, rendering this protein the most significant one produced quantitatively, in the culture supernatant, (3) avoiding disrupting the general metabolism of the producing strain, (4) obtaining a protein that will be processed by the post-traductional modification system, (5) finally showing that all foreign DNA, with the exception of the enzyme coding genes have been removed during the strain construction process.

The inserted sequences are amplified by PCR and sequenced in order to check that no sequence of *E. coli* remains in the production strain.

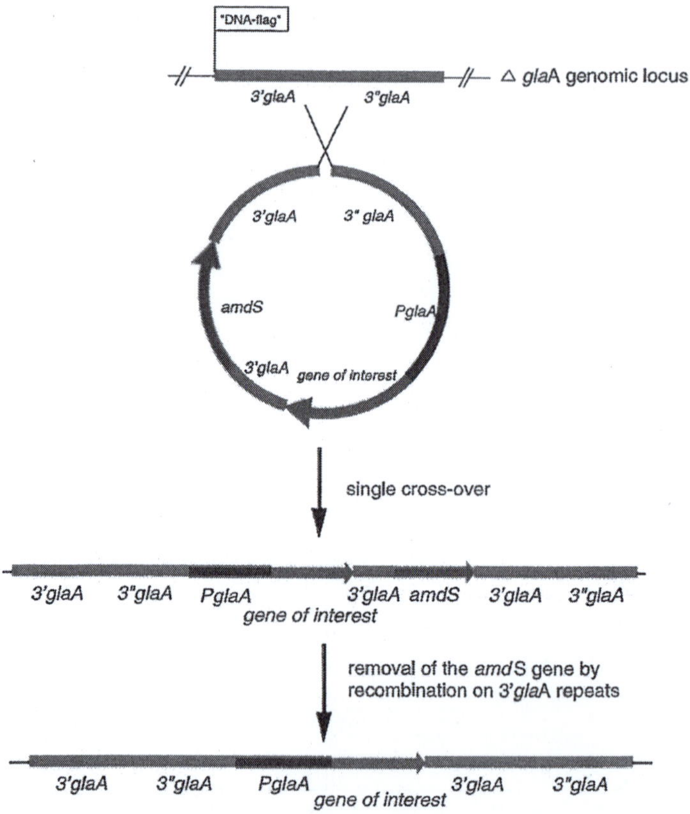

Figure 3.3 Example of the marker-gene free insertion of an expression unit. The expression unit, a linear piece of DNA, is integrated into one of the seven marked loci by homology of the 30 regions of the loci. By varying the conditions of transformation of the expression units multiple copies of the gene of interest arranged in tandem can be integrated in a marked locus. Current practice is that both the gene of interest and the amdS selection marker are on different expression units. So the transformation is carried out with a mixture of expression units carrying the gene of interest and expression units carrying the amdS marker. By selection on agar plates containing acetamide as sole carbon source the transformants are selected. By counter-selection on agar plates containing fluoro-acetate variants can be selected from these transformants which have lost the amdS marker but which have retained (multiple copies of) the gene of interest (van Dijck et al, 2003).

This approach was designed for the construction of enzymes used as food processing aids and was carried out following very strict safety assessment guidelines. The idea was to have strains produce enzymes which no longer needed to be tested on animals to demonstrate product safety. Irrespective and, at least in Europe, the present legislation still obliges companies to test the enzyme preparations in animals to evaluate for different types of toxicity (e.g., subchronical toxcity for 9 weeks). These tests must be completed prior to permission being granted for use in food preparations.

3.4.4 The revolution of CrispR-Cas9 editing for the construction of fungal strains that hyperproduce enzyme

The Clustered Regularly Interspaced Short Palindromic Repeats (CRISPR)/CRISPR-associated (Cas) systems, are systems containing *cas* genes and CRISPR array(s) being naturally derived from the adaptive immune system of prokaryotes. The *cas* genes are short sequences that originated from foreign genetic material, called spacers, interspaced with identical palindromic repeats (Jinek et al., 2012). The *cas* genes code proteins responsible for acquisition of new foreign sequences into the CRISPR array(s) as well as for disruption of exogenous DNA through the activity of Cas proteins bearing endonuclease activity, such as the Cas9, Cas12, or Cas 13 proteins (Mota et al., 2020). Over the past decade, the ability to reprogram Cas proteins by creating crRNAs (guide RNAs) designed to bind and/or cleave specific DNA or RNA sequences has made them useful in a variety of applications, including genome editing and modulation, specific control of gene expression and nucleic acid detection (Mota et al., 2020) (Fig. 3.4)

In the CRISPR/Cas9 system, the Cas9 protein forms a complex with two RNA molecules: CRISPR RNA (crRNA), encoded by the random spacers found in the CRISPR array(s), and transactivating CRISPR RNA (tracrRNA). It has been shown, that these two RNAs form a dual-tracrRNA:crRNA, which can also be designed as a single guide RNA (sgRNA) for genome editing purposes (Jinek et al., 2012). The Cas9-sgRNA complex is guided to a target DNA via homology of the crRNA to the protospacer region of the target sequence (Fig. 3.5).

The Cas9 nuclease subsequently binds the target DNA through interaction with the protospacer adjacent motif (PAM) immediately downstream of the protospacer sequence. If the crRNA part of the sgRNA sequence successfully pairs with the target DNA, the Cas9 nuclease will perform a double strand break (DSB) three nucleotides upstream of the PAM motif, which can be repaired. The simple design and construction of a single guide RNA for precise genome editing has rapidly expanded the toolbox of molecular methods, such as Zinc Finger Nucleases (ZFN) and Transcription Activator-Like Effector Nucleases (TALENs) (Kim and Kim, 2014).

CRISPR/Cas9 is a cost-efficient and simple platform to perform genetic manipulations with a single enzymatic activity guided by a predesigned

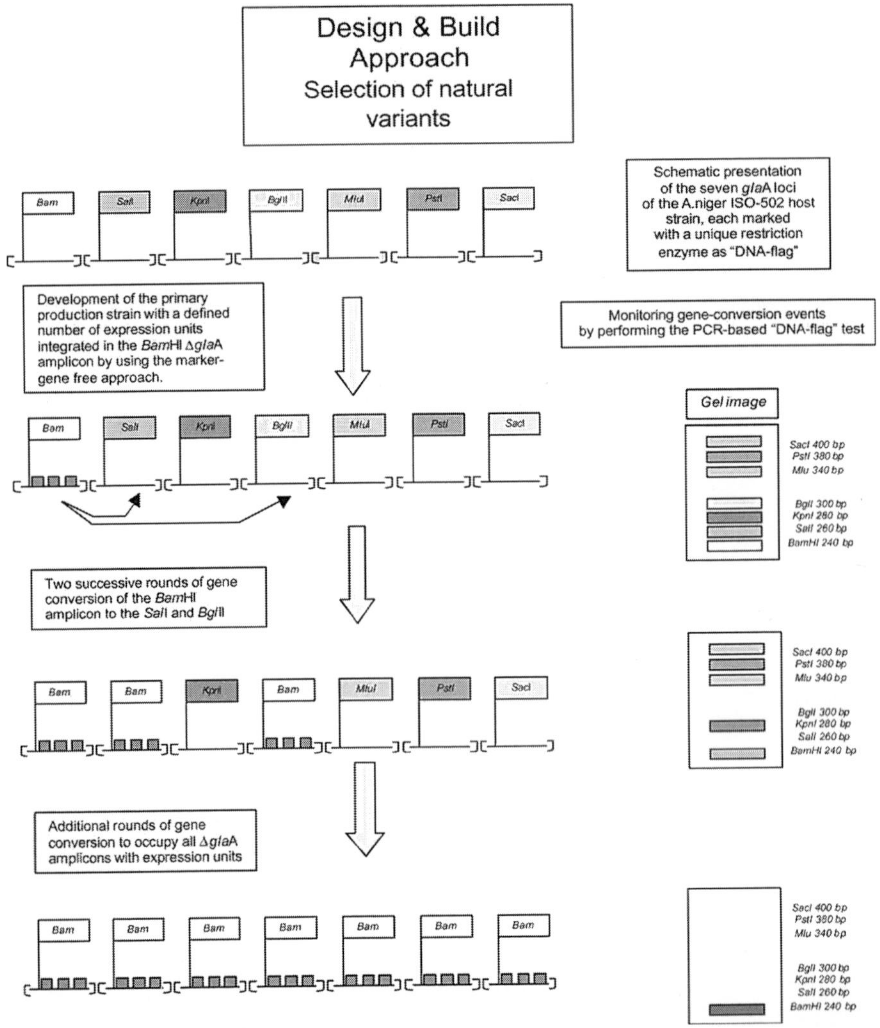

Figure 3.4 Selection of natural variants. The natural process of gene conversion is used to select for variants that have acquired a set of multiple gene copies of the gene of interest in any the remaining empty marked loci. The filling-up process in the several selection rounds can be followed by analysis in a gel electrophoretic system (van Dijck et al, 2003; Mota et al., 2020).

sgRNA molecule. Therefore, it has become a common genome editing method in a variety of organisms and has been reported to work in numerous filamentous fungi. In filamentous fungi, this system has been used to target genes involved in enzyme production (Liu et al., 2017a,b; Katayama

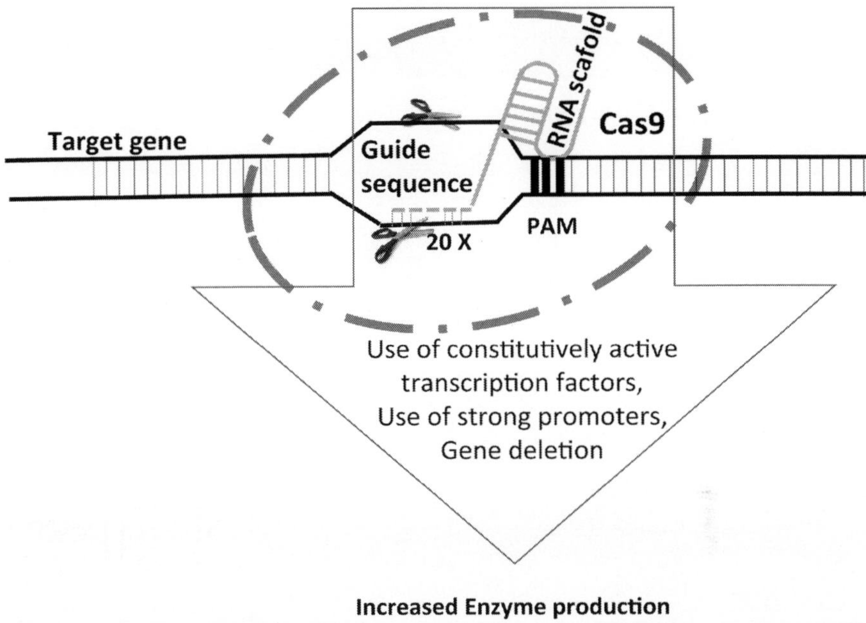

Figure 3.5 Utilization of the CRISPR/Cas9 system in filamentous fungi for increased enzyme production. The artificial sgRNA is composed of a guide sequence *(in gray dotted line)* and a scaffold sequence *(in grey solid line)*, which are the corresponding parts of the naturally occurring crRNA and tracrRNA, respectively. The sgRNA forms a complex with the Cas9 nuclease *(within the circle)*, which will be directed to a targeted locus of a gene of interest. In case there is sequence homology between the guide sequence and the target sequence upstream of a protospacer adjacent motif (PAM) *(Bold Black Steps)*, a double strand break (DSB) will be performed. The DSB can be repaired in a targeted manner using repair templates, which can result in precise gene editing (Kun et al., 2019).

et al., 2018; Liu et al., 2015). The system can also be used to efficiently change functional properties of the proteins encoded by the target gene by on site genomic mutation (e.g., in *Aspergillus niger*) (Kun et al., 2020).

Once the overproducing strain is obtained a validation of the strain stability is necessary. Strains with a high number of copies of a gene are often not very stable. It seems that there is an economy of the cell that leads to the removal or inactivation of non-compulsory genes. The high number of copies is not necessary and some are often deleted after a limited number of generations. Irrespective, it is important to set up a proper strain stock management of the efficient strain, in order to keep the best activity obtained.

3.5 Growth requirements of microorganisms

3.5.1 Media development

The knowledge of the strain growth requirements and regulation systems are prerequisites to controlling the over-expression of enzyme production. Sugar and nitrogen substrates are quantitatively used during the process. By-products of the food industry such as molasses, whey powder or corn steep liquors can be convenient, efficient and more affordable substrates than purer substrates such as pure sugars or protein hydrolysates (peptones or tryptones). Yeast extracts are also commonly used. However, the complexity of the medium can be a drawback for the future purification of the enzymatic activity at the end of the fermentation process.

Another substrate that should be provided in significant amounts is oxygen. The oxygen transfer from gas to microorganisms can be a limiting step especially when filamentous fungi are used. This will be discussed later.

3.5.2 Nutritional requirements

3.5.2.1 Effect of carbon sources

Apart from their influence on growth rates the availability of carbon sources has a significant metabolic regulation effect. It starts by the expression of the substrate uptake, but may also influence the pathway expression. Some of the enzyme activities are associated with growth and, in this case, easy to use sugars such as sucrose, fructose or maltose can be efficient for enzyme production. However, the availability of glucose, even if very efficient on cell growth is also one of the most repressive substrates for a large number of genes and as a consequence limits the production of many enzymes. So, the use of other sugars such as lactose or maltose or the use of glycerol can be a better choice to develop a full expression of the strain potentiality.

Arora et al. (2018) have shown the diversity of by-products that could be used for the production of enzymes by solid phase fermentation (see 3.5). Substrates as diverse as apple or tomato pomace, different types of bran, linseed oil cake, palm straw, citrus pulp or peel, etc. can be used and offer the advantage of being cheap substrates as well as being substrates easily used by fungi.

3.5.2.2 Effect of nitrogen sources

Nitrogen is often added in the mineral form with salts such as ammonium sulfate or ammonium nitrate. Organic nitrogen can also be used with simple substrates such as urea but peptone or tryptone of different proteins, even if they are significantly more expensive are also very efficient. Other efficient complex sources of nitrogen are used for industrial processes such as sugar slop or spent or corn steep liquor which are also rich in nitrogen. However, a complete analysis of the flow sheet costs should be done to take the right economical decision because the more complex the medium the more difficult and costly could be the purification process.

Yeast extract is providing nitrogen, but it is also providing a large diversity of vitamins especially B vitamins.

3.5.2.3 Effect of salts and oligoelements

Phosphate and sulfate sources are usually added to the culture medium as phosphorus and sulfur are significant constituent atoms of living cells.

Metal ions and halogens can be efficient biochemical mediators and necessary oligo-elements for efficient enzyme production (e.g., sodium, potassium, calcium, manganese, magnesium, iron, copper, molybdenum, cobalt, iodine, chloride,...) however some contaminant metals must be absent in enzyme preparations for food applications. It is the case for lead, cadmium, arsenic, mercury, and the concentration of these metals must be controlled regularly for these applications.

3.6 Process design and optimization

Automated liquid fermentation equipment available on the market nowadays, enables organizations to control temperature, pressure, pH and weight during industrial scale production.

Sensors are also available for biomass measurement but are mainly adapted for monitoring bacterial or yeast growth rather than filamentous fungi. Such sensors are not adapted to measuring the biomass of fungi because the fungi usually form pellets when the liquid medium is agitated, a step that is necessary for providing oxygen to the microorganisms. Special geometry of reactors called "air lift" are favoring oxygen transfer without increasing the shearing stress associated to high oxygen transfer with propellers (Fig. 3.6).

It can be interesting to produce enzymes by solid state fermentation (SSF). A recent review (Arora et al., 2018) showed the diversity of enzymes

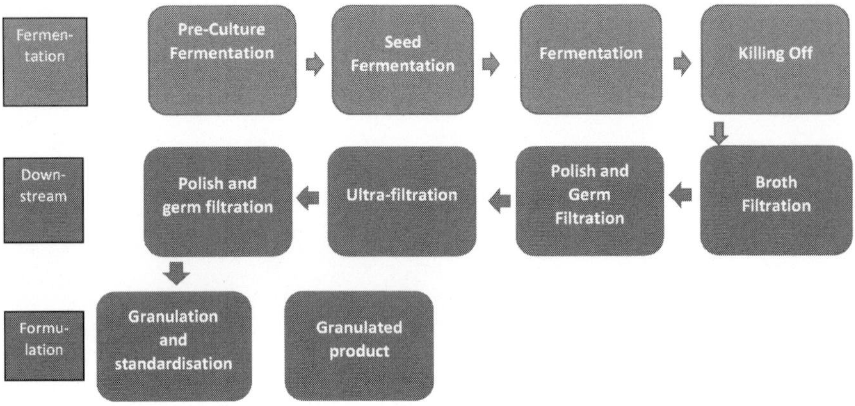

Figure 3.6 Example of a process sheet used for food enzyme production.

and efficiency when using such a method for producing enzymes. Different types of solid state bioreactors are used: tray bioreactors, packed bed bioreactor, air pressure pulsation bioreactors, and mixed SSF bioreactors. A diverse range of enzymes has been produced with such reactors. The best results were obtained for enzymes such as amylase, pullulanase, and xylanase.

In this case the enzymes are more concentrated in the solution that can be obtained by pressing the solid substrate (such as sugar beet pulps). However such equipment is less common than the liquid fermentation ones. The control of sterility, relative humidity and temperature can be an issue when scaled up for industrial production. However, this technique is used by several companies.

3.7 Downstream processes

3.7.1 Concentration and purification of enzymes

At the end of fermentation, the purification of enzymes is facilitated by the titer and can be hindered by the complexity of the medium. Most often enzymes in the culture supernatant are concentrated by tangential ultrafiltration and then precipitated using salts such as ammonium sulfate. It is followed by a desalination process, using membrane techniques.

As for the use in food, the regulations stipulate that there are maximum permissible limits for heavy metals and there must be an absence of pathogenic micro-organisms (such as *Salmonella*, *Staphylococcus aureus*, sulfito-reducting *Clostridiae*,...).

3.7.2 Product formulation and standardization

The activity of the concentrated form is measured in order to standardize the number of enzyme units per pack. Some enzymes are sensitive to oxidation and they should be protected by antioxidants such as sulfites or ascorbic acid. They are often mixed with carriers agents such as dry starch or dextrines and granulated in order to facilitate their use.

3.8 Conclusion

The screening methods are still prerequisites for selecting overproducing strains but the recent progress in genome knowledge and genome expression have made genetic engineering and automated protein engineering easier. It has contributed to the improvement of specific qualities and quantities of enzymes produced including producing them at reasonable prices. The revolutionary CRISPR/Cas gene technology also helps these operations, however in some cases the insertion sites are not as specific as expected.

One of the critical steps is to keep the built strain efficiency for a long duration. The number of copies of the interesting gene can be deleted when the number of generations become high. As far as we know the systems regulating the excision of genes with low interest for the producing cells are not completely elucidated and seem to be difficult to inactivate.

The production of enzymes in liquid medium are now quite controlled. However, production of enzymes by filamentous fungi, requires large amounts of oxygen. Oxygen transfer presents technical problems such as shear stress damaging the mycelia and energy cost that can be critical at large scale. New techniques of solid state fermentation of agro industry by-products are now at an industrial scale. Even if they are more difficult to control than liquid fermentation, obtaining high enzyme concentrations is possible. Moreover, solid state fermentations save water and has shown to reduce sterilization, downstream processes, and water treatment costs.

Downstream processes can be very specific depending on the type of enzyme and its sensitivity to environmental factors. Levels of purity is sometime a significant factor for some applications and secondary enzymatic activities present in the preparation can be a problem.

The simultaneous development of these techniques has helped to develop a large number of applications first for cleaning powders, then for food and feed applications, and more recently for the lignocellulose breakdown and the production of specific chemicals.

References

Archer, D.B., Peberdy, J.F., 1997. The molecular biology of secreted enzyme production by fungi. Crit. Rev. Biotechnol. 17, 273–306.

Arora, S., Rani, R., Ghosh, S., 2018. Bioreactors in solid state fermentation technology: Design, applications and engineering aspects. J. Biotechnol. 269, 16–34.

Baker, S.E., 2018. Protein hyperproduction in fungi by design. Appl. Microbiol. Biotechnol. 102, 8621–8628.

Cai, D., Rao, Y., Zhan, Y., Wang, Q., Chen, S., 2019. Engineering *Bacillus* for efficient production of heterologous protein: current progress, challenge and prospect. J. Appl. Microbiol. 126, 1632–1642.

Fiedler, M.R.M., Barthel, L., Kubisch, C., Nai, C., Meyer, V., 2018. Construction of an improved *Aspergillus niger* platform for enhanced glucoamylase secretion. Microb Cell Fact. 17, 95.

Goddard, J.-P., Reymond, J.-L., 2004. Recent advances in enzyme assays. Trends Biotechnol. 22, 363–370.

Jinek, M., Chylinski, K., Fonfara, I., Hauer, M., Doudna, J.A., Charpentier, E., 2012. A programmable dual-RNA-guided DNA endonuclease in adaptive bacterial immunity. Sci. 337, 816–821.

Katayama, T., Nakamura, H., Zhang, Y., Pascal, A., Fujii, W., Maruyama, J., 2018. Forced recycling of an AMA1-based genome-editing plasmid allows for efficient multiple gene deletion/integration in the industrial filamentous fungus *Aspergillus oryzae*. Appl. Environ. Microbiol. 85 (3), 1–16, e01896-18.

Kim, H., Kim, J.-S., 2014. A guide to genome engineering with programmable nucleases. Nat. Rev. Genet. 15, 321–334.

Kun, R.S., Gomes, A.C.S., Hilden, K.S., Cerezo, S.S., Makela, M.R., de Vries, R.P., 2019. Developments and opportunities in fungal strain engineering for the production of novel enzymes and enzyme cocktails for plant biomass degradation. Biotechnol. Adv. 37, 1–16, (Doi: 10.1016/j.biotechadv.2019.02.017).

Kun, R.S., Meng, J., Salazar-Cerezo, S., Makela, M.R., de Vries, R.P., Garrigues, S., 2020. CRISPR/Cas9 facilitates rapid generation of constitutive forms of transcription factors in *Aspergillus niger* through specific on-site genomic mutations resulting in increased saccharification of plant biomass. Enzyme Microb. Technol. 136, 109508. doi: 10.1016/j.enzmictec.2020.109508.

Liu, R., Chen, L., Jiang, Y., Zhou, Z., Zou, G., 2015. Efficient genome editing in filamentous fungus *Trichoderma reesei* using the CRISPR/Cas9 system. Cell Discov. 1, 15007–15017, (Doi: 10.1038/celldisc.2015.7).

Liu, R., Chen, L., Jiang, Y., Zou, G., Zhou, Z., 2017b. A novel transcription factor specifically regulates GH11 xylanase genes in *Trichoderma reesei*. Biotechnol. Biofuels. 10, 194, (Doi: 10.1186/s13068-017-0878-x).

Liu, Q., Gao, R., Li, J., Lin, L., Zhao, J., Sun, W., Tian, C., 2017a. Development of a genome-editing CRISPR/Cas9 system in thermophilic fungal *Myceliophthora* species and its application to hyper-cellulase production strain engineering. Biotechnol. Biofuels. 10 (1), (Doi: 10.1186/s13068-016-0693-9).

Mota, D.S., Marques, J.M., Guimarães, J.M., Mariúba, L.A.M., 2020. Research article CRISPR/Cas Class 2 systems and their applications in biotechnological processes. Genet. Mol. Res. 19 (1), gmr18478. doi: 10.4238/gmr18478.

Selten, G.C.M., van Gorcom, R.F.M., Swinkels, B.W., 1995. Selection marker gene free recombinant strains: a method for obtaining them and the use of these strains. *Eur. Patent Appl.* EP 00635574 A1.

van Dijck, P.W.M., Selten, G.C.M., Hempenius, R.A., 2003. On the safety of a new generation of DSM *Aspergillus niger* enzyme production strains. Reg. Toxicol. Pharmacol. 38, 27–35.

CHAPTER FOUR

Recovery and purification of industrial enzymes

Sumit Kumar, Sunil K. Khare
Indian Institute of Technology, Delhi, India

4.1 Introduction

Enzymes are the catalysts of biological origin, having the remarkable property of high-catalytic power and substrate/reaction specificity. They have the upper hand because catalysis involving them leads to an efficient generation of pure stereoisomers, fewer side reactions, easier separation of products, and lower environmental burden. Because of their unique properties, enzymes found wide-scale applications in industrial sectors, namely agriculture, environment, medical, energy, food, and nutrition (Benbelgacem et al., 2019; Tanokura et al., 2015; Vingiani et al., 2019). The above applications require the enzyme in a purified form for which purification is carried out to get rid of impurities. Varied applications of enzymes require different degree of purified forms; for example, samples for kinetic characterization can be of less purity, while for the X-ray crystallization study highest degree of enzyme purity is required.

The cost of enzymes is a major factor influencing their industrial applications, making the price of production a prime concern. In view of the critical role that downstream processing plays in the overall cost of the enzyme/protein, biochemists are slowly becoming aware of the need for the development of purification protocols that are less costly, more efficient, and easily scaleable (Nadar et al., 2017). The emphasis is shifted to separation protocols with less number of steps to minimize product loss, thus making the whole protocol more efficient. Usually, to purify enzymes from fermentation broth or cell lysate, a preliminary fractionation step such as ultrafiltration, ammonium sulfate, and acetone precipitation is employed. Further, a combination of chromatography steps, such as affinity, ion-exchange, hydrophobic, size-exclusion chromatography, etc., are employed to get pure preparations. The enzyme and microbial biochemistry laboratory has been engrossed to purify a range of industrially important enzymes. These enzymes were mainly sourced from

Enzymes: Novel Biotechnological Approaches for the Food Industry.
http://dx.doi.org/10.1016/B978-0-12-800217-9.00004-6

extremophiles and are termed extremozymes for their capability to carry out catalytic activities under harsh conditions (Dumorné et al., 2017; Raddadi et al., 2015; Suriya et al., 2016). Extremophiles are a class of organisms that can thrive in environments considered unsuitable for the existence of life. They possess an inherent ability to function and carry out metabolic activities under extreme conditions (Brininger et al., 2018; Saralov, 2019). We primarily worked on enzymes from two groups of extremophilic microbes, that is, solvent-tolerant and halophiles. Halophiles, a class among extremophiles, have the unique ability to thrive in environments rich in high salt concentrations as their metabolic and physiological activities are adapted to function under such conditions (Ghosh et al., 2019; Ma et al., 2010; Oren, 2002; Yin et al., 2015). Their enzymes are deemed suitable for diverse industrial applications such as peptide synthesis and biofuel production (Białkowska et al., 2017; DasSarma and DasSarma, 2015; Karan et al., 2012).

In the present book chapter, an overview of purification processes employed to purify industrial enzymes from extremophiles isolated by us are presented. In general, combinations of chromatographic processes have been utilized to get a pure form of the enzyme. While in other cases, a single-step purification process or nonconventional method such as three-phase partitioning worked well.

4.2 Proteases

Proteases represent one of the major groups of industrial enzymes, accounting for approximately 60% of the total worldwide enzyme sales. The proteolytic enzymes hydrolyze the peptide links of proteins and peptides to form smaller peptides and amino acids. Their application encompasses a wide range of bioprocesses, namely baking, brewing, detergents, leather processing, pharmaceuticals, meat tenderization, cosmetics, peptide synthesis, medical diagnosis, and silver recovery (Contesini et al., 2018; da Silva, 2017; Gupta et al., 2002a,b; Gurumallesh et al., 2019). The choice of protease as target enzyme is based on the fact that (1) these are the largest selling industrial enzymes, (2) the demand for proteases functional at high salt concentrations and alkaline pH is increasing for detergent application, and (3) the requirement of organic solvents stable protease is being foreseen for applications in peptide synthesis (Gupta and Khare, 2009).

Keeping in mind the importance of organic solvent-tolerant protease, a protease from *Pseudomonas aeruginosa* was purified and characterized. The protease was purified by a combination of ion exchange and hydrophobic interaction chromatography using the Q-Sepharose and Phenyl Sepharose

6 Fast Flow matrix, respectively. The process led to 11.6-fold protease purification with a 60% recovery. The protease was stable in a range of organic solvents having a log P value above 3.2 for at least up to 72 h. Added to this; it was also stable and active in an array of detergents and surfactants (Gupta et al., 2005a,b). In another study, a polyextremophilic protease from *Pseudomonas putida* SKG-1 was purified 21.4-folds with 57.2% recovery by ultrafiltration using a 50 kDa membrane and Phenyl Sepharose 6 Fast Flow column hydrophobic interaction chromatography. The protein was remarkably solvent-, detergent-, thermo-, and alkali-stable and can find prospective biotechnological applications (Singh et al., 2013).

We have been extensively involved in research on halophilic enzymes. The purification of halophilic enzymes has been tedious to achieve as they tend to lose their activity during the course of purification if salt concentration is reduced. Their native structure is lost in the absence of high salt because they need salt concentration matching their native environment for maintaining structural integrity. Incorporation of high salt during the purification often disturbs chromatographic procedures/separation processes. Hence, specific protocols need to be developed for purification of the halophilic enzyme. Interestingly, hydrophobic interaction chromatography on Phenyl Sepharose 6 Fast Flow matrix has proved an ideal matrix for purification of haloalkaline protease sourced from halophiles. A salt- and solvent-stable protease from *Geomicrobium* sp. EMB2 was purified 22.6-fold with 51% yield by a combination of ultrafiltration and Phenyl Sepharose 6 Fast Flow hydrophobic chromatography. The 38 kDa protease was most active at 50°C and stable in the pH range 6.0–12.0. Its unique feature of salt, detergent, surfactant, and solvent (log $P > 2$) stability could be of potential biotechnological interest (Karan and Khare, 2010). In other studies, also haloalkaliphilic proteases from *Bacillus* sp. and haloalkaliphilic bacterium sp. AH-6 were purified by single-step purification employing Phenyl Sepharose 6 Fast Flow column chromatography. *Bacillus* sp. protease was purified 10-fold with an 82% yield and was marked for detergent application (Gupta et al., 2005a,b). Whereas, the protease from haloalkaliphilic bacterium sp. AH-6 was purified 23.42-fold with a 28% yield and was salt and pH stable (Dodia et al., 2008). Another protease with industrially important properties from halophilic *Bacillus* sp. EMB9 was purified 489-fold with a 2.8% recovery by 80% ammonium sulfate precipitation and CM cellulose chromatography. The protease required Ca^{2+} ion to maintain structure, while Na^+ and K^+ ions were needed for activity and assisted in refolding. It was stable in a range of surfactants as well as compatible with commercial detergents (Sinha and Khare, 2013). Ultrafiltration, DEAE ion exchange, and Sephadex

G-75 gel filtration chromatography was employed to purify protease from *Virgibacillus* sp. EMB13. The protease was organic solvent stable and can be useful for non-aqueous synthesis. Additionally, on account of compatibility with commercial detergents and the ability to remove blood stain, *Virgibacillus* protease was potentiated for additive in detergent formulations (Sinha and Khare, 2012). In short, hydrophobic chromatography on Phenyl Sepharose 6 Fast Flow matrix was reasonably efficient for purification of organic solvent-tolerant and halophilic proteases, while for others, a combination of chromatographic procedures worked well.

4.3 Lipases

Lipases are one of the prominent enzymes because of their novel activities and multifold biotechnological applications. The diverse sectors having applicability of lipases are detergent, textile, food, pharmaceutical, cosmetics, and biorefinery. They catalyze three major reactions: hydrolysis, esterification, and transesterification (Gaur and Khare, 2011a,b; Rodrigues et al., 2019). The latter two reactions predominate if the nonaqueous medium and keeping this in mind a solvent-tolerant lipase from *P. aeruginosa* was investigated in my laboratory. Following initial ultrafiltration through the 30 kDa molecular weight cut-off (MWCO) membrane, the lipase was purified by a single-step by Sephadex G-100 gel exclusion chromatography. The 60 kDa protein was purified 8.6-fold with 51.6% yield and was stable in a range of polar and nonpolar solvents such as ethanol, benzene, hexane, decane, and tetradecane. Lipase exhibited substrate preference toward longer carbon chain (C16) fatty acid ester over the shorter ones and showed random positional specificity for triolein hydrolysis. Solvent and surfactant stability of the lipase mark it for possible application in non-aqueous systems and detergent formulations (Gaur et al., 2008). Additionally, this lipase was used for palm oil hydrolysis in an isooctane-aqueous biphasic system. The optimization of conditions by response surface methodology (RSM) led to 89% palm oil hydrolysis in 2 h at 30°C (Gaur and Khare, 2011b). This lipase, when produced under solid-state fermentation (SSF) could also be purified following the above purification procedure. Structural studies using circular dichroism and intrinsic fluorescence spectra revealed that the secondary and tertiary structure of lipase remains intact in organic solvents (Joshi and Khare, 2013).

In another research with the objective to bioremediate waste cooking oil, a lipase producing *Penicillium chrysogenum* strain was isolated from waste grease. The lipase was purified by 90% ammonium sulfate precipitation followed by DEAE-cellulose ion-exchange chromatography (Kumar et al., 2012a,b). Further work on this lipase revealed that this was stable in a wide range of polar and nonpolar organic solvents except ethyl butyrate at 50% (v/v) concentrations for 72 h. Structural studies were carried out on this lipase to get the perspective for solvent stability. The gained knowledge from this system could be used for rational engineering of other lipases for synthesis application in non-aqueous media (Sadaf et al., 2018a,b).

While decoding properties of lipases from halophilic bacteria a halo-alkaline lipase from *Marinobacter* sp. EMB5 lipase (LipEMB5) was found to form catalytically active aggregates. The aggregate form of lipase was purified by ultrafiltration (10 kDa) and Sephacryl S200 gel filtration chromatography. It was surfactant and organic solvent stable. The property of solvent stability was correlated with secondary and tertiary structural studies (Hemamalini and Khare, 2016). The aggregated form of LipEMB5 (172.35 kDa) could be disaggregated by the addition of 70% 2-propanol. The catalytically active aggregate form was 24 times more active with 1000-fold less *K*m compared to the disaggregated form (Hemamalini and Khare, 2018).

4.4 α-Amylase

α-Amylase is another important industrial enzyme that catalyzes the hydrolysis of α-1,4-glycosidic linkages in starch, and related polysaccharides. It plays a central role in carbohydrate metabolism, and on account of this, it is ubiquitously present among microbes, plants, and animals. α-Amylase ranks among the largest selling industrial enzymes and has found potential applications in many industrial processes such as starch hydrolysis, biofuel production, and the paper industry. It is also used in the pharmaceutical industry as a digestive aid. Demand for α-amylase is on the continuous rise due to their versatile use for diverse industrial processes in food, textile, detergent, analytical chemistry, fine-chemicals, pharma, and beverage industries (Janecek et al., 2014; Kumar et al., 2016a,b; van der Maarel et al., 2002). Our laboratory has worked on α-amylase sourced from halophilic microorganisms. Halophilic amylases have an advantage over mesophilic one as they exhibit activity and stability in the presence of varying concentrations of salt. Amylases from halophiles have been least explored, although they offer the benefit of working better toward saline samples. It is interesting to mention that many halophilic amylases exhibit polyextremophilic properties of industrial relevance. Some common properties encountered in this class of amylases are stability in alkaline pH, high temperature, and low water activity (presence of organic solvents) conditions (Kumar et al., 2016a,b).

Keeping in mind the above advantages of amylases, especially the halophilic one, the α-amylase from moderately halophilic *Marinobacter* sp. EMB8 was studied. The moderately halophilic amylase producing isolate of *Marinobacter* sp. EMB8 was isolated from the seawater of Kozhikode, Kerala, India (Kumar et al., 2012a). The colony characteristics, cell morphology, amylase activity, and purification method employed for amylase are depicted in Fig. 4.1. The isolate was rod-shaped, and the amylase activity was checked by the zone of hydrolysis on starch agar plate as well as produc-

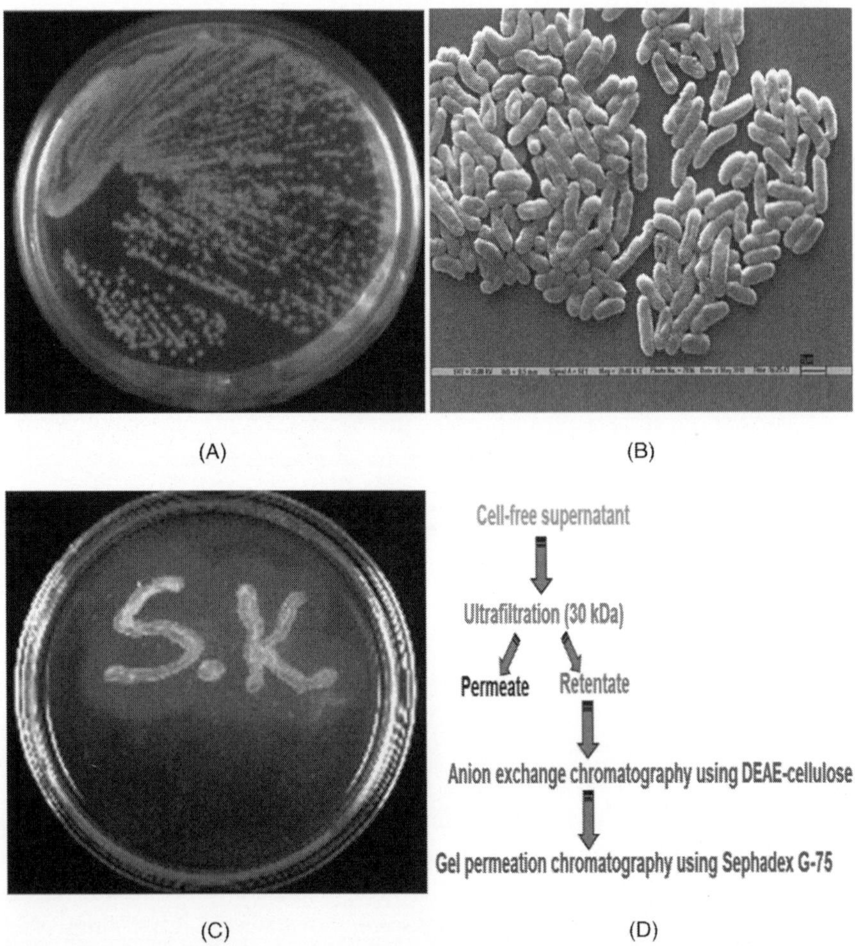

Figure 4.1 Amylase producer moderately halophilic *Marinobacter* sp. EMB8: (A) isolate EMB8 streaked on nutrient agar plate containing 5% (w/v) NaCl (B) scanning electron microscopy (SEM) of *Marinobacter* sp. EMB8 (scale bar 1 μm with 20000x magnification) (C) amylase activity (zone formation on starch agar plate containing 5% (w/v) NaCl) (D) procedure employed for α-amylase purification.

tion in a liquid nutrient broth medium containing starch. The amylase was purified to homogeneity by ultrafiltration, DEAE-cellulose, and Sephadex G-75 column chromatography with a 52% yield and 76-fold purification. It exhibited a temperature optimum at 45°C and a pH optimum at 7.0 and had many novel features, namely stability up to 20% (w/v) NaCl, 80°C temperature, and 6.0–11.0 pH range. α-Amylase in the present study was stable in 25% (v/v) concentration of various organic solvents up to 24 h. Solvent stability has been observed as a generic and unique trait among the halophilic enzymes. This is attributed to the fact that they work at high salt surroundings, which is akin to low water medium. *Marinobacter* α-amylase produces maltotriose and maltotetraose upon starch hydrolysis. The advantage of the solvent stability of amylase was harvested to enhance maltotetraose yield during starch hydrolysis. The starch hydrolysis carried out in the presence of 10% (v/v) ethanol led to a three time increase in maltotetraose concentration (Kumar and Khare, 2012). Further, structural studies using circular dichroism (CD) and fluorescence spectroscopy were carried out to decipher the basis of salt and solvent stability of *Marinobacter* amylase. Salt was necessary for optimum catalysis and structure stabilization as, in its absence, tertiary structure was lost. The α-amylase was active in the presence of nonpolar solvents, namely hexane and decane but inactivated by ethanol. The structural elucidation correlated well with the loss of tertiary structure in the presence of ethanol while retained in hexane and decane (Kumar et al., 2016b). *Marinobacter* amylase also showed the property of chloride ion activation commonly encountered in eukaryotic amylases. Phylogenetic analysis revealed that the amylase was closer to eukaryotes such as humans and pigs and might have been acquired by *Marinobacter* through horizontal gene transfer (Kumar and Khare, 2015).

4.5 Purification of other potential enzymes

4.5.1 *Glucoamylase and β-amylase*

Affinity precipitation using alginate was found to be an effective purification method for carbohydrate-active enzymes. This works as a single-step procedure in which the enzyme is precipitated by alginate as macroaffinity ligand and afterwards eluted with 1.0 M maltose. Glucoamylase from *Aspergillus niger* was purified employing this process to obtain a 28-fold purification with an 81% yield. This protocol also worked well to purify *Bacillus amyloliquefaciens* glucoamylase. The *Bacillus* glucoamylase was purified efficiently with 38-fold purification and 78% recovery (Teotia et al., 2001a,b). β-Amylase having industrial relevance in maltose syrup production was also purified by this process. Sweet potato β-amylase was purified by affinity precipitation with alginate with 44-fold purification and 80% activity yield. The purified β-amylase by this process could safely be used in food industries (Teotia et al., 2001b).

4.5.2 *Xylanase*

Xylanases are another important industrial enzyme that hydrolyzes xylan a plant cell wall polysaccharide. Xylanases are used in pulp and paper bioleaching and food industries (Walia et al., 2017). The xylanase from *Melanocarpus albomyces* was purified in a single step by Phenyl Sepharose 6 Fast Flow hydrophobic interaction chromatography. Purification led to the recovery of two major isoenzymes having a molecular mass of 18.6 and 25.4 kDa. Ethylene glycol enhanced the thermal stability of this xylanase as, in its presence, thermal stability was improved from 2.2 h to almost 40 h at 55°C (Gupta et al., 2002).

4.5.3 *Aminopeptidase*

Studies to carry out purification and characterization of the aminopeptidase from *P. aeruginosa* PseA was also carried out in my laboratory. Aminopeptidases are exopeptidases that catalyze the hydrolysis of amino acids from the N-terminal of peptides and proteins. They are sought after in the food and pharmaceutical industries (Nandan and Nampoothiri, 2017). Research on PseA aminopeptidase was carried out with the viewpoint that solvent-tolerant aminopeptidases could be useful for peptide/ester synthesis in a non-aqueous medium. The 56 kDa enzyme was purified by Q-Sepharose anion-exchange chromatography with 11.9-fold purification and 38% recovery. The purified preparation showed noteworthy stability in hydrophilic and hydrophobic solvents such as dimethyl sulfoxide (DMSO), methanol, cyclohexane, hexane, decane, tetradecane, etc. Cloning and analysis of the aminopeptidase gene revealed that 536 amino acids long peptide is encoded by a 1611-bp open reading frame. Conserved domain search revealed that PseA aminopeptidase belonged to the zinc-metallopeptidase family. The presence of a large excess of hydrophobic amino acid residues on the surface of the enzyme was thought to impart it the property of solvent-stability (Gaur et al., 2010).

4.5.4 *β-Lactamase*

β-Lactamases enzyme that confers antimicrobial resistance (AMR) to microorganisms was also purified and characterized by our research group. The β-lactam ring containing antibiotics such as penicillins, cephalosporins, monobactams, and carbapenems are hydrolyzed and rendered inactive by this class of enzyme. Consequently, microorganisms producing this enzyme are resistant to a β-lactam ring containing antibiotics (Khan et al., 2017; Pratt, 2016). It is thought that a large excess of antibiotics present in the ecosystem due to its rampant use, imparts antibiotic resistance to microbes. In this scenario, β-lactamases can come handy to degrade antibiotics present in the environment. The β-lactamase of *Bacillus cereus* EMB20 was purified by ultrafiltration through a 10 kDa MWCO membrane followed by three-phase partitioning (TPP), and Sephadex G-100 gel filtration chromatography. The 43 kDa protein was purified 30-fold with a 15% recovery. Ampicillin was found to be the best substrate for it. As generally encountered among lactamases, this enzyme was not a metalloenzyme but was activated by Mg^{2+} and Mn^{2+} ions. The lactamase was stable in the presence of inhibitors and denaturing reagents such as urea and guanidinium chloride. The improved activity of EMB20 lactamase was correlated with structural studies using CD and fluorescence spectroscopy (Sadaf et al., 2018a,b).

4.6 Application of purified enzyme

The purified enzymes have been used for various applications as summarized in Table 4.1. The protease of *P. aeruginosa* PseA was stable in organic solvents, detergent, and surfactants. The unique stability characteristics of this enzyme mark its applications in detergent formulations and peptide synthesis (Gupta et al., 2005). *Bacillus* EMB9 protease was marked for application in detergents as it exhibited effective proteinaceous blood stain removal (Sinha and Khare, 2013). It was also utilized to efficiently hydrolyze protein-laden defatted soy flour and for complete recovery of silver from used/waste X-ray films (Sinha et al., 2014). In general, proteases studied by us were surfactant and detergent stable and were potentiated for use in the detergent industry.

PseA lipase was utilized for palm oil hydrolysis and was potentiated for applications in bioconversion and bioremediation of oil-rich industrial waste (Gaur and Khare, 2011b). Additionally, the solvent stability of this lipase was harnessed for the efficient synthesis of ethyl butyrate in the non-aqueous hexane medium (Joshi and Khare, 2013). *P. chrysogenum* SNP5 lipase was utilized for bioremediation of waste cooking oil (Kumar et al., 2012). The *P. chrysogenum* strain was also effectively used to remediate waste grease with the concomitant production of economically useful fatty acids (Kumari et al., 2017). *Marinobacter* α-amylase was producing maltotriose-

Table 4.1 Purification strategies and characteristics of industrial enzymes from extremophilic sources.

Microorganism/ Source	Purification procedure	Fold purification and recovery (%)	Characteristics	Additional properties	Reference(s)
(A) Protease					
P. aeruginosa PseA	Q-Sepharose ion exchange and Phenyl Sepharose 6 Fast Flow hydrophobic interaction chromatography	11.6; 60	Mw 35 kDa; pHopt 8.0; Topt 60°C; pH stability 6.0–9.0; *Km* 2.7 mg/mL and *Vmax* 3.0 μmol/min	Stable in detergents, surfactants, and solvents	Gupta et al., 2005a
Bacillus sp.	Single step Phenyl Sepharose 6 Fast Flow hydrophobic interaction chromatography	10; 82	Mw 29 kDa; pHopt 10.0–11.0; Topt 37°C; pH stability 8.5–12.0; *Km* 2 mg/mL and *Vmax* 289.8 μg/min	Stable in detergents and surfactants	Gupta et al., 2005b
Geomicrobium sp. EMB2	Ultrafiltration and Phenyl Sepharose 6 Fast Flow hydrophobic chromatography	22.6; 51.2	Mw 38 kDa; pHopt 10.0; Topt 50°C; pH stability 6.0–12.0; *Km* 0.10 mM and *Vmax* 526 U/min	Stable in salt, detergent, surfactant and solvent	Karan and Khare (2010)
Bacillus sp. EMB9	Ammonium sulfate precipitation and CM cellulose chromatography	489; 2.8	Mw 29 kDa; pHopt 9.0; Topt 55°C; pH stability 5.0–12.0; $t_{1/2}$ 190 min at 60°C; *Km* 2.22 mg/ mL, and *Vmax* 1111.11 U/mL	Bloodstain removal; detergent application	Sinha and Khare (2013)

(B) Lipase					
P. aeruginosa PseA	Ultrafiltration and Sephadex G-100 gel exclusion chromatography	8.6; 51.6	Mw 60 kDa; pHopt 8.0; Topt 40°C; pH stability 6.5–8.5; *Km* 70.4 mM and *Vmax* 2.24 μmol/(min. mg) toward p-nitrophenyl palmitate (pNPP)	Solvent and surfactant stable	Gaur et al. (2008)
P. chrysogenum SNP5	Ammonium sulfate precipitation and DEAE-cellulose ion-exchange chromatography	26.28; 39.6	Mw 40 kDa; pHopt 8.0; pH stability 7.0–9.0; *Km* 0.4 mM and *Vmax* 47.61 U/mL toward pNPP	Waste cooking oil bioremediation	Kumar et al., 2012b
Marinobacter sp. EMB5	Ultrafiltration and Sephacryl S200 gel exclusion chromatography	28.2; 5.6	Mw 172 kDa (aggregate); pHopt 9.0; Topt 50°C; pH stability 8.0–10.0; *Km* 1.43 μM and *Vmax* 1.43 μmoles/mL/min toward p-nitrophenyl octanoate (PNPO)	Solvent and surfactant stable	Hemamalini and Khare (2016)
(C) Other Enzymes					
α-Amylase					
Marinobacter sp. EMB8	Ultrafiltration, DEAE-cellulose and Sephadex G-75 column chromatography	76; 52	Mw 72 kDa; pHopt 7.0; Topt 45°C; pH stability 6.0–11.0; *Km* 4.6 mg/mL and *Vmax* 1.3 mg/min/mL toward soluble starch	Maltooligosaccharides synthesis; salt and solvent stable	Kumar and Khare (2012)

(*Continued*)

Table 4.1 Purification strategies and characteristics of industrial enzymes from extremophilic sources. (*Cont.*)

Microorganism/ Source	Purification procedure	Fold purification and recovery (%)	Characteristics	Additional properties	Reference(s)
β-Amylase					
Sweet potato	Affinity precipitation with alginate	44; 80	Mw 50 kDa for the single subunit of tetrameric protein	Potential application in food industries	Teotia et al., 2001a
Xylanase					
M. albomyces	Single-step Phenyl Sepharose 6 Fast Flow hydrophobic interaction chromatography	7.3; 61	Mw 18.6 and 25.4 kDa for two major isozymes	Thermal stability enhanced in the presence of ethylene glycol	Gupta et al., 2002a
Aminopeptidase					
P. aeruginosa PseA	Q-Sepharose anion-exchange chromatography	11.9; 38	Mw 56 kDa; pHopt 8.0; Topt 60°C; pH stability 6.0–8.0; *Km* 3.02 mM and *Vmax* 6.71 μmol/mg/min toward L-Leu-*p*-nitroanilide	Solvent- and thermal-stable; application in nonaqueous enzymology	Gaur et al. (2010)
β-Lactamase					
B. cereus EMB20	Ultrafiltration, TPP, and Sephadex G-100 gel filtration chromatography	30; 15	Mw 43 kDa; pHopt 7.0; Topt 35°C; pH stability 7.0–11.0; *Km* 150.0 μM and *Vmax* 66.0 μM/mL/min toward ampicillin	Stable in inhibitors and denaturing agents	Sadaf et al., 2018b

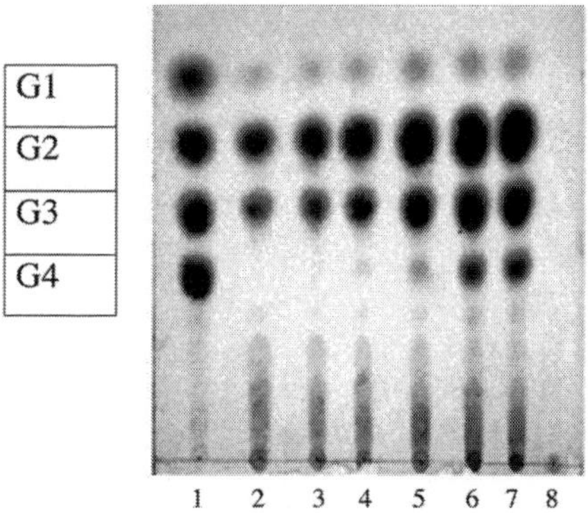

Figure 4.2 Thin layer chromatogram of the maltooligosaccharides formed from starch hydrolysis by *Marinobacter* sp. α-amylase. Lane 1, maltooligosaccharide standards; G1-Glucose, G2-Maltose, G3-Maltotriose, G4-Maltotetraose; Lane 2, 1% starch; Lane 3, 2% starch; Lane 4, 3% starch; Lane 5, 5% starch; Lane 6, 8% starch; Lane 7, 10% starch; Lane 8, starch.

and maltotetraose-rich maltooligosaccharides (MOS) upon starch hydrolysis. Maltotriose and maltotetraose-rich MOS is more desirable in bread-making industries due to their better antistaling properties thus, this enzyme was marked for use in bread industries. The thermal stability of this amylase could also be exploited for starch hydrolysis. It can hydrolyze a high concentration of starch and formed maltose, maltotriose, and maltotetraose as major hydrolysis products (Fig. 4.2) (Kumar and Khare, 2012). β-Lactamase studied by us can also be made use to biodegrade antibiotics present in the environment. As a generic feature, the enzymes from the solvent-tolerant and halophile class of extremophiles were stable in organic solvents and can be employed in synthesis in nonaqueous enzymology.

4.7 Conclusions

The book chapter presents the studies carried out for purification and recovery of industrial enzymes from extremophilic sources mainly from solvent-tolerant and halophilic microbes. A range of industrially useful enzymes such as proteases, lipase, amylase, aminopeptidase, xylanase, β-lactamase, etc., were purified and characterized. Though, a single purification protocol cannot work out for all enzymes; however, a combination of chromatographic procedures preceded by preliminary fractionation steps such as ultrafiltration and ammonium sulfate can be tried for a successful recovery. Phenyl Sepharose 6 Fast Flow hydrophobic interaction chromatography has worked well during the purification of extremophilic enzymes and in some cases, enzymes were purified in single step employing this chromatography. It is thought that solvent-tolerant and halophilic enzymes having hydrophobic patches on their protein surface can be effectively purified utilizing this hydrophobic column chromatography. Carbohydrate active enzymes have also been purified in single step employing affinity precipitation by alginate. Single-step purification protocol saves a lot of reagents, chemicals, and time and is economically viable. Compared to the multistep process, better enzyme recovery is obtained in the single-step process. The purified enzymes were used for varied industrial applications as well as potentiated for future use in diverse sectors. The experiences gained through purification of these enzymes will come in handy for developing successful purification procedures for other proteins.

Acknowledgments

The financial support by the Department of Biotechnology (Government of India), Indian Institute of Technology Delhi, and MHRD, Government of India is gratefully acknowledged. Author SK is grateful to the Council of Scientific and Industrial Research (CSIR, Govt. of India) for Senior Research Associateship (Scientists' Pool Scheme).

References

Benbelgacem, F.F., Bellag, O.A., Mohd Salleh, H.N., Batcha, I.A., 2019. Functional diversity of biocatalysts: Whether to bioprospect or bioengineer? Biol. Natural Res. Engineer. J. 2, 64–84.

Białkowska, A.M., Morawski, K., Florczak, T., 2017. Extremophilic proteases as novel and efficient tools in short peptide synthesis. J. Indust. Microbiol. Biotechnol. 44, 1325–1342.

Brininger, C., Spradlin, S., Cobani, L., Evilia, C., 2018. The more adaptive to change, the more likely you are to survive: Protein adaptation in extremophiles. Seminars Cell Develop. Biol. 84, 158–169.

Contesini, F.J., Melo, R.R., Sato, H.H., 2018. An overview of *Bacillus* proteases: From production to application. Crit. Rev. Biotechnol. 38, 321–334.

da Silva, R.R., 2017. Bacterial and fungal proteolytic enzymes: Production, catalysis and potential applications. Appl. Biochem Biotechnol. 183, 1–19.

DasSarma, S., DasSarma, P., 2015. Halophiles and their enzymes: Negativity put to good use. Curr. Opin. Microbiol. 25, 120–126.

Dodia, M.S., Rawal, C.M., Bhimani, H.G., Joshi, R.H., Khare, S.K., Singh, S.P., 2008. Purification and stability characteristics of an alkaline serine protease from a newly isolated *Haloalkaliphilic bacterium* sp. AH-6. J. Indust. Microbiol. Biotechnol. 35, 121–131.

Dumorné, K., Córdova, D.C., Astorga-Eló, M., Renganathan, P., 2017. Extremozymes: A potential source for industrial applications. J. Microbiol. Biotechnol. 27, 649–659.

Gaur, R., Grover, T., Sharma, R., Kapoor, S., Khare, S.K., 2010. Purification and characterization of a solvent stable aminopeptidase from *Pseudomonas aeruginosa*: Cloning and analysis of aminopeptidase gene conferring solvent stability. Process Biochem. 45, 757–764.

Gaur, R., Gupta, A., Khare, S.K., 2008. Purification and characterization of lipase from solvent tolerant *Pseudomonas aeruginosa* PseA. Process Biochem. 43, 1040–1046.

Gaur, R., Khare, S.K., 2011a. Solvent tolerant Pseudomonads as a source of novel lipases for applications in non-aqueous systems. Biocatal. Biotransform. 29, 161–171.

Gaur, R., Khare, S.K., 2011b. Statistical optimization of palm oil hydrolysis by *Pseudomonas aeruginosa* PseA lipase. Asia-Pacific J. Chem. Engineer. 6, 147–153.

Ghosh, S., Kumar, S., Khare, S.K., 2019. Microbial diversity of saline habitats: An overview of biotechnological applications. In: Giri, B., Varma, A. (Eds.), Microorganisms in Saline Environments: Strategies and Functions. Springer International Publishing, Champlain, IL, pp. 65–92.

Gupta, R., Beg, Q.K., Lorenz, P., 2002b. Bacterial alkaline proteases: Molecular approaches and industrial applications. Appl. Microbiol. Biotechnol. 59, 15–32.

Gupta, A., Khare, S.K., 2009. Enzymes from solvent-tolerant microbes: Useful biocatalysts for non-aqueous enzymology. Crit. Rev. Biotechnol. 29, 44–54.

Gupta, A., Roy, I., Khare, S.K., Bisaria, V.S., Gupta, M.N., 2002a. One-step purification of xylanase from *Melanocarpus albomyces* and ethylene glycol as a novel soluble additive for enhancing its thermal stability. Biotechnol. Lett. 24, 2005–2009.

Gupta, A., Roy, I., Khare, S.K., Gupta, M.N., 2005a. Purification and characterization of a solvent stable protease from *Pseudomonas aeruginosa* PseA. J. Chromatogr. A. 1069, 155–161.

Gupta, A., Roy, I., Patel, R.K., Singh, S.P., Khare, S.K., Gupta, M.N., 2005b. One-step purification and characterization of an alkaline protease from haloalkaliphilic *Bacillus* sp. J. Chromatogr. A. 1075, 103–108.

Gurumallesh, P., Alagu, K., Ramakrishnan, B., Muthusamy, S., 2019. A systematic reconsideration on proteases. Int. J. Biol. Macromol. 128, 254–267.

Hemamalini, R., Khare, S.K., 2016. Purification and characterization of active aggregates of an organic solvent tolerant lipase from *Marinobacter* sp. EMB5. Insights Enzym. Res. 1, 1.

Hemamalini, R., Khare, S.K., 2018. Halophilic lipase does forms catalytically active aggregates: Evidence from *Marinobacter* sp. EMB5 lipase (LipEMB5). International. J. Biol. Macromol. 119, 172–179.

Janecek, S., Svensson, B., MacGregor, E.A., 2014. α-Amylase: An enzyme specificity found in various families of glycoside hydrolases. Cell. Mol. Life Sci. 71, 1149–1170.

Joshi, C., Khare, S.K., 2013. Purification and characterization of *Pseudomonas aeruginosa* lipase produced by SSF of deoiled Jatropha seed cake. Biocatal. Agric. Biotechnol. 2, 32–37.

Karan, R., Khare, S.K., 2010. Purification and characterization of a solvent-stable protease from *Geomicrobium* sp. EMB2. Environ. Technol. 31, 1061–1072.

Karan, R., Kumar, S., Sinha, R., Khare, S.K., 2012. Halophilic microorganisms as sources of novel enzymes. In: Satyanarayana, T., Johri, B.N. (Eds.), Microorganisms in Sustainable Agriculture and Biotechnology. Springer, Dordrecht, The Netherlands, pp. 555–579.

Khan, A.U., Maryam, L., Zarrilli, R., 2017. Structure, genetics and worldwide spread of New Delhi metallo-β-lactamase (NDM): A threat to public health. BMC Microbiol. 17, 101.

Kumar, S., Grewal, J., Sadaf, A., Hemamalini, R., Khare, S.K., 2016a. Halophiles as a source of polyextremophilic α-amylase for industrial applications. AIMS Microbiol. 2, 1–26.

Kumar, S., Karan, R., Kapoor, S., Singh, S.P., Khare, S.K., 2012a. Screening and isolation of halophilic bacteria producing industrially important enzymes. Brazilian J. Microbiol. 43, 1595–1603.

Kumar, S., Khan, R.H., Khare, S.K., 2016b. Structural elucidation and molecular characterization of *Marinobacter* sp. α-amylase. Prep. Biochem. Biotechnol. 46, 238–246.

Kumar, S., Khare, S.K., 2012. Purification and characterization of maltooligosaccharide-forming α-amylase from moderately halophilic *Marinobacter* sp. EMB8. Bioresour. Technol. 116, 247–251.

Kumar, S., Khare, S.K., 2015. Chloride activated halophilic α-amylase from *Marinobacter* sp. EMB8: Production optimization and nanoimmobilization for efficient starch hydrolysis. Enzym. Res. 2015, e859485.

Kumar, S., Mathur, A., Singh, V., Nandy, S., Khare, S.K., Negi, S., 2012b. Bioremediation of waste cooking oil using a novel lipase produced by *Penicillium chrysogenum* SNP5 grown in solid medium containing waste grease. Bioresour. Technol. 120, 300–304.

Kumari, A., Ahmad, R., Negi, S., Khare, S.K., 2017. Biodegradation of waste grease by *Penicillium chrysogenum* for production of fatty acid. Bioresour. Technol. 226, 31–38.

Ma, Y., Galinski, E.A., Grant, W.D., Oren, A., Ventosa, A., 2010. Halophiles 2010: life in saline environments. Appl. Environ. Microbiol. 76, 6971–6981.

Nadar, S.S., Pawar, R.G., Rathod, V.K., 2017. Recent advances in enzyme extraction strategies: A comprehensive review. Int. J. Biol. Macromol. 101, 931–957.

Nandan, A., Nampoothiri, K.M., 2017. Molecular advances in microbial aminopeptidases. Bioresour. Technol. 245, 1757–1765.

Oren, A., 2002. Molecular ecology of extremely halophilic Archaea and Bacteria. FEMS Microbiol. Ecol. 39, 1–7.

Pratt, R.F., 2016. β-Lactamases: Why and how. J. Medicinal Chem. 59, 8207–8220.

Raddadi, N., Cherif, A., Daffonchio, D., Neifar, M., Fava, F., 2015. Biotechnological applications of extremophiles, extremozymes and extremolytes. Appl. Microbiol. Biotechnol. 99, 7907–7913.

Rodrigues, R.C., Virgen-Ortiz, J.J., Dos Santos, J.C.S., Berenguer-Murcia, A., Alcantara, A.R., Barbosa, O., Ortiz, C., Fernandez-Lafuente, R., 2019. Immobilization of lipases on hydrophobic supports: Immobilization mechanism, advantages, problems, and solutions. Biotechnol. Adv. 37, 746–770.

Sadaf, A., Grewal, J., Jain, I., Kumari, A., Khare, S.K., 2018a. Stability and structure of *Penicillium chrysogenum* lipase in the presence of organic solvents. Prep. Biochem. Biotechnol. 48, 977–983.

Sadaf, A., Sinha, R., Khare, S.K., 2018b. Structure and functional characterisation of a distinctive β-lactamase from an environmental strain EMB20 of *Bacillus cereus*. Appl. Biochem. Biotechnol. 184, 197–211.

Saralov, A.I., 2019. Adaptivity of archaeal and bacterial extremophiles. Microbiol. (Russian Federation) 88, 379–401.

Singh, S.K., Singh, S.K., Tripathi, V.R., Garg, S.K., Khare, S.K., 2013. Downstream processing, characterization, and structure-function relationship of solvent-, detergent-, psychro-, thermo-, alkalistable metalloprotease from metal-, solvent-tolerant psychrotrophic *Pseudomonas putida* SKG-1 isolate. Biotechnol. Progr. 29, 99–108.

Sinha, R., Khare, S.K., 2012. Isolation of a halophilic *Virgibacillus* sp. EMB 13: Characterization of its protease for detergent application. Indian J. Biotechnol. 11, 316–326.

Sinha, R., Khare, S.K., 2013. Characterization of detergent compatible protease of a halophilic *Bacillus* sp. EMB9: differential role of metal ions in stability and activity. Bioresour. Technol. 145, 357–361.

Sinha, R., Srivastava, A.K., Khare, S.K., 2014. Efficient proteolysis and application of an alkaline protease from halophilic *Bacillus sp.* EMB9. Prep. Biochem. Biotechnol. 44, 680–696.

Suriya, J., Bharathiraja, S., Krishnan, M., Manivasagan, P., Kim, S.K., 2016. Extremozymes from marine actinobacteria. Adv. Food Nutr. Res. 79, 43–66.

Tanokura, M., Miyakawa, T., Guan, L., Hou, F., 2015. Structural analysis of enzymes used for bioindustry and bioremediation. Biosci. Biotechnol. Biochem. 79, 1391–1401.

Teotia, S., Khare, S.K., Gupta, M.N., 2001a. An efficient purification process for sweet potato β-amylase by affinity precipitation with alginate. Enzym. Microb. Technol. 28, 792–795.

Teotia, S., Lata, R., Khare, S.K., Gupta, M.N., 2001b. One-step purification of glucoamylase by affinity precipitation with alginate. J. Mol. Recogn. 14, 295–299.

van der Maarel, M.J., van der Veen, B., Uitdehaag, J.C., Leemhuis, H., Dijkhuizen, L., 2002. Properties and applications of starch-converting enzymes of the alpha-amylase family. J. Biotechnol. 94, 137–155.

Vingiani, G.M., De Luca, P., Ianora, A., Dobson, A.D., Lauritano, C., 2019. Microalgal enzymes with biotechnological applications. Marine Drugs 17, 459.
Walia, A., Guleria, S., Mehta, P., Chauhan, A., Parkash, J., 2017. Microbial xylanases and their industrial application in pulp and paper biobleaching: A review. 3 Biotech. 7, 11.
Yin, J., Chen, J.C., Wu, Q., Chen, G.Q., 2015. Halophiles, coming stars for industrial biotechnology. Biotechnol. Adv. 33, 1433–1442.

CHAPTER FIVE

Chemical stabilization of enzymes

Yusuf Chisti
School of Engineering, Massey University, Palmerston North, New Zealand

5.1 Introduction

This chapter is concerned with stabilization of enzymes to extend their life during storage and use. Enzyme structure is discussed as being directly relevant to understanding the causes of activity loss or denaturation and approaches to stabilization. Factor affecting stability, and the kinetics of thermally induced activity loss, are discussed. Methods of enhancing stability through chemical modification of the native enzyme molecule, are focused on. Enzyme use in nonpolar hydrophobic solvents and other unconventional media, is discussed. Stabilization may permit use of an enzyme at an elevated temperature, for example, and this can greatly enhance the productivity of an industrial process.

An enzyme is a giant protein molecule that catalyzes a chemical reaction, or speeds up its rate. The smallest enzymes are around 10 kDa in molecular weight, or molar mass. Some of the largest enzymes may exceed 240 kDa in molar mass. Enzymes generally function in a mild environment (e.g., temperature, pH, water activity, ionic strength) that is able to support life. Environmental conditions that deviate a little from the physiological norms, can easily damage enzymes so that they lose their catalytic activity, or bioactivity. Relatively stable and robust enzymes are needed for industrial use, as the conditions in practicable industrial operation are necessarily more extreme. For example, the substrate concentration may be high, the temperate may be relatively high, and the pH may be extreme so as to minimize microbial contamination of a bioreactor.

An enzyme catalyzed reaction may involve the conversion of a reactant (a chemical, also known as a substrate) to some product, or complete degradation of the substrate to small molecules. An enzyme catalyzes a reaction but is not itself consumed in the reaction. An enzyme molecule can be chemically modified to improve its stability, or other attributes, but such modifications cannot be used to alter the reaction thermodynamics (Woodley, 2013). Enzymatic processes are generally believed to be more

Enzymes: Novel Biotechnological Approaches for the Food Industry.
http://dx.doi.org/10.1016/B978-0-12-800217-9.00005-8

environmentally sustainable compared to conventional chemical processes (Jegannathan and Nielsen, 2013).

Stability considerations of particular relevance to industrial use of enzymes are thermostability, stability in unnatural solvents (e.g., anhydrous hydrophobic organic solvents), stability in media with extremes of pH and high ionic strength, stability at elevated concentrations of substrates and products, and tolerance to mechanical forces (or hydrodynamic shear forces) encountered during processing, especially the forces at gas-liquid and solid-liquid interfaces. Stability considerations are important also in enzyme production and recovery processes (Weijers and Van't Riet, 1992). Although the focus here is on enzymes, the interest in stabilizing proteins extends to the many biopharmaceutical proteins and diagnostic antibodies (Balcão and Vila, 2015; Tamer and Chisti, 2001).

5.2 Commercial applications of enzymes

Enzymes are used to catalyze reactions in wide-ranging industrial applications. Areas of application include: specialty syntheses and biotransformations in biochemical and pharmaceutical processes (Banerjee et al., 1992; Patel, 2000); analyses and biosensing (Schaertel and Firstenberg-Eden, 1988; Vaidya and Annapure, 2019; Yunus, 2019); medicine (Banerjee et al., 2004; Rouf et al., 1996; Vellard, 2003); processing of foods (Fernandes, 2010; Raveendran et al., 2018; Sirisansaneeyakul et al., 2007; Wee et al., 2011), beverages, animal feeds, crops, fibers, agricultural residues, and animal hides; production of industrial chemicals and fuels; bioremediation and biological treatment of wastes; and as additives in detergents. New applications of enzymes continue to be found especially in production of high-value chemicals where cost is less important and the needs are for high specificity, excellent activity under mild conditions, an absence of toxicity, and biodegradability.

Enzymes are generally highly enantioselective: in a mixture of enantiomers, they are able to convert one particular enantiomer to a product while the other enantiomers are unaffected. Therefore, enzymes are of particular interest in enantioselective, or asymmetric, biotransformations (Patel, 2000). Enzyme-based biosensors are commonly used in some applications, but wider use is limited by the poor stability of enzymes (Reyes-de-Corcuera et al., 2018).

Orally administered enzymes are used in treating certain disease in which the activity of the enzyme is required in the gastrointestinal tract.

Stabilizing such therapeutic enzymes to better withstand the denaturing and proteolytic environment of the human digestive system is discussed by Fuhrmann and Leroux (2014). Among emerging applications, large quantities of extremely cheap enzymes are essential to economic viability of processes being developed to breakdown lignocellulosic biomass to fermentable sugars mainly for production of alcohol biofuels (Illanes et al., 2012; Karimi and Chisti, 2017). The diverse applications of enzymes have been extensively reviewed (Buchholz et al., 2012; Choi et al., 2015; Patel et al., 2016; Singh et al., 2016; van Beilen and Li, 2002).

The global enzyme market in 2016 was worth more than US$5 billion. Food and beverage sector was one of the largest users (>US$1.8 billion) of the enzymes. Nearly 50% of the market value was contributed by carbohydrate processing enzymes (e.g., cellulases, amylases, invertases). Animal feed, pulp and paper, textiles, and detergents were the other main users. Bulk of the enzymes were used in the soluble form on a single-use basis.

Most commercial applications use enzymes in their native form (Schmid, 1979) without any modification. Production of microbial enzymes and their use in bioreactors for various enzyme-catalyzed processes, have been discussed by others (Buchholz et al., 2012; Chisti, 2010; Chutmanop et al., 2008,b; Chisti and Moo-Young, 1999, 2002; Headon and Walsh, 1994). Industrial enzymes have been reviewed (Aehle, 2004; Choi et al., 2015; Polaina and MacCabe, 2007). Overall, the global enzyme market is small: high cost (Iyer and Ananthanarayan, 2008), limited availability, and poor relative stability of enzymes constrain their role in commercial processes. Hence, the focus is on stabilizing enzymes.

5.3 Sources of industrial enzymes

All lifeforms are potential sources of enzymes. Most industrial enzymes are obtained from microorganisms that have been selected, or genetically modified, to overproduce them. Other sources include extraction from animal and plant material. Among microbial sources, microfungi provided more than 50% of the commercial enzymes and bacteria provide more than 33% (Illanes et al., 2012). Higher animals and plants provide around 12% (Illanes et al., 2012). The great preponderance of commercial microbial enzymes is simply because microorganisms can generally be grown extremely rapidly and cheaply compared to higher animals and plants. Recombinant DNA technology has allowed production of many proteins and enzymes in microbial hosts (Manderson et al., 2006a,b;

Mattanovich et al., 2012; Overton, 2014; Rosano and Ceccarelli, 2014) and this has greatly reduced the cost of production. In addition, many microbial enzymes are naturally released from the cell into the extracellular culture medium and this simplifies recovery. Microbial processes for producing a great diversity of enzymes have been described (Chutmanop et al., 2008; Garrido et al., 1994; Ghoshal et al., 2012; Jain et al., 2012; Patil et al., 2017; Qureshi et al., 2012, 2016, Salehmin et al., 2013, 2014; Singh et al., 2011; Sirisansaneeyakul et al., 2007).

Advantages of microbial production notwithstanding, there is a strong potential for many enzymes (Biesgen et al., 2002; Hood, 2002; Obembe et al., 2011; Park et al., 2016; Yoshida et al., 2004) and other bioactive proteins (Cardi et al., 2010; Daniell et al., 2009; Obembe et al., 2011; Prado et al., 2014; Yoshida et al., 2004) to be produced cost effectively in transgenic plants. If enzymes can be produced cheaply, the need for prolonged use, or recovery and reuse, may be reduced.

5.4 Enzyme structure and stability

An understanding of enzyme structure is essential for understanding the loss of stability in a denaturing environment and developing methods to counter the activity loss. An enzyme is made up of one or more polypeptide chains (Fig. 5.1). Each polypeptide chains is comprised of many amino acid residues. Around 20 different types of amino acid residues may be present in each chain. Amino acid residues are formed by the reaction between an α amino group (i.e., the $-NH_2$ group on the carbon closest to the carboxylic acid group) on one amino acid and the carboxylic acid group ($-COOH$) of another amino acid. This results in the formation of a peptide bond ($R_1-C(O)-NH-R_2$) and elimination of a water molecule. The sequence of amino acid residues in a polypeptide chain constitutes the primary structure of the chain. Free functional groups derived from the unreacted side chains of the various amino acids occur throughout the polypeptide chain. In addition, each polypeptide chain has N- and C-termini (Fig. 5.1) with free amino ($-NH_2$) and carboxylic acid ($-COOH$) functional groups, respectively.

The polypeptide chains contain regions of helical folding (the α-helices) and β-pleated sheets (Fig. 5.1). These are generated through hydrogen bonding between the $-NH$ and $-C{=}O$ groups on the main chain. These helical folds and the pleated sheets constitute the secondary structure of the enzyme. A polypeptide chain is further folded into a globular structure

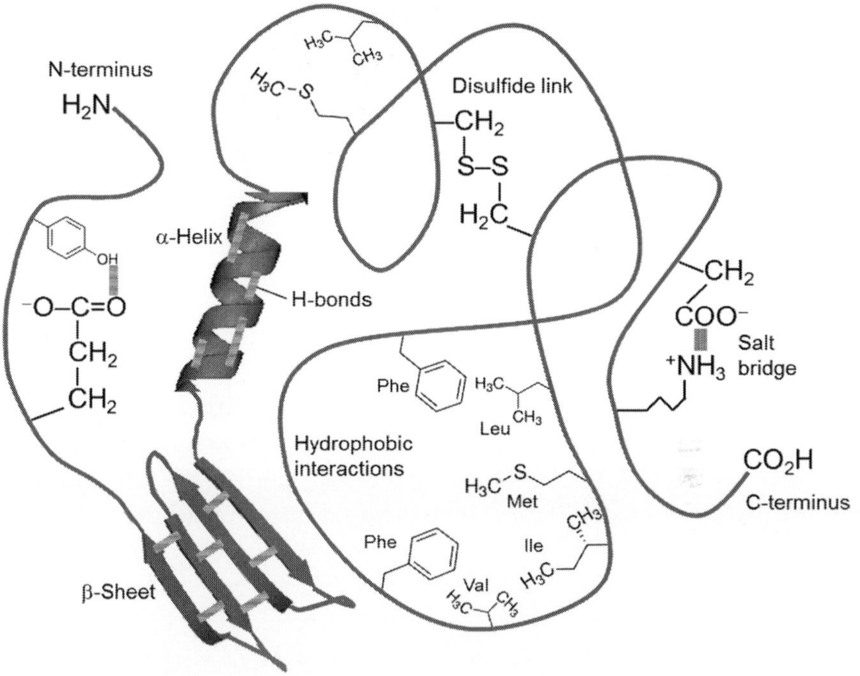

Figure 5.1 Schematic structure of a monomeric enzyme. Shown are: the C- and N-termini; the α-helical coiled regions; the β-pleated sheets; the hydrogen bonds; the hydrophobic interaction zones; the ionic interaction zones, or salt bridges; and the covalent disulfide links. Hydrogen bonding, hydrophobic interactions, ionic interactions, and the disulfide links among the various side chain functional groups of the amino acid residues of the polypeptide contribute to stabilizing the globular tertiary structure (Phe, phenylalanine; Met, methionine; Leu, leucine; Ile, isoleucine; Val, valine). Based on Janeček, 1993.

(Fig. 5.1). This highly specific folding is determined by the interactions among the side groups on the amino acid residues in the polypeptide. These interactions include: (1) hydrogen bonding between the free –OH (e.g., on the amino acids tyrosine, serine, threonine) and the free –COOH groups (e.g., on the amino acids aspartic acid and glutamic acid); (2) hydrophobic interactions among the hydrophobic side chains of the amino acids such as phenyl alanine, valine, leucine, and isoleucine; (3) formation of salt bridges, or ionic interactions, between the ionizable side chain functional group of amino acids such as arginine, lysine, aspartic acid, and glutamic acid; and (4) formation of covalent disulfide bonds (–S–S–) between the side groups of two cysteine residues. The globular folding of the polypeptide chain constitutes the tertiary structure of an enzyme. The secondary and tertiary structures are ultimately influenced by the primary structure of the polypeptide molecule.

An enzyme comprised of a single polypeptide chain is monomeric. Other enzymes may be made of multiple folded polypeptide chains, that is, they may be multimeric. The interactions described above for a single polypeptide chain, occur also between two or more folded polypeptide chains and help to hold them together. The arrangement of the distinct multimers in the molecule of an enzyme constitutes its quaternary structure. Only multimeric enzymes have a quaternary structure. In addition, an enzyme may have complexed metal ions and covalently bound carbohydrates, or lipids, that may be important for its stability and biological activity.

Most enzymes are glycosylated, that is, they have carbohydrate moieties attached to the polypeptide chain. Glycosylation does not alter the tertiary structure of the enzyme, but often enhances its thermal stability and resistance to proteolysis (Khan et al., 2003; Meldgaard and Svendsen, 1994; Rouf et al., 1996; Solá and Griebenow, 2009; Wang et al., 1996). This notwithstanding, cases are known where the thermal stability was barely affected by glycosylation (Schülke and Schmid, 1988).

Folding into a globular shape minimizes the surface area (Erickson, 2009) and hence the surface energy of an enzyme molecule. In their typically aqueous hydrophilic natural milieu, the polypeptide chains fold such that the hydrophilic functional groups are exposed to the surface and the hydrophobic functional groups are forced to the interior of the molecule. The globular enzyme molecule is tightly packed without substantial internal spaces and almost no water molecules within the globule (Erickson, 2009). The density of the molecule is in the range of 1370–1429 kg m^{-3} (Erickson, 2009). The volume (V, nm^3) of the roughly spherical molecule depends on its molecular weight M (kg $kmol^{-1}$, or Da), as follows (Erickson, 2009):

$$V = 1.212 \times 10^{-3}\ M \tag{5.1}$$

Therefore, the diameter d_p (nm) of the molecule can be approximated, as follows (Erickson, 2009):

$$d_p = 0.132\ M^{1/3} \tag{5.2}$$

In Eq. (5.2), d_p is the minimum diameter of a sphere capable of containing the given molecular mass (Erickson, 2009). The estimated diameters (Eq. 5.2) of some industrial enzymes are shown in Table 5.1. The actual diameter tends to be a little larger than the d_p calculated using Eq. (5.2).

Table 5.1 Estimated molecular diameters (Eq. 5.2) of some industrial enzymes.

Enzyme	Source	Molar mass (Da)	Molecular diameter, d_p (nm)
α-Amylase (EC 3.2.1.1)	*Aspergillus oryzae*	51,000–51,700	~4.9
β-Galactosidase (EC 3.2.1.23)	*Aspergillus oryzae*	112,000	6.4
Invertase (EC 3.2.1.26)	*Saccharomyces cerevisiae*	270,000	8.5
Lysozyme (EC 3.2.1.17)	Hen egg white	14,400	3.2
Pepsin A (EC 4.3.3.1)	Porcine mucosa	34,600	4.3

The circumference (C_s, nm) of a spherical enzyme molecule of molar mass M (kg kmol^{-1}) may be estimated using the following equation:

$$C_s = 0.4147\ M^{1/3} \tag{5.3}$$

The forces contributing to stability do so by making unfolding more difficult, or by rigidifying the enzyme molecule. Stability contribution of a salt bridge is up to 20 kJ/mol per bridge located in the interior of the enzyme molecule (Weijers and Van't Riet, 1992). Stability contribution is reduced to around 4–10 kJ/mol per bridge, if located in the exterior (Weijers and Van't Riet, 1992). Intramolecular disulfide links contribute significantly to enzyme stability (Wong and Wong, 1992). Each disulfide bond contributes to the tune of 15–20 kJ per bond/mol (Weijers and Van't Riet, 1992). Enzymes can be structurally modified to enhance various attributes including the substrate specificity and range; the pH and temperature for optimal activity; and stability under various denaturing environments.

5.5 Inherently stable enzymes

Most of the industrial enzymes used in large scale processes are relatively thermostable and have been obtained from thermophilic microorganisms (optimal growth temperature of ~50°C), but not from hyperthermophiles (Haki and Rakshit, 2003; Janeček, 1993). Hyperthermophilic microorganisms are able to grow at temperatures of >80°C (Vieille and Zeikus, 2001). Enzymes of such microorganisms are generally more heat resistant compared to other microbial enzymes but are made of the same amino acids as the other enzymes. The higher thermostability is apparently a consequence of the stronger intramolecular forces keeping the enzyme from unfolding under thermal stress (Mozhaev and Martinek, 1984).

Many subtle structural differences contribute to the higher thermostability of thermophilic enzymes compared to their mesophilic analogs (Gianfreda and Scarfi, 1991; Hancock, 1957; Iyer and Ananthanarayan, 2008; Mozhaev and Martinek, 1984; Ó'Fágáin, 1995). For example, the thermophilic enzymes generally contain more internal hydrophobic regions and fewer external hydrophobic regions compared to mesophilic enzymes (Gianfreda and Scarfi, 1991). In addition, thermophilic enzymes may have extra salt bridges that link residues in the vicinity of the N-terminus to others near the C-terminus (Gianfreda and Scarfi, 1991). For example, the thermostable ruberodoxin of the hyperthermophile *Pyrococcus furiosus* appears to have an N-terminus fixed by a salt bridge that does not occur in the less thermostable ruberodoxins of mesophiles (Janeček, 1993). Compared to their mesophilic counterparts, thermophilic enzymes may contain additional metal ions (Schmid, 1979).

The sequence of amino acids in the polypeptide chains of hyperthermophilic enzymes and the folding of the chains have probably evolved to permit stronger intramolecular interactions to prevent denaturation. Some hyperthermophiles produce polyhydroxy compounds such as 2-O-β-mannosylglycerate which have been shown to protect enzymes against thermal denaturation in vitro (Ramos et al., 1997). The protein-synthesizing machinery of thermophiles is more thermostable compared to that in mesophiles (Gianfreda and Scarfi, 1991).

For homologous enzymes, irrespective of the microbial source, whether mesophilic or thermophilic, the structure of the active sites and the mechanisms of action are highly conserved (Feller, 2010), but active sites are less flexible in thermostable enzymes. Homologous enzymes from different microbial sources are believed to retain comparable levels of flexibility at their optimal temperatures. Thus, a mesophilic enzyme with an optimal temperature of 37°C, for example, has the same level of flexibility as a counterpart enzyme functioning at 80°C in an extremothermophile. Compared at the same temperature of 37°C, the extremophilic enzyme would be much less flexible than its mesophilic counterpart (Feller, 2010).

As a consequence of the more interactions (e.g., hydrophobic interactions, hydrogen bonds, disulfide links, electrostatic interactions) contributing to the folded structure of a thermophilic enzyme, such enzymes are more rigid and tightly packed than mesophilic enzymes (Feller, 2010; Iyer and Ananthanarayan, 2008). A drawback of enhanced rigidity is a reduced flexibility, and this may result in the thermophilic enzyme being catalytically less efficient at moderate temperatures than a mesophilic counterpart (Iyer and

Ananthanarayan, 2008). Thus, a thermophilic enzyme may be more stable and active at elevated temperatures but would be barely active at room temperature (Feller, 2010). Therefore, thermophilic enzymes are merited only for processes requiring a high temperature (Iyer and Ananthanarayan, 2008).

Enzymes of certain halophilic archaea are highly salt tolerant, requiring a high-salt (1–4 M) environment for activity and stability (Mevarech et al., 2000). These enzymes have a higher proportion of the acidic amino acid residues in their structure compared to mesophilic enzymes. The resulting high negative surface charge on these enzymes keeps them in solution in a high-salt environment that would precipitate the nonhalophilic proteins (Mevarech et al., 2000). Halophilic enzymes are potentially useful in certain industrial processes (Iyer and Ananthanarayan, 2008).

Although extremophiles may be difficult to culture because of the extreme conditions required for their isolation and growth, their enzymes may be produced in easily grown recombinant microorganisms (Cowan and Fernandez-Lafuente, 2011). Hyperthermophilic and other extremophilic enzymes are further discussed by others (Ellueche et al., 2014, 2015; Vieille and Zeikus, 2001). Stabilizing features of thermophilic enzymes relative to their mesophilic counterparts, are discussed by Mozhaev and Martinek (1984).

5.6 Protein engineering for stability enhancement

Protein engineering (Bommarius and Paye, 2013; Böttcher and Bornscheuer, 2010; Ó'Fágáin, 1995; Ó'Fágáin, 2003; Singh et al., 2013; Woodley, 2013) through rational design (Lutz, 2010), random mutagenesis and selection, and directed evolution, is being used to improve enzymes and develop ones with new traits. Protein engineering for stability improvement may require: introduction of internal disulfide links; modifications to increase internal hydrogen bonding; enhancements to internal hydrophobic interactions through improved internal packaging of the hydrophobic groups; and introduction of more, or stronger, surface salt bridges (Janeček, 1993). Nondisulfide covalent links formed between novel side groups on genetically encoded noncanonical amino acid residues in a polypeptide chain, have also been used to improve protein stability (Moore et al., 2017).

In rational design, specific outcomes are targeted by knowledge-driven changes to the enzyme structure, including modification of the primary amino acid sequence and substitution of specific amino acids with others. The outcome is not necessarily predictable. Therefore, after each change, the enzyme is sequenced to confirm the change and catalytic properties are characterized in a process that is tedious and expensive (Chen, 2001). Computational design and bioinformatics-driven strategies that may be used to predict the effects of structural changes on stability and catalytic activity of an enzyme, have been extensively reviewed (Damborsky and Brezovsky, 2014; Eijsink et al., 2004; Lutz, 2010; Suplatov et al., 2015; Wijma et al., 2013), although not much used in practice. Rational design has been extensively reviewed (Chen, 2001; Dambrosky and Brezovsky, 2014; Eijsink et al., 2004; Hammer et al., 2017; Lutz, 2010; Wijma et al., 2013).

Directed evolution does not depend on a knowledge of how structure might affect function (Chen, 2001). Instead, randomly mutated enzyme genes are created, cloned, and expressed. High throughput screening is used to identify mutant enzymes with improved properties. Directed evolution has been further enhanced by using in vitro recombination, or DNA shuffling, to rapidly combine the beneficial mutations generated through random mutagenesis (Chen, 2001). Directed evolution has been reviewed (Chen, 2001; Cole and Gaucher, 2011; Eijsink et al., 2005; Gershenson and Arnold, 2000; Hammer et al., 2017; Socha and Tokuriki, 2013).

Of the various protein engineering approaches, directed evolution (Gershenson and Arnold, 2000; Illanes et al., 2012) has proven more successful than rational design, in improving enzymes (Woodley, 2013). This is presumably because rational design requires substantial prior knowledge. For example, the enzyme structure and the mechanism of action must be known (Chen, 2001) and a sensitive assay of enzyme activity must be available (Chen, 2001).

5.7 Understanding enzyme catalysis

An enzyme-catalyzed reaction may be represented as follows:

$$aA + bB \overset{\text{Enzyme}}{\Leftrightarrow} cC + dD \tag{5.4}$$

the reactants A and B are converted to products C and D. The reactants are also known as substrates. The stoichiometric coefficients a, b, c, and d are the relative molar amounts of the species involved in the reaction. The reaction is reversible and can occur both in forward and reverse directions. The position of the equilibrium is established by the equilibrium constant, K_e:

$$K_e = \frac{[C]^c [D]^d}{[A]^a [B]^b} \tag{5.5}$$

where, the square brackets denote the active molar concentrations of the species within brackets at equilibrium. A large value of the equilibrium constant at the conditions of the reaction generally implies that the products *C* and *D* are thermodynamically favored and the reaction tends towards the products. Enzymes do not alter the equilibrium constant. The equilibrium constant depends on temperature and may be affected by other factors such as the pH, the ionic strength of reaction medium, and the concentrations of the reactants and products.

As a catalyst, the enzyme is not consumed by the reaction, although it may lose activity if it is being denatured by the conditions used in the reaction. An apparent loss of activity may occur if the enzyme is being inhibited by the product of the reaction. Activity of an enzyme is typically quantified as specific activity, or activity per unit quantity (mass, or mole) of the enzyme, under the conditions of the reaction. These include the temperature, the pH, the ionic strength, the concentrations of the reactants and products, and the presence organic solvents.

Rate of a chemical reaction roughly doubles for each 10°C rise in temperature. Therefore, if an enzyme with an optimal reaction temperature of 25°C in vivo could somehow be used at 75°C in an industrial process without being denatured, the rate of reaction would be around 100-fold greater than in nature (Iyer and Ananthanarayan, 2008). This explains the quest for thermostable enzymes and the reason for stabilizing them.

If the specific activity of an enzyme is high, a lower concentration will be required for a desired outcome and the cost would be low. A low specific activity would mean a larger quantity of the enzyme for a given reaction outcome. Because of cost considerations, industrial processes generally require enzymes with a high specific activity. An industrial enzyme needs to be able to function at high concentrations of the reactants and products. Concentrations may range between 60 and 400 g L^{-1} (Woodley, 2013). A stable enzyme retains a high-specific activity for a longer period under the conditions of the reaction. Possible inhibitory effects of high concentrations of substrates on enzymes may be managed by using a fed-batch reactor (Chisti and Moo-Young, 2002), for example. A packed bed immobilized enzyme reactor (Chisti and Moo-Young, 1999) may be optimal if the substrate inhibition is low. Continuous removal of an inhibitory product from the reaction system may be used to reduce enzyme inhibition.

Different enzymes require different conditions for optimal activity. Therefore, in a multienzyme reaction, involving two enzymes for example, one of the two enzymes may have to be modified via protein engineering,

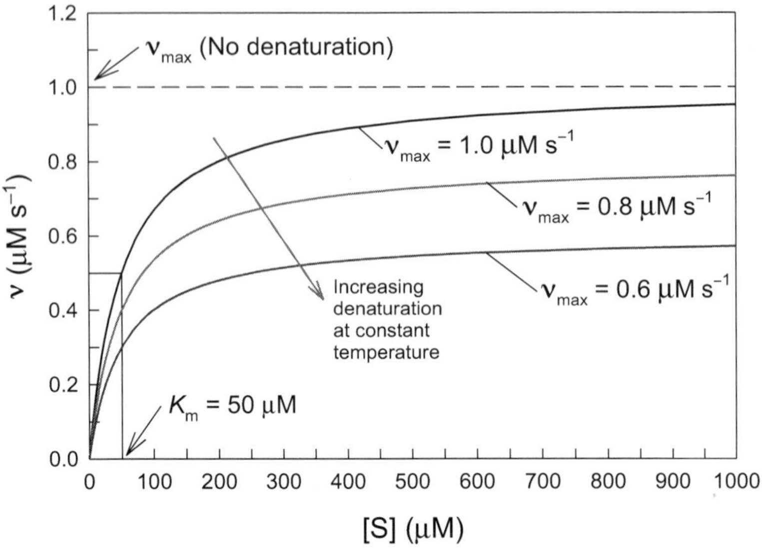

Figure 5.2 Dependence of the reaction rate v on the substrate concentration [S] for a reaction obeying Michaelis-Menten kinetics. For the undenatured enzyme $v_{max} = 1.0\ \mu M\ s^{-1}$ and $K_m = 50\ \mu M$. As the enzyme denatures at constant temperature, the v_{max}-value decreases, but the K_m is not affected. The reaction rate declines for a given concentration of the substrate, as shown by the curves.

or other methods, to alter its optimal activity conditions to match those of the other enzyme. Alternatively, both enzymes may be modified to work optimally at a common set of optimal reaction conditions.

Many enzymes obey Michaelis-Menten kinetics (Briggs and Haldane, 1925; Michaelis and Menten, 1913). Thus, the rate of reaction ($v = -d[S]/dt$) depends on the concentration of the substrate [S], as follows:

$$v = \frac{v_{max}[S]}{K_m + [S]} \tag{5.6}$$

In the above equation, the kinetic constant v_{max} (kmol m^{-3} s^{-1}) declines with time t if the enzyme is being denatured at constant temperature, but the Michaelis-Menten constant K_m (kmol m^{-3}) is not affected. The rate of reaction (v, kmol m^{-3} s^{-1}) declines with time, as shown in Fig. 5.2.

5.8 Enzyme denaturation and its causes

Heat-induced unravelling of the tertiary and quaternary structures results in a loss of the biological activity at an elevated temperature. Suppression of thermal denaturation is probably of greatest importance in industrial processing. In addition to heat, other factors can inactivate, or denature, enzymes. The denaturing factors include: the pH; the ionic strength; removal of essential metal ions by chelating reagents, or replacement of one metal ion by a different one; proteolytic damage to the polypeptide chain because of the presence of contaminating proteases; oxidation of disulfides and other damage because of the presence of thiol contaminants. Mechanical forces associated with high pressure (Section 5.8.3), fluid shear (Section 5.8.4), and ultrasound can damage enzymes. Although ultrasound can be damaging to enzymes (Chisti, 2003), sonication regimes can sometimes be found to actually enhance the rate of an enzymatic reaction relative to the nonsonicated control. Enzyme catalyzed processes have been successfully enhanced using ultrasound (Chisti, 2003; Gumel et al., 2012, 2013a,b; Sulaiman et al., 2013).

Thermostable enzymes are desired for several reasons: a high temperature enhances the rate of reaction and reduces the likelihood of microbial contamination; the need for expensive cooling is reduced; the viscosity of media is lowered for improved mixing; diffusional limitations are reduced; and in most cases the solubility of substrates and products is enhanced (Iyer and Ananthanarayan, 2008; Janeček, 1993).

5.8.1 Thermal denaturation

Thermal denaturation is a loss of activity as a consequence of the heat-induced unravelling of the tertiary and quaternary structures of the enzyme. Enzymes are highly susceptible to thermal denaturation. Temperature of a substance is measure of the vibrational motion within its molecule. Dynamic molecular models reveal that an enzyme molecule at room temperature is pulsating all the time because of the vibrations of its atoms and bonds. The frequency and magnitude of pulsing increase with temperature. A complete cessation of molecular vibrations would imply an unattainable state of zero absolute temperature. As temperature increases, the forces holding the polypeptide chain in the biologically active folded configuration are gradually overcome, resulting in unravelling of the folded structure and loss of biological activity. In addition to temperature, the thermal denaturation rate of an enzyme depends on other factors including the pH, the ionic strength of the medium, and the presence of the substrate.

5.8.1.1 Thermodynamic versus kinetic stability

An elevated temperature results in unfolding of the biologically active native enzyme to an inactive unfolded form. Thermodynamic stability of an enzyme is the difference between the free energies of the folded (i.e., active) state and the unfolded (or denatured) state (Stepankova et al., 2013). Thermodynamic stability is measured in terms of the following parameters (Stepankova et al., 2013): ΔG_u, Gibbs free energy for unfolding, that is, the difference in Gibbs free energies of the folded and unfolded states; K_u, the unfolding equilibrium constant, that is, the concentration of the denatured (unfolded) enzyme divided by the concentration of the active (folded) enzyme at equilibrium; T_m, the melting temperature (Section 5.8.1.2), or the equilibrium temperature at which half of the enzyme molecules are in an unfolded state; $C_{1/2}$, or the concentration of a denaturant needed to unfold half of the enzyme at equilibrium.

Thermodynamic stability is an intrinsic characteristic of the enzyme but is of little value in determining the time-dependent stability, or kinetic stability. Kinetic stability is a measure of how quickly the enzyme unfolds (Balcão and Vila, 2015). Kinetic stability is important for practical purposes and is characterized in terms of the following parameters (Stepankova et al., 2013): k_d, the denaturation rate constant; $t_{1/2}$, the enzyme half-life, or the time required to reduce activity by 50%; T_{50}, the temperature at which the activity is reduced to 50% after a set incubation period; C_{50}, the concentration of a denaturant needed to reduce the activity by 50% after a defined period; T_{opt}, the optimum temperature, or the temperature for maximum activity; T_{n}, or the total turnover number, that is the number of moles of the substrate converted by a mole of the enzyme over its lifetime.

5.8.1.2 Melting temperature

If the temperature of an enzyme solution is increased, the enzyme molecules unfold. The temperature at which the concentrations of the fully folded active enzyme and the unfolded denatured enzyme are equal is known as the melting temperature, T_m, of the enzyme (Iyer and Ananthanarayan, 2008). Melting temperature is probably the best measure of thermostability. T_m of enzymes can range from 30°C to more than 100°C, but in aqueous solutions it tends to be between 60°C and 80°C (Iyer and Ananthanarayan, 2008). Melting temperature is greater for thermophilic enzymes, compared with their psychrophilic and mesophilic counterparts (Feller, 2010).

The typical thermal denaturation curve for an enzyme is shown in Fig. 5.3 (Zoldák et al., 2004). The melting temperature is determined from the curve (Fig. 5.3) as shown in the figure. Only when the melting is fully reversible, is the melting temperature directly related to conformational stability and the enthalpy of unfolding can be estimated from the melting

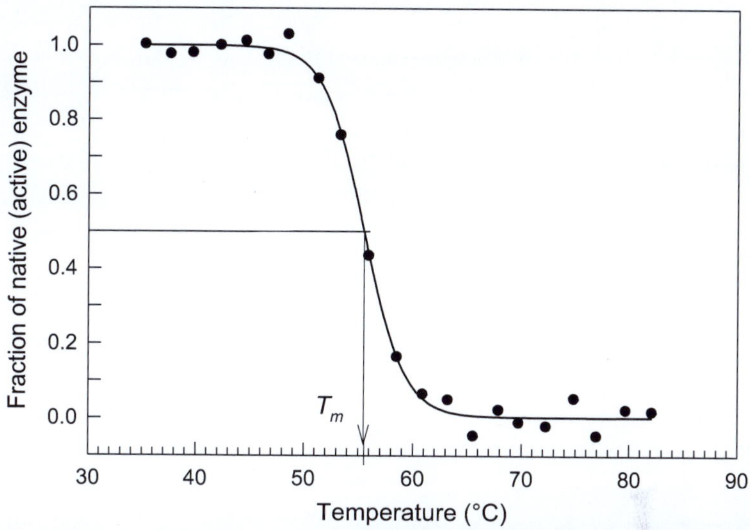

Figure 5.3 Thermal denaturation curve of glucose oxidase from *Aspergillus niger*. The melting temperature (T_m) of the enzyme is indicated. Based on data of Zoldák et al. (2004).

curve. Enthalpy of unfolding is typically in the range of 200–540 kJ mol^{-1} for most enzymes.

5.8.1.3 Monitoring stability loss through unfolding

Circular dichroism (CD) spectroscopy is used to characterize changes in secondary and tertiary structures of an enzyme (Kelly and Price, 2000). CD spectra can be used to compare the conformational stability of a protein subjected to different environments (temperature, pH, stabilizers, and buffers). A conformational change in the molecule results in spectral changes. The observed changes are an average effect of the molecules in solution.

Changes to secondary structure can be monitored in the spectral wavelength range of 190–250 nm, that is, the far-ultraviolet region of the spectrum. The part of the protein molecule absorbing in this region, or the chromophore, is the peptide bond in the native folded structure. Changes in the environment of the peptide bond because of unfolding, change the absorption. The typical required sample size for the CD measurements is 20 to 200 μL of solution containing between 50 and 1000 μg protein per mL. Care is required to ensure that the other components (e.g., buffers, stabilizers) in solution do not absorb in the spectral range of interest.

Changes to some features of the tertiary structure can be monitored in the CD spectral range of 250–350 nm. This is the near-ultraviolet region of the spectrum. The absorption in this region is attributed to the phenylalanine residues (250–270 nm); the tyrosine residues (270–290 nm); and tryptophan residues (280–300 nm). Disulfide bonds also produce weak broad absorption bands in near-UV wavelength range. Absorption in the near-UV range is weaker than in the far-UV range and, therefore, a more concentrated (protein concentration of 250–2000 μg mL^{-1}) enzyme solution is required for the measurements.

CD spectra can be used to characterize thermal stability either by comparing spectra measured at different time intervals for a sample incubated at a fixed temperature, or by monitoring the sample continuously at a specific wavelength as the temperature is raised. Subtle differences in conformation generally do not result in a detectable difference in the far-UV CD spectra, but differences may be seen in the near-UV CD spectra.

5.8.1.4 Kinetics of thermal denaturation

5.8.1.4.1 First-order denaturation kinetics

In a simplified analysis, thermal denaturation of an enzyme at a fixed temperature follows first-order kinetics (Ó'Fágáin, 2003; Sadana, 1988):

$$E \xrightarrow{k_d} D \tag{5.7}$$

where, E is the active enzyme, D is the denatured enzyme and k_d (s^{-1}) is the first-order denaturation rate constant. For first-order denaturation, the rate of denaturation depends on the molar concentration of the active enzyme remaining to be denatured; thus,

$$-\frac{d[E]}{dt} = k_d[E] \tag{5.8}$$

where $[E]$ is the molar concentration of the active enzyme at time t (s). Eq. (5.8) can be integrated using the following limits: at $t = 0$ the concentration of the active enzyme is $[E_0]$ and at $t = t$, the concentration of the active enzyme is $[E]$:

$$-\int_{E_0}^{E} \frac{dE}{[E]} = k_d \int_0^t dt \tag{5.9}$$

or

$$-\ln\frac{[E]}{[E_0]} = k_d t \tag{5.10}$$

The rate constant k_d can be estimated from the slope of a semilog (natural log) plot of $\ln[E]$ versus time. The concentration of the active enzyme at any time can be calculated using a rearranged form of Eq. 5.10, as follows:

$$[E]=[E_0]e^{-k_d t} \tag{5.11}$$

As the activity of the enzyme is proportional to its active molar concentration, the concentrations $[E]$ and $[E_0]$ in the Eqs. 5.10 and 5.11 may be replaced with activity.

The kinetic half-life ($t_{1/2}$) of an enzyme is the time required to reduce the activity to 50% of the original value under specified conditions. For first-order denaturation kinetics, at $t = t_{1/2}$, $E = 0.5\ E_0$. Therefore, from Eq. (5.11), the denaturation rate constant k_d and the kinetic half-life are related as follows:

$$t_{1/2} = \frac{\ln 2}{k_d} \tag{5.12}$$

The denaturation rate constant k_d depends on the activation energy of thermal denaturation (E_a, kJ mol^{-1}) and the absolute temperature T (K), as follows:

$$k_d = Ae^{-E_a/RT} \tag{5.13}$$

where, A (s^{-1}) is the Arrhenius frequency factor, or the preexponential factor and R is the universal gas constant (= 8.314×10^{-3} kJ mol^{-1} K^{-1}). For any given absolute temperature T (K), A may be calculated using the following equation:

$$A = \frac{k_B T}{h} \tag{5.14}$$

In Eq. 5.14, k_B (= 1.38×10^{-23} J K^{-1}) is the Boltzmann constant and h (= 6.626×10^{-34} J s) is Planck constant.

The activation energy for thermal denaturation of enzymes is generally around 200–300 kJ mol^{-1} and the native configuration is only marginally stable. A few highly thermostable enzymes have high activation energies of denaturation. For example, the highly thermostable extracellular α-amylase of the aquatic anaerobic hyperthermophilic archeal microbe *P. furiosus* has an activation energy of 310 kJ mol^{-1} (Brown et al., 2013). This enzyme is unusually stable, with a half-life of 19.1 min at 117°C (Brown et al., 2013).

The substrate concentration profiles can be calculated for a reaction obeying some particular kinetics and being performed at a temperature that causes a simultaneous first-order denaturation of the enzyme. For example, for a reaction obeying Michaelis-Menten kinetics (Eq. 5.6), the substrate is consumed and its concentration declines with time, as follows:

$$v = -\frac{d[S]}{dt} = \frac{k_{\text{cat}}[E][S]}{K_{\text{m}} + [S]} \tag{5.15}$$

In the above expression, the constant $v_{\max}$ of Eq. (5.6) has been replaced with $k_{\text{cat}}[E]$ where k_{cat} (s^{-1}) is the catalytic constant of the enzyme and $[E]$ is the normally fixed molar concentration of the active enzyme in the reaction system. If the enzyme is being denatured at the constant temperature of the reaction, its active concentration will decline with time. For a first-order denaturation process, the active concentration of the enzyme at any time will depend on its initial concentration $[E_0]$ and the denaturation rate constant k_d, as shown in the previously derived Eq. (5.11). Substitution of Eq. (5.11) in Eq. (5.15) results in the following:

$$-\frac{d[S]}{dt} = \frac{k_{\text{cat}}[E_0]e^{-k_d t}[S]}{K_{\text{m}} + [S]} \tag{5.16}$$

The above can be rearranged to:

$$-\left(\frac{K_m + [S]}{[S]}\right)d[S] = k_{\text{cat}}[E_0]e^{-k_d t}dt \tag{5.17}$$

or,

$$-K_m\frac{d[S]}{[S]} - d[S] = k_{\text{cat}}[E_0]e^{-k_d t}dt \tag{5.18}$$

Eq. (5.18) can be integrated using the following limits: at $t = 0$, the substrate concentration is $[S_0]$ and at any time t, the substrate concentration is $[S]$; thus,

$$-K_m\int_{[S_0]}^{[S]}\frac{d[S]}{[S]} - \int_{[S_0]}^{[S]}d[S] = k_{\text{cat}}[E_0]\int_0^t e^{-k_d t}dt \tag{5.19}$$

Solving the above results in the following equation:

$$K_m \ln\frac{[S_0]}{[S]} - [S] + [S_0] = -\frac{k_{\text{cat}}[E_0]}{k_d}\left(e^{-k_d t} - 1\right) \tag{5.20}$$

Known values of the various kinetic constants (i.e., K_m, k_{cat}, k_d) and the initial concentrations (i.e., $[S_0]$, $[E_0]$), may be used in Eq. (5.20) to calculate the substrate concentration remaining at any time t during the reaction performed at such a temperature that there is denaturation of the enzyme.

5.8.1.4.2 More complex denaturation kinetics

The first-order denaturation kinetics are a simplification and, often, the thermal denaturation kinetics are more complex (Daniel and Danson, 2013; Feller, 2010; Iyer and Ananthanarayan, 2008; Ó'Fágáin, 2003; Sadana and Henley, 1986). In practice, denaturation may involve a two-step mechanism (Feller, 2010; Sadana, 1988; Weijers and Van't Riet, 1992) in which the native active enzyme E first reversibly, or partly, unfolds to a denatured inactive form D. A subsequent irreversible step transforms D to the permanently inactivated form I; thus,

$$\underbrace{E}_{Active} \rightleftarrows \underbrace{D}_{Denatured} \rightarrow \underbrace{I}_{Inactivated} \tag{5.21}$$

The native state E is the conformation of minimal free energy. The native state is only marginally thermodynamically stable with a low free energy of stabilization of around 40–120 kJ mol^{-1} (Weijers and Van't Riet, 1992). Irreversible unfolding, that is, the D to I reaction, is driven by kinetics and may be viewed as kinetic stability (Feller, 2010). The reversible denaturation, that is, the E to D reversible step, is associated with thermodynamic stability (Feller, 2010).

Most of the different deactivation mechanisms proposed can be reduced to the following simpler biexponential equation for concentration of the active enzyme at any time during the reaction (Ó'Fágáin, 2003):

$$[E]=[E_0]\left\{\theta e^{-\alpha t}+(1-\theta)e^{-\beta t}\right\} \tag{5.22}$$

where α, β, and θ are complex functions of the various rate constants. If $\theta = 0$, the above expression reduces to the first-order form (see Eq. 5.11) (Ó'Fágáin, 2003). When $\beta = 0$, some residual activity remains after the denaturation treatment (Ó'Fágáin, 2003). The above three-parameter model (Eq. 5.22) applies to many deactivation processes (Ó'Fágáin, 2003).

In a dilute solution of an enzyme, the thermally induced unfolding may be fully reversible by lowering the temperature. In other cases, thermal denaturation is followed by protein aggregation and precipitation, preventing a return to the active folded state (Ó'Fágáin, 1995). Aggregation is driven by the need to minimize the thermodynamically unfavorable interactions between the solvent and the exposed hydrophobic residues (Balcão and Vila, 2015).

Folding of a polypeptide chain is driven by entropy changes (Shaw and Bott, 1996). An unfolded polypeptide chain has many more degrees of freedom compared to a folded chain. Thus, folding considered alone, would seem to reduce entropy. In practice, an unfolded chain has a much higher degree of hydration compared to a folded chain and folding increases the entropy of the solvent. A net effect of folding is an increased entropy of the system (Shaw and Bott, 1996).

Whereas entropic factors derive folding, the free energy, or enthalpic factors contribute to stabilizing the folded molecule, or preventing it from unfolding. Free energy, or internal energy, of an enzyme molecule is lowered by the attractive interactions (e.g., hydrogen bonding, hydrophobic interactions, salt bridges, disulfide bridges) between various regions of the folded polypeptide chains. These enthalpic factors contribute to stability. Both entropic and enthalpic factors contribute to the overall stability and can be manipulated for stability enhancement (Shaw and Bott, 1996).

5.8.2 Chemicals and other environmental factors affecting stability

The chemical environment of an enzyme affects its stability. The chemical factors include the pH, the ionic strength, and the presence of other chemicals including proteolytic enzymes (proteases). Thermal stability is also influenced by the chemical environment including the pH, the ionic strength, and the hydrophobicity of the solvent.

An enzyme may be acidic, basic, or neutral, depending on the whether it contains more of acidic or basic amino acids in its structure. At neutral pH, arginine, lysine and histidine have basic side chains whereas aspartic acid and glutamic acid have acidic side chains. Changes in pH affect the ionization of the functional groups on the enzyme molecule and this in turn influences stability by affecting the ease of unfolding (Gianfreda and Scarfi, 1991). If the pH is less than its isoelectric pH (or pI), the enzyme molecule has a net positive charge. The net charge is negative at pH > pI. The pH may also affect the ionization status of the functional groups at the active site, to strongly influence activity. The rigidity of tertiary and quaternary structures depends on pH as well as the ionic strength of the surrounding medium.

Chemicals such as urea and acrylamide (and other amides) are well-known enzyme denaturants. Nonetheless, these chemicals have been shown to have a protective effect on enzymes against thermal denaturation

(Martinek et al., 1975). Denaturation by urea has multiple mechanisms including direct binding to the enzyme and indirect alteration of the solvation environment (Bennion and Daggett, 2003; Gianfreda and Scarfi, 1991).

Oxidation is another damaging factor. Although in vivo some enzymes are known to protect an organism against oxidative damage, certain enzymes are deactivated by oxygen and hydrogen peroxide (Gianfreda and Scarfi, 1991) both in vivo and in vitro. Damage by oxygen or peroxide is generally associated with oxidation of functional groups on the enzyme molecule (Davies, 2003; Gianfreda and Scarfi, 1991). The availability of oxygen in a medium may be reduced by reducing its solubility by adding salt and also by operating at an elevated temperature (Fig. 5.4).

Intense light, especially ultraviolet light, damages enzymes (Kerwin and Remmele, 2007) and accentuates the damaging effects of oxygen. Amino acid residues particularly susceptible to photooxidation include tryptophan, tyrosine, phenylalanine, and cysteine/cystine (Kerwin and Remmele, 2007).

Compounds present in the reaction mixture may act as suicide inhibitors (or suicide inactivators, or mechanism-based inactivators) of the enzyme. A suicide inhibitor binds to the active site and the ensuing reaction results in intermediates that covalently attach to functional groups of the active site, making it irreversibly inactive.

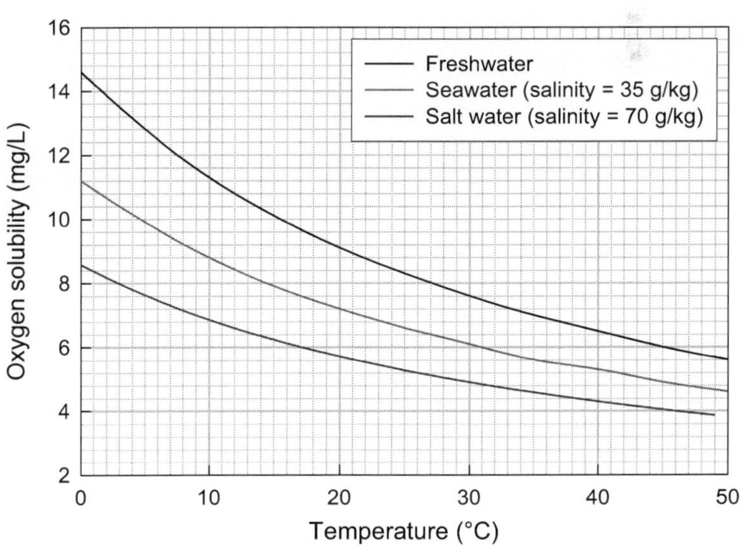

Figure 5.4 Solubility of oxygen in water at various temperatures and salinities (g salt kg^{-1} solution). The data shown are for an absolute pressure of 1 atmosphere.

5.8.3 Pressure effects on stability and activity

High pressures (100–200 MPa) have been found to dissociate some multimeric enzymes (Fernandez-Lafuente, 2009). Pressures of >300 MPa generally cause irreversible enzyme denaturation at room temperature (Hendrickx et al., 1998). Denaturation is attributed to pressure-induced changes in conformational volume and, therefore, the secondary structure of the enzyme. The mechanics of the pressure effect are discussed by Silva and Weber (1993). Industrial processes for enzyme production and use typically operate at much lower pressures (e.g., 0.1–0.2 MPa) than the damaging threshold, but higher pressures are encountered in operations such as high-pressure homogenization (typical 50–200 MPa operating pressure) used in breaking microbial cells to release intracellular enzymes (Chisti and Moo-Young, 1986). Enzyme damage is known to occur during high-pressure homogenization, but this is not the effect of pressure per se, but of the intense shear rates (Chisti, 2010) generated during homogenization.

Some enzymes, for example those of the deep-sea creatures, actually function best under high pressure. At the deepest point in the ocean, the depth is nearly 11 km. The hydrostatic pressure at a depth of 3 km in the ocean is around 30 MPa, rising to nearly 111 MPa at a depth of 11 km. Such pressure activated or piezophilic enzymes are quite similar in their amino acid sequence to their normal pressure counterparts; therefore, the pressure adaptation of deep-sea enzymes cannot be attributed to their primary structure (Ohmae et al., 2013). The local physiological environment may contribute to pressure tolerance and under identical conditions of the surrounding medium, the deep-sea enzymes may be no more pressure tolerant than their terrestrial equivalents (Ohmae et al., 2013). The pressure activation of a piezophilic enzyme is likely associated with the pressure-linked changes in the internal cavity volume and the state of hydration. Piezophilic enzymes are further discussed by Ohmae et al. (2013).

5.8.4 Damage by high shear rates in the fluid

Certain flow processes subject the enzymes in solution to intense shear rates with the potential to damage the protein (Chisti, 2010). Examples of such high-shear operations include mechanical disruption of microbial cells in high-pressure homogenizers and bead mills, as commonly used for recovering intracellular enzymes (Chisti and Moo-Young, 1986). In addition, enzymes tend to unfold at gas-liquid interfaces and also at some solid-liquid interfaces, and such unfolded enzymes are particularly susceptible to damage by mechanical forces such as the fluid velocity gradients (or hydrodynamic shear forces) encountered during flow (pumping, mixing) in processing equipment. Effects of hydrodynamic forces on enzymes in solution have been discussed elsewhere (Chisti, 2010). Damage is generally mitigated by a redesign of the process equipment, changes to the operational regime, avoidance of gas-liquid interfaces and foam formation, a reduced processing temperature, and rapid processing to minimize exposure to a damaging environment.

5.9 Storage stability

Storage stability is important, as after production the enzyme must be transported to the point of use and stored until used. Storage conditions may be optimized to reduce loss of activity over a relatively long period and may be quite different than the conditions of use. During storage and use, enzymes need to be protected from light as some functional groups on the polypeptide chain are susceptible to photooxidation and other photochemical damage (Schmid, 1979). Enzymes are generally more stable in concentrated solutions (Chisti, 2007). A dilute solution (<1 g L^{-1}) is more susceptible to inactivation and activity may be lost as a consequence of binding to the container walls (Silva et al., 2018). Therefore, an inactive "filler" protein such as bovine serum albumin may be added to bring the total protein concentration to around 2–6 g L^{-1} (Silva et al., 2018). Storage temperature is typically 4°C and freezing is avoided as thawing is time consuming and freeze-thaw cycles tend to damage enzymes (Section 5.9.1). Protective additives are commonly used to enhance storage stability as discussed in Section 5.9.2.

5.9.1 Storage temperature

Freeze-dried enzyme powders kept in the dark are highly stable at 4°C. More often, enzymes in solution need to be stored. Storage temperature is typically 4°C. Freezing is avoided, as freeze-thaw cycles damage enzymes. The damage is associated with the formation of ice crystals and ice-liquid interfaces during freeze and thawing (Cao et al., 2003). Freezing of an aqueous solution can be prevented by adding glycerol, propylene glycol, dimethyl sulfoxide (DMSO), and other cryoprotectants. The concentration of cryoprotectant needed in a solution depends on the temperature to which freezing protection is desired. For example, 50% by weight glycerol in an aqueous solution will prevent freezing down to −20°C. Similarly, 50%

by volume of propylene glycol in an aqueous solution will prevent freezing down to −30°C and 45% by weight of DMSO in water will effectively prevent freezing down to −20°C. Other cryoprotectants may be used, but the ones mentioned here are essentially nontoxic. (For example, DMSO is generally considered nontoxic to many live animal cells at a concentration of 10% by volume. In practice, a cryopreserved enzyme solution would be used so diluted that the concentration of the cryoprotectant in the final reaction medium as well as the product would be low. Similarly, unlike the highly toxic ethylene glycol (also a cryoprotectant), propylene glycol is generally recognized as safe, in most scenarios. Glycerol is essentially nontoxic.) Normal freezing points (i.e., at standard atmospheric pressure) of aqueous solutions of some cryoprotectants are shown in Fig. 5.5.

If freezing becomes necessary, a slow freezing (a freezing rate about 1°C min^{-1}) and rapid thawing (a thawing rate >10°C min^{-1}) cycle is recommended as more activity is recovered compared to a cycle involving fast freezing followed by slow thawing (Cao et al., 2003). Rapid freezing results in smaller ice crystals being formed and therefore a relatively large ice-liquid interface. Recrystallization during thawing causes further damage to enzyme because of the mechanical forces at the solid-liquid interface. The buffer used during freezing also affects the extent of enzyme damage (Cao et al., 2003).

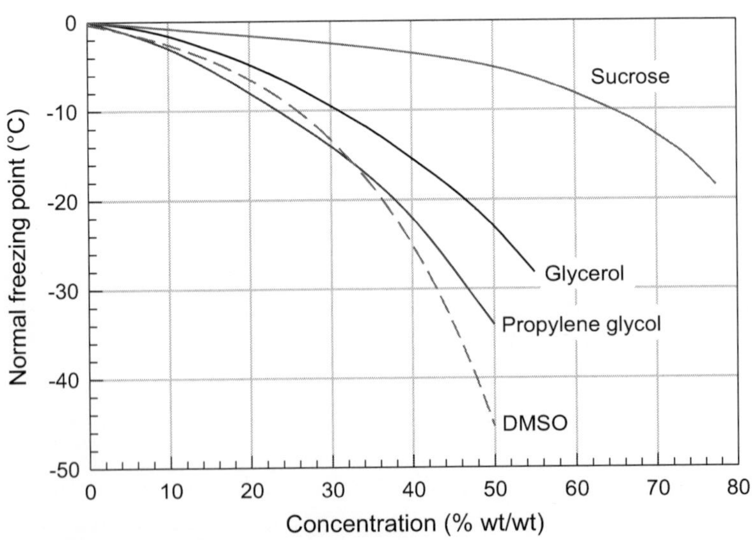

Figure 5.5 Normal freezing points of aqueous solutions of some cryoprotectants.

5.9.2 Protective additives

Additives are used to enhance the storage stability of enzymes. Enzyme solution preparations sold commercially commonly contain protective additives. Specific additives serve specific needs, as discussed in the following sections.

Protective effects of some additives are at least partly linked with their impact on the water activity in solution and the degree of structural organization of water around the enzyme (Combes et al., 1987). For example, deuterium oxide (D_2O, heavy water) has a higher degree of molecular structural organization than water and, consequently, at 60°C invertase dissolved in pure D_2O is nearly 3.6-fold more stable compared with invertase dissolved in water (Combes et al., 1987). In water at 60°C, the half-life of invertase is around 14 min. This could be increased by nearly 770-fold by dissolving the enzyme in D_2O supplemented with sorbitol to a concentration of 2.55 M (Combes et al., 1987). Thus, substantial increases in stability are possible through a combination of stabilizing mechanisms. In practice, D_2O is not used as a stabilizing agent because of its high cost. Whether an additive is protective, or destabilizing, depends on the nature of the enzyme, particularly on its hydrophilic or hydrophobic character (Combes et al., 1987). Enzyme stabilization with additives is further discussed in the literature (Balcão and Vila, 2015; Gianfreda and Scarfi, 1991; Iyer and Ananthanarayan, 2008; Schmid, 1979; Torchilin and Martinek, 1979). The protective mechanisms attributed to various additives are discussed by Schmid (1979), Balcão and Vila (2015), and Silva et al. (2018).

5.9.2.1 Substrates

Presence of the substrate, or substrate analogs, may stabilize an enzyme in solution. For example, the thermal stability of α-amylase of *Bacillus amyloliquefacienes* is enhanced in the presence of soluble starch at a concentration of 200 g L^{-1} (Tomazic, 1991). Binding of the substrate to the active site increases an enzyme's resistance to unfolding; that is, the melting temperature (Section 5.8.1.2) of the enzyme is increased by substrate or ligand binding.

In addition to storage stability, the operational stability (i.e., the retention of activity during the conditions of use) may also be enhanced in the presence of the substrate. Presence of the substrate is known to enhance thermostability of many, but not all enzymes (Combes et al., 1987; Schmid, 1979). For example, the extracellular invertase of *Saccharomyces cerevisiae* is most active in the temperature range of 50°C–60°C and has an optimum temperature of 55°C in dilute sucrose solutions (Tananchai, 2011). In the absence of the substrate (sucrose), nearly all the activity is lost within 15 min at 65°C

(Schülke and Schmid, 1988); however, in concentrated solutions of sucrose the enzyme is stable up to 70°C (Tananchai, 2011).

5.9.2.2 *Water soluble polymers*

Polyethylene glycol (PEG; Fig. 5.6) is a widely used protective additive. It is believed to act by enhancing the hydration of the enzyme molecule (Chisti, 2007; Silva et al., 2018). Stabilizing effect PEG increases both with its concentration and molecular mass (Balcão and Vila, 2015). Other stabilizing water soluble polymers include the polysaccharide dextran and the cationic polyethylenimine (PEI; Fig. 5.6). For example, PEI (50 kDa molar mass) at concentrations between 0.3 and 0.5 mM, had a stabilizing effect on glucose dehydrogenase during use in an electrochemical reactor (Obón et al., 1997). Both low (2 kDa) and high (600 kDa) molecular weight PEIs (0.1–10 g L^{-1}) have been found to be equally effective in stabilizing certain multimeric enzymes (Iyer and Ananthanarayan, 2008).

5.9.2.3 *Water miscible organic solvents*

Water miscible organic solvents such as ethanol, methanol and acetone added at concentrations of up to 200 g L^{-1} stabilize some enzymes (Tomazic, 1991), apparently by reducing the concentration of water in the vicinity of the enzyme.

5.9.2.4 *Polyhydroxy alcohols (polyols) and sugars*

High concentrations (10–300 g L^{-1}) of polyhydroxy alcohols, or polyols (e.g., glycerol, sorbitol) and sugars such as sucrose, glucose, lactose, and trehalose, enhance stability of many enzymes during storage (Chisti, 2007; Tomazic, 1991). Presence of these compounds in the medium results in the enzyme molecule folding tighter and this enhances its resistance to denaturation (Iyer and Ananthanarayan, 2008). Formation of hydrogen bonds between water

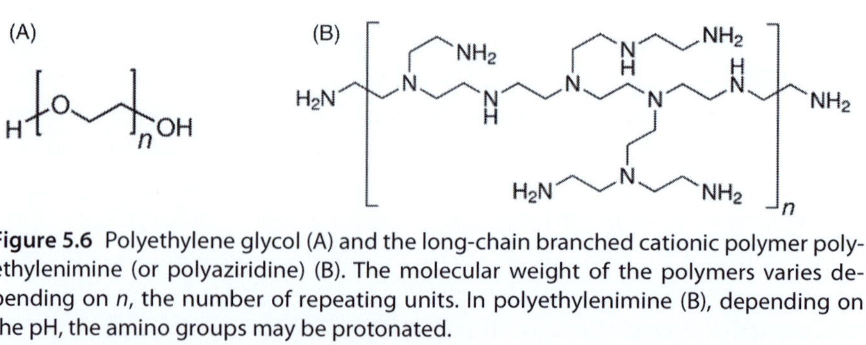

Figure 5.6 Polyethylene glycol (A) and the long-chain branched cationic polymer polyethylenimine (or polyaziridine) (B). The molecular weight of the polymers varies depending on *n*, the number of repeating units. In polyethylenimine (B), depending on the pH, the amino groups may be protonated.

and the additive molecules reduces water activity and, therefore, may protect against microbial degradation of protein in aqueous solutions. In addition, these compounds rigidify enzyme molecules by enhancing the hydrophobic interactions among nonpolar surface functional groups on the molecule (Stepankova et al., 2013). Furthermore, the viscosity of the medium surrounding the enzyme molecule is increased because of the dissolved solute, restricting movement of the surface regions of the enzyme molecules.

5.9.2.5 Antioxidants

Disulfide bonds in enzymes are highly susceptible to oxidation. Amino acid residues particularly susceptible to oxidation are cysteine (the thiol group), tryptophan (indole ring), and methionine (Weijers and Van't Riet, 1992). Antioxidants such as vitamin E and ascorbic acid are used to improve stability (Chisti, 2007; Tomazic, 1991; Weijers and Van't Riet, 1992). Thiols readily react with the disulfides and, therefore, should be excluded from enzyme solutions. Low concentrations of thiol compounds (e.g., mercaptoethanol, dithiothreitol) are often used to suppress inactivation of sulfur-containing enzymes during recovery and purification, but high concentrations of these compounds lead to reduction (breakage) of the disulfide bond (Schmid, 1979).

5.9.2.6 Neutral salts

High concentrations ($\geq$0.1 M) of neutral salts of Na^+, NH_4^+, Cl^-, and SO_4^-, have a generic stabilizing effect on enzymes (Tomazic, 1991; Iyer and Ananthanarayan, 2008). Ammonium sulfate and potassium phosphate are commonly used as stabilizers in solution (Chisti, 2007). In some cases, presence of nonmetal ions such as sulfate and phosphate may be necessary for stability. For example, the dimeric structure of starch phosphorylase of *Corynebacterium callunae* is stabilized by phosphate and sulfate (Fernandez-Lafuente, 2009). The enzyme is comprised of two 90 kDa subunits each containing a pyridoxal 5'-phosphate as an active-site cofactor (Fernandez-Lafuente, 2009). The quaternary structure of industrially important tetrameric β-glactosidases of *Escherichia coli* and *Kluyveromyces lactis* is strongly stabilized by the presence of Mn^{2+} and Mg^{2+} (Fernandez-Lafuente, 2009).

5.9.2.7 Specific divalent metal ions

Specific divalent metal ions (Ca^{2+}, Mn^{2+}, Zn^{2+}, Mg^{2+}) used in low concentrations ($\leq$0.1 M) stabilize specific enzymes, apparently by forming ionic bridges that increase the rigidity of the tertiary structure (Chisti, 2007; Fernandez-Lafuente, 2009; Iyer and Ananthanarayan, 2008; Tomazic, 1991).

Other stabilizing mechanisms may be involved. This effect typically occurs with enzymes that require the relevant metal ion for catalytic activity. In such cases, stability may depend on the presence of the relevant metal ions in the medium. In other instances, metal ions may be involved in the catalytic role of the enzyme. For example, the extracellular α-amylase of *Aspergillus oryzae* is a metaloenzyme that requires Ca^{2+} for activity and stability.

5.9.2.8 Protease inhibitors

Proteases inactivate enzymes by proteolysis (Schmid, 1979; Weijers and Van't Riet, 1992). The conformation, size, charge, and structural features of the enzyme influence its susceptibility to proteolysis. Acid proteins and larger molecules are generally degraded more rapidly compared to neutral and basic proteins and smaller molecules (Schmid, 1979; Weijers and Van't Riet, 1992). Some proteases are susceptible to autolysis (Stoner et al., 2004). Protease inhibitors can be added to enzyme solutions to protect against proteolytic damage. Either a cocktail of protease inhibitors, or a specific type of inhibitor, may be used, depending on the contaminating proteases present. Inhibitors targeting serine, cysteine, and aspartic acid proteases, and aminopeptidases are commercially available. Phenylmethylsulfonyl fluoride (PMSF; 0.1–1.0 mM), for example, inhibits serine proteases (chymotrypsin, trypsin, and thrombin) and also cysteine proteases such as papain. Aprotinin (1–2 $\mu g\ mL^{-1}$) and benzamidine (15 $\mu g\ mL^{-1}$) may be used to inhibit serine proteases. Leupeptin (1–2 $\mu g\ mL^{-1}$) inhibits cysteine and serine proteases. Metal ion chelating agents such as EDTA (1–10 mM) may be used to inhibit metalloproteases.

5.9.2.9 Surfactants

Depending on type and concentration, surfactants can both destabilize and stabilize enzymes (Imamura et al., 2014; Kaspersen et al., 2017; Madsen et al., 2015; Ó'Fágáin, 2003; Otzen, 2011, 2015; Schmid, 1979; Silva et al., 2018). Certain surfactants have been effectively used to stabilize specific enzymes during storage (Ó'Fágáin, 2003; Schmid, 1979; Silva et al., 2018). Interactions between surfactants and enzymes are further discussed by others (Holmberg, 2018; Imamura et al., 2014).

5.10 Immobilized enzymes and stability enhancement by immobilization

Enzymes are generally soluble in aqueous media and are commonly used as such. An enzyme in solution is lost after a single use. Immobilization arose as a method of making the enzyme insoluble by attaching it to a solid support, or confining it within a gel matrix, so that it could be recovered easily and reused (Cowan and Fernandez-Lafuente, 2011). Immobilization was found to commonly enhances enzyme stability. In some forms of immobilization, a soluble enzyme may be attached to a soluble inert polymer to increase the molecular weight of the enzyme-polymer entity that remains soluble but is relatively easily retained in a homogeneous reaction system by using ultrafiltration membranes (Chisti, 2007b) with a pore size that is smaller than the size of the enzyme-polymer complex. An enzyme may be immobilized also by linking several of its molecules together to produce a larger protein complex that remains in solution but is easier to retain by ultrafiltration membranes. Alternatively, so many molecules of the enzyme may be linked together that it comes out of solution as a bioactive suspended particle made almost entirely of the enzyme of interest. Enzymes naturally immobilized within viable and inactivated microbial cells are also used in bioprocesses (Pai et al., 2014; Patil et al., 2013; Pawar et al., 2012; Singh et al., 2012). This approach eliminates the expense of recovery and purification of the enzyme from the cells but allows prolonged use as the enzyme is more stable compared to the same purified enzyme in solution.

Immobilized enzymes are commonly used in industrial processes particularly those involving soluble substrates with a low molecular weight. Although, the ease of recovery and reuse drove the early development of immobilized enzymes, other benefits of immobilization were soon discovered. These included improved thermal stability and an altered optimal pH relative to the free enzyme (Cowan and Fernandez-Lafuente, 2011; Klibanov, 1979; Lee et al., 2017; Mateo et al., 2007) if the immobilization matrix carried charged functional groups. Multiple enzymes may be immobilized in a matrix in close proximity to facilitate transformations requiring several enzymes (Jia et al., 2014).

Immobilization may extend the half-life of an enzyme by thousands of fold relative to the soluble native enzyme (Mateo et al., 2007). The thermostability enhancing effect of immobilization is explained by the resistance to unfolding imparted by the confining solid matrix, or by the multipoint covalent binding to the solid support (Klibanov, 1979). In multipoint attachment to a solid support (Fig. 5.7), the support acts as a rigid multifunctional crosslinker (Cowan and Fernandez-Lafuente, 2011). In addition, an entrapping matrix may protect the enzyme against other denaturants including proteases, the adverse pH values, the high concentrations of damaging chemicals, drops of a hydrophobic solvent, and gas bubbles (Cowan and Fernandez-Lafuente, 2011).

A shortcoming of immobilization on or within a solid matrix is an inevitable reduction in the substrate binding capacity of the enzyme (Mozhaev

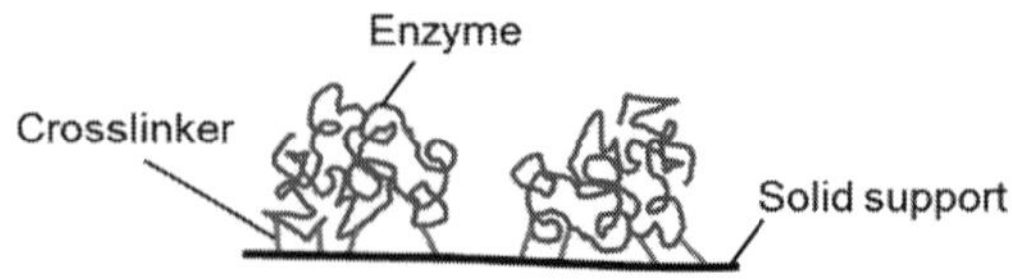

Figure 5.7 Multipoint attachment of an enzyme to a solid support rigidifies the enzyme. The degree of rigidification depends on the number of enzyme surface points attached to the solid surface, the surface area of the enzyme molecule involved in the attachment, the length of the linker molecule, and possible other factors. Based on Cowan and Fernandez-Lafuente (2011).

et al., 1990) because of steric interferences. Furthermore, an enzyme bound to solid support my be completely unusable if the substrate is insoluble (e.g., cellulose, starch), because the active site is almost inaccessible to the substrate (Gianfreda and Scarfi, 1991) and the diffusional mobility of both the enzyme and the substrate are severely limited (Chisti, 2010).

Immobilization methods and immobilized enzymes have been extensively reviewed in the literature (Balcão and Vila, 2015; Bernal et al., 2018; Hanefeld et al., 2009; Homaei et al., 2013; Iyer and Ananthanarayan, 2008; Lee et al., 2017; Sheldon, 2007; Singh et al., 2013; Sirisha et al., 2016). Stability considerations for enzymes immobilized in membranes are discussed by Yabuki (2017). Two special forms of immobilized enzymes not involving an inert solid support, or a soluble inert polymer, are the crosslinked enzyme crystals and the crosslinked enzyme aggregates, as discussed next.

5.11 Crosslinked enzyme crystals (CLEC) and crosslinked enzyme aggregates (CLEA)

Enzymes may be immobilized by crosslinking their microcrystals (Fernandez-Lafuente, 2009; Govardhan, 1999; Ó'Fágáin, 2003). Individual microcrystals range in size from 1 to 200 μm (Govardhan, 1999). Channels between the crosslinked crystals are of the order of 2–10 nm in diameter (Govardhan, 1999), allowing easy diffusion of small molecules. Bifunctional reagents such as glutaraldehyde (Fig. 5.8) have been used for crosslinking. Such crosslinked enzyme crystals (CLEC) tend to be stable against diverse denaturants including anhydrous organic solvents, elevated temperature, autolysis and proteolysis. High volumetric activities are achieved with CLECs, but their disadvantages include a high cost because the enzyme needs to be purified and crystalized; poor utilization of the enzyme as only a fraction of the total enzyme is actually active; and diffusional limitations for large substrate molecules.

Figure 5.8 Glutaraldehyde, or pentanedial. The distance between the reactive ends is around 0.496 nm.

Another approach for support-free crosslinking is to use crosslinked enzyme aggregates (CLEA) (Sheldon, 2011) that are generally much cheaper than CLEC. In preparing CLEA, a crude enzyme is precipitated from solution and the resulting particles are crosslinked with a bifunctional reagent such as glutaraldehyde to prevent redissolution in the absence of the precipitating reagent (Fernandez-Lafuente, 2009). If a reaction scheme requires sequential use of several enzymes, they could all be coprecipitated and crosslinked for use in a one-pot reaction. CLEAs of many commercially important enzymes have been produced (Illanes et al., 2012; Sheldon, 2011). Like CLECs, CLEAs tend to be generally more stable compared to native enzymes.

5.12 Immobilization on nanoparticles and other nanocarriers

Enzymes are increasingly being immobilized on nanoparticles and other nanocarriers (Hwang and Gu, 2013; Illanes et al., 2012; Kim et al., 2006; Lee and Au-Duong, 2018). The advantage of nanosupports is a high surface-to-volume ratio compared to the conventionally immobilized enzymes. As with conventional immobilization, enzymes attached to nanostructures are often more stable compared to their counterparts in solution (Hwang and Gu, 2013; Lee and Au-Duong, 2018). Enzyme nanoparticles in which individual enzyme molecules are surrounded by a porous network of organic or inorganic molecules, have been reported for improving enzyme stability (Kim et al., 2006). The confining network around the enzyme molecule is thinner than a few nanometers (Kim et al., 2006). Such stabilized enzymes may be further immobilized in conventional ways for ease of recovery, or use (Kim et al., 2006).

5.13 PEGylation and attachment to other water-soluble polymers

Polyethylene glycol (PEG; Fig. 5.6), and its derivatives, can be covalently and noncovalently linked to enzymes in a process known as PEGylation. PEGylation is a method of surface modification of enzymes.

PEG is a water-soluble polymer with the general formula shown in Fig. 5.6. PEG has both hydrophilic and hydrophobic properties and PEGylated enzymes are soluble in hydrophobic organic solvents such as toluene (Stepankova et al., 2013). PEGylation improves enzyme stability in organic media and may affect other properties such as catalytic activity and enantioselectivity (Stepankova et al., 2013). Low molecular weight PEG (e.g., 400 Da) is typically used for surface modification. PEGylation may result in a substantial increase in the hydrodynamic volume of the enzyme (Silva et al., 2018) and this may have associated mass transfer effects especially if the substrate is a large molecule. The reactions used in protein PEGylation have been discussed by others (Baslé et al., 2010; Dozier and Distefano, 2015). Unbound PEG may be used as a protective additive as discussed in Section 5.9.2.2.

Multipoint covalent attachment to various other water-soluble polymers has been used to enhance enzyme stability (Ó'Fágáin, 2003; Schmid, 1979). Covalent attachment of an enzyme to a polysaccharide such as dextran has nearly always improved stability (Iyer and Ananthanarayan, 2008).

5.14 Chemical modifications for enhancing enzyme stability

The stability of a native enzyme may be enhanced by chemically modifying the functional groups on its surface so that the rigidity of the molecule and its resistance to unfolding are improved. Chemical modification has certain advantages over immobilization because the stabilized enzyme remains soluble and the immobilization matrix does not interfere with the binding of the substrate. Furthermore, the chemically modified enzyme is generally close to its native size and therefore, the severe mass transfer limitations encountered with immobilized enzymes especially if the substrate is a macromolecule, do not generally occur.

Chemical modification involves reacting the surface functional groups of the enzyme with a modifying reagent. As multiple reactive functional groups (e.g., $-NH_2$, $-COOH$) of a type may be typically available (Fig. 5.9), chemical modification is generally not site specific, unless a unique functional group can be targeted by the chemistry involved (Cowan and Fernandez-Lafuente, 2011). For example, the thiol group of cysteine may be specifically modified in an enzyme with only a single surface cysteine residue. Chemical modification can be used to increase the variety of the functional groups available on an enzyme molecule (Cowan and Fernandez-Lafuente, 2011).

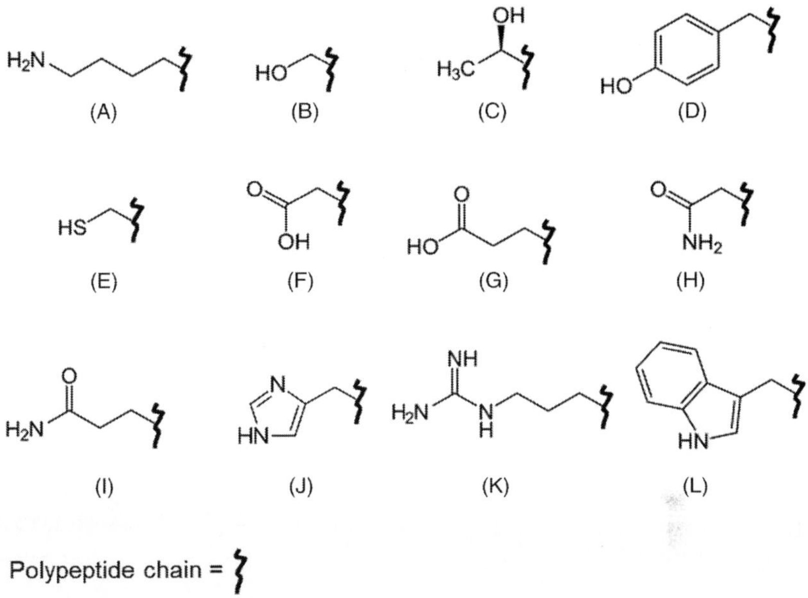

Figure 5.9 Reactive functional groups available on an enzyme molecule for crosslinking: (A) primary amine of lysine; (B) primary alcohol of serine; (C) secondary alcohol of threonine; (D) aryl alcohol of tyrosine; (E) thiol of cysteine; (F) carboxyl groups of aspartic and (G) glutamic acids; (H) amides of asparagine and (I) glutamine; (J) the imidazole ring of histidine; (K) the guanidino group of arginine; and (L) the indole group of tryptophan.

The ε-amino groups of lysine residues are the most abundant and accessible reactive functional groups on an enzyme.

Covalent binding of modifying agents to an enzyme may involve one or more of the following approaches (Janeček, 1993; Mozhaev et al., 1990): (1) use of bifunctional molecules to crosslink two functional groups on the enzyme surface; (2) strengthening of the hydrophobic interactions by modifying some of the hydrophilic surface groups with hydrophobic monofunctional modifiers; (3) enhancing ionic interactions, or hydrogen bonding, by replacing some of the hydrophobic functional groups by hydrophilic or ionic groups using monofunctional reagents; (4) surface hydrophilization to improve compatibility with the aqueous medium. Chemical modifications aiming to modify the surface functional groups without forming crosslinks, and those intended to crosslink the surface functional groups to build a network to physically constrain the tendency to unfold, are discussed separately in the following sections.

As many functional groups are potentially available on the enzyme surface (Fig. 5.9) for reacting with a modifying reagent and the reactions involved are not highly specific, not all the surface modified and crosslinked enzyme molecules are exactly the same (Davis, 2003; Silva et al., 2018). Instead, a mixture of variously crosslinked or surface modified enzyme molecules is produced (Silva et al., 2018). Chemical modifications of specific enzymes have been reviewed (Davis, 2003; DeSantis and Jones, 1999; Qi et al., 2001; Tann et al., 2001). Modifications have targeted various catalytic attributes and not just stability.

Although the various forms of protein engineering also involve changes to the chemical structure of the enzyme, they are not considered chemical modifications per se as the native enzyme is not directly modified through one or more chemical reactions.

5.14.1 Enzyme surface modification without crosslinking

Surface of an enzyme molecule is predominantly hydrophilic because the folding of the polypeptide chain into a globular structure tends to bury the hydrophobic side chains of the various amino acid residues in the interior of the globule (Balcão and Vila, 2015). More than 95% of the charged functional groups are also located on the surface (Balcão and Vila, 2015).

Surface hydrophilicity/hydrophobicity and charge can be modified by modifying the surface functional groups by reaction with a suitable monofunctional (i.e., noncrosslinking) reagent. The surface may be hydrophilized either by converting the hydrophobic surface functional groups to hydrophilic ones, or by modifying the nearby groups to enhance their hydrophilicity so that the hydrophobic groups are better shielded from the aqueous surroundings (Janeček, 1993; Mozhaev et al., 1990). The hydrophilization functional groups on the most promising monofunctional reagents are $-OH$, $-COOH$, $-NH_2$, and $-N^+(CH_3)_3$ (Mozhaev et al., 1990). For example, the relatively less hydrophilic surface $-NH_2$ functional groups may be transformed to the more hydrophilic $-NHCH_2COOH$ to enhance hydrophilicity (Gianfreda and Scarfi, 1991). The principal reactive functional groups commonly available on the surface of an enzyme molecule are shown in Fig. 5.9.

If immobilization on supports is excluded, hydrophilization of the surface functional groups has been found to be the most effective method of stabilizing enzymes by chemical modifications (Janeček, 1993). Hydrophilization improves the compatibility of the enzyme surface with the surrounding aqueous medium.

If necessary, the surface hydrophilicity can be reduced by reacting some of the hydrophilic functional groups on the surface with a monofunctional reagent having a hydrophobic nonreactive end. This converts a hydrophilic group to a hydrophobic one. For example, the free amino groups (E–NH_2; E = enzyme) on the surface may be acetylated (E–NH–C(O)CH_3), or alkylated (E–NH–CH(OH)CH=CH_2) (Tananchai, 2011). This way an enzyme may be made more compatible with a solvent that is relatively less hydrophilic than water.

The ionic character of the surface can be modified, for example, by succinylation (E–NH–C(O)–$(CH_2)_2$COOH) of the free amino groups (E–NH_2) (Tananchai, 2011). Succinylation increases the number of carboxylic acid (–COOH) groups on the enzyme surface at the expense of the amino groups.

Surface modifications have been used to improve resistance to heat-induced unfolding (Tananchai, 2011). Changes in optimal pH can be achieved through surface modification and the solvent-compatibility of an enzyme may be altered by this method. Examples of some monofunctional reagents for surface modification are shown in Table 5.2.

5.14.2 Enzyme surface modification involving crosslinking

A crosslinker is a molecule with two reactive ends capable of forming covalent bonds with one or more kinds of functional groups on the surface of a protein molecules. If the reactive ends of the crosslinker are identical, it is said to be a homobifunctional crosslinker. A heterobifunctional crosslinker has different reactive functional groups at its two ends. Crosslinking of two

Table 5.2 Some monofunctional reagents for surface modification of enzymes.

Reagent	Chemical formula	Enzyme (E) surface functional group modified
Butyl isocyanate	$H_3C(CH_2)_3N=C=O$	E–NH_2 at pH> ~7; E–S^- at pH 6–8; E–Ph–OH (Ph = phenyl ring) at pH >5; E–COO^- at pH ~5; E–Im (Im = imidazole ring in histidine) at pH ~8
Ethyl acetimidate	$H_3CC(=NH)OC_2H_5$	Reacts with primary E–NH_2 groups to form an amidine bond (ε-amine in lysine and N-terminal amines). Reaction occurs at alkaline pH (pH 7.5–10). The amidine bond is rapidly hydrolyzed at room temperature and pH 7–9. At pH > 11, the amidine bond is unstable. Around neutral pH, the amidine is protonated and, therefore, has a positive charge.
Primary amine	R–NH_2	E–COOH is modified to E–C(=O)NHR

functional groups on the surface of the same enzyme molecule is intramolecular crosslinking. If one functional group on an enzyme molecule is linked to a group on a different enzyme molecule, the crosslinking is said to be intermolecular. The distance between the reactive ends of a cross-linker molecule determines the maximum distance between two functional groups that may be crosslinked on the surface of the enzyme.

Factors affecting the choice of a crosslinker include: its length, that is the distance between the its reactive ends; the nature and specificity of the reactive groups on the crosslinker and the enzyme; and solubility of the crosslinker in an aqueous medium. The conditions of a crosslinking reaction must be mild to minimize damage to the native structure of the enzyme. The enzyme concentration in solution must be sufficiently low to favor intramolecular crosslinking over the formation of intermolecular linkages. The number of the functional groups of a given type available on the enzyme surface is an important consideration. If many target groups are available, a lower molar ratio of the cross-linker-to-enzyme may be used. Only the free functional groups located on the surface of the enzyme molecule can be crosslinked, as the crosslinking reagent does not penetrate to the interior of the molecule.

The nature of the functional groups available for crosslinking and their frequency of occurrence on the surface, depend on the enzyme and its molecular diameter. During crosslinking, a consideration of the functional groups at the active site of the enzyme is important, as binding of the cross-linker to these functional groups will permanently inactivate the enzyme. Furthermore, care is necessary so that crosslinks do not hinder access of the substrate to the active site.

The distance between the two surface functional groups being targeted for crosslinking must be a little less than, or equal to, the effective length of the crosslinker molecule. If the distance is too large, no crosslinking will occur, although the target functional groups on the surface may both react with one reactive end of two crosslinker molecules. In addition, if the crosslinker is much longer than the distance between the targeted surface functional groups, crosslinking may occur but is unlikely to be effective in reducing any thermally induced unravelling of the enzyme molecule.

The crosslinking reaction conditions depend mainly on the crosslinking reagent and the enzyme surface groups being crosslinked. Mild reaction conditions are desired to prevent damage to the enzyme. The reaction conditions should be selected with regard to the stability of the enzyme. For example, some enzymes are quite stable at fairly acidic pH, but others are less stable. Ideally, the reaction should be rapid especially if the conditions used differ significantly from the conditions for optimal stability of the native enzyme. Depending on the intended use of the stabilized enzyme, the reagents used may have to be either nontoxic, or any residual reagent may have to be removed completely.

Chemical crosslinking may stabilize an enzyme, or have no effect on stability (Iyer and Ananthanarayan, 2008), or, in some cases, the enzyme may be destabilized (Iyer and Ananthanarayan, 2008). Crosslinking of the surface functional groups typically enhances an enzyme molecule's resistance to conformational change, or unfolding (Cowan and Fernandez-Lafuente, 2011; Gianfreda and Scarfi, 1991). A priory prediction of the outcome of a crosslinking strategy is difficult. A single crosslink between specific surface functional groups may be highly effective in conferring stability, but not all crosslinks are equally effective. Crosslinking reagents are generally not particularly selective and may react with many surface functionals of a given type, or several types, quite randomly. Intramolecular crosslinking has been used to significantly stabilize both monomeric and multimeric enzymes (Mozhaev et al., 1990). Intramolecular crosslinking can protect an enzyme against proteases by sterically hindering the attacking enzyme (Schmid, 1979).

The mole ratio of the reactive functional groups on the crosslinker relative to the reactive functional groups available on the surface of an enzyme molecule in a reaction mixture, is also likely to influence the frequency of intramolecular crosslinking. For example, for a 0.1 M concentration of a bifunctional crosslinker, the total concentration of the reactive functional groups derived from the crosslinker would be 0.2 M. Further, if the enzyme concentration in the reaction mixture is 0.1 mM and number of reactive functional groups per molecule is 30, the concentration of the enzyme reactive groups would be 3 mM. For intramolecular crosslinking to be statistically favored, the concentration of the enzyme surface reactive functional groups should be at least twice as high as the concentration of the reactive groups derived from the bifunctional crosslinker. Thus, for a 0.1 mM enzyme solution, the maximum concentration of the reagent in the mixture should not exceed 1.5 mM. Failing this, only one end of the crosslinker molecule would be likely to react with the enzyme and this will reduce to probability of intramolecular crosslinks forming.

An estimate of the distance between the adjacent enzyme molecules in solution is needed to ensure that the solution is sufficiently dilute that the bifunctional cross-linking reagent preferentially crosslinks the functional

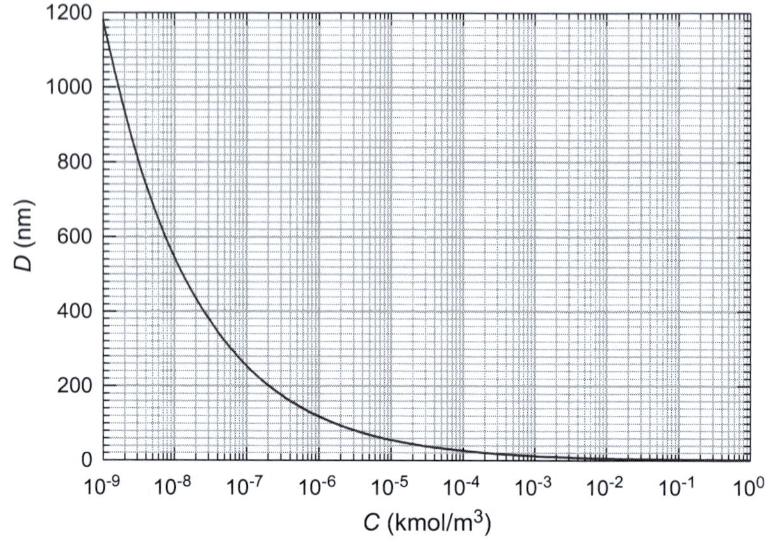

Figure 5.10 Average center-to-center intermolecular distance D between enzyme molecules in solution with a molar enzyme concentration C.

groups on the surface of the same enzyme molecule instead of forming intermolecular crosslinks. The average center-to-center distance D (nm) between adjacent molecules depends on the molar concentration C (kmol m^{-3}) of the enzyme, as follows (Erickson, 2009):

$$D = 1.18\ C^{-1/3} \tag{5.23}$$

A plot of Eq. (5.23) is shown in Fig. 5.10. If the length of the bifunctional crosslinker is well below the distance D, the likelihood of intermolecular crosslinks forming would be small.

In another approach to intramolecular crosslinking, some of the native surface functional groups of the enzyme are first modified by reacting with a short heterobifunctional reagent. The free functional groups of the newly attached reagent are then reacted with other monomers to form a confining polymer network around the enzyme (Fig. 5.11) (Schmid, 1979). Polyfunctional crosslinkers with multiple reactive functional groups on each molecule may also be used to crosslink multiple surface sites on an enzyme molecule (Cowan and Fernandez-Lafuente, 2011).

The recognition that rigidification through crosslinking enhances stability, suggests that a more rational and targeted crosslinking might be possible if the most flexible regions of the enzyme molecule are first identified.

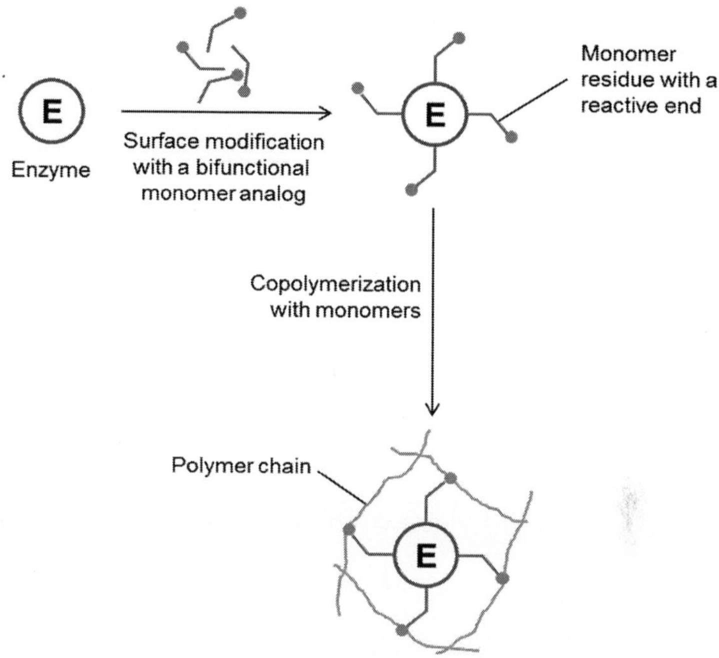

Figure 5.11 Confinement within a polymer network formed on the surface of the enzyme (E) by multipoint covalent binding to polymer chains. Based on Schmid (1979).

Both experimental (Kamerzell and Middaugh, 2008) and computer-aided molecular modeling methods may be used to identify flexible regions (Yu and Huang, 2014).

5.14.2.1 Crosslinking agents

Glutaraldehyde and diimidoesters are some of the most frequently used crosslinking agents (Schmid, 1979). Diisocyantes are also commonly used.

5.14.2.1.1 Glutaraldehyde

Glutaraldehyde ($H(O{=})C(CH_2)_3C({=}O)H$) (Fig. 5.8) has been extensively used for crosslinking (Schmid, 1979; Silva et al., 2018; Torchilin and Martinek, 1979), mostly because it is readily available and inexpensive. In aqueous media, glutaraldehyde occurs not as a dialdehyde, but in hydrated and polymeric forms. In crosslinking with glutaraldehyde, a typical protocol is as follows: A sufficiently dilute protein solution is used so that intermolecular crosslinking is minimized (see Section 5.14.2). Typically, the enzyme is dissolved in a buffer at pH 7.5–8. Phosphate and HEPES buffers (20 mM) are suitable. Tris buffers should not be used and the buffer should be free of amines. Around 5 mL of a freshly made solution of glutaraldehyde (25 g L^{-1}) is added dropwise per 100 mL protein solution. The reaction mixture is incubated at 37°C for 2–5 minutes. The reaction is quenched by adding Tris-HCl buffer (1 M, pH 8, 10 mL). Glutaraldehyde crosslinks the surface amino groups on the enzyme.

5.14.2.1.2 Imidoesters

The imidoesters react with the primary amino functional groups to form amidine bonds. Alkaline conditions (pH 10) provide specificity for primary amines. The buffer solution used in the crosslinking reaction must be free of primary amines; therefore, Tris and glycine buffers should not be used, except to quench the reaction. Imidoester solution must be freshly made as the reactive functional group is easily hydrolyzed in solution. The maximum distance between the reactive functional groups of some diimidoesters (DMA, DMP, DMS) is shown in Table 5.3. Crosslinking of ε-amino groups of two lysine residues on the surface of an enzyme with the homobifunctional diimidoester dimethyl adipimidate (DMA) is shown in Fig. 5.12.

In a typical crosslinking protocol, the enzyme solution is prepared in the crosslinking buffer. The crosslinker dissolved in water, or the buffer, is added to achieve a 10-fold molar excess of the crosslinker relative to enzyme, if the

Table 5.3 Distance between reactive groups of some homobifunctional crosslinker molecules.

Reagent	Chemical formula	Backbone length (×100 nm)
1,4-Diisocyanatobutane	$O{=}C{=}N(CH_2)_4N{=}C{=}O$	39.6
1,6-Diisocyanatohexane	$O{=}C{=}N(CH_2)_6N{=}C{=}O$	64.7
1,8-Diisocyanatooctane	$O{=}C{=}N(CH_2)_8N{=}C{=}O$	90.3
Dimethyl adipimidate (DMA)	$H_3COC(=NH)(CH_2)_4C(=NH)OCH_3$	86.0
Dimethyl pimelimidate (DMP)	$H_3COC(=NH)(CH_2)_5C(=NH)OCH_3$	92.0
Dimethyl suberimidate (DMS)	$H_3COC(=NH)(CH_2)_6C(=NH)OCH_3$	110.0
Hydrazine	H_2NNH_2	14.6
1,2-Diaminoethane	$H_2N(CH_2)_2NH_2$	15.4
1,4-Diaminobutane	$H_2N(CH_2)_4NH_2$	38.5
1,6-Diaminohexane	$H_2N(CH_2)_6NH_2$	63.6
1,8-Diaminooctane	$H_2N(CH_2)_8NH_2$	90.3
1,10-Diaminodecane	$H_2N(CH_2)_{10}NH_2$	115.8
1,12-Diaminododecane	$H_2N(CH_2)_{12}NH_2$	141.1

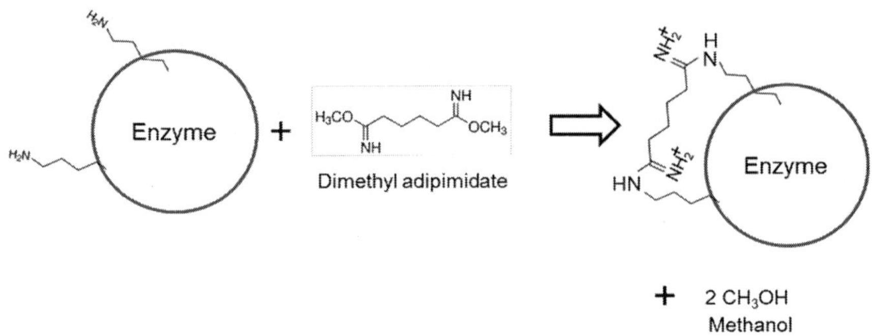

Figure 5.12 Reaction between the ε-amino groups of lysine residues on an enzyme and the diimidoester dimethyl adipimidate (DMA).

enzyme concentration is >5 mg mL^{-1}. For enzyme concentrations below 5 mg mL^{-1}, a 20- to 30-fold molar excess of the crosslinker is used. The reaction mixture is incubated at room temperature for 30–60 minutes. The reaction is stopped by adding 1 mL of glacial acetic acid for each 4 mL of the reaction mixture. Alternatively, the reaction may be quenched by adding Tris or glycine buffer to a final concentration of 20–50 mM.

5.14.2.1.3 Isocyanate

The reactive functional group of diisocyanates is cyanate, –N=C=O. Depending on the pH, the cyanate functional group can react with many of the functional groups on the enzyme molecule (Table 5.2) (Brown et al., 1987). These include the amino, the thiol, the imidazole, the tyrosyl, and the carboxyl groups (Fig. 5.9). A more stable product is formed by reaction with the amino group, compared to the other functional groups. Reaction occurs at neutral pH and room temperature. In comparison with the amino groups, the thiols react more rapidly with the cyanate, but this reaction is reversible (Tananchai, 2011). Crosslinking of ε-amino groups of two lysine residues on the surface of the enzyme with the homobifunctional reagent 1,4-diisocyanatobutane is shown in Fig. 5.13. The distance between the reactive functional groups of some diisocyanates is shown in Table 5.3.

Crosslinking of a yeast invertase with 1,4-diisocyanatobutane (Table 5.3) reduced the first-order thermal denaturation rate constant at 60°C from 1.567 min^{-1} for the native enzyme, to 0.437 min^{-1} for the stabilized enzyme (Tananchai and Chisti, 2010). The crosslinking resulted in an increase in the activation energy for thermal denaturation, from 391 kJ mol^{-1} for the native enzyme to 466 kJ mol^{-1} for the stabilized enzyme (Tananchai and Chisti, 2010).

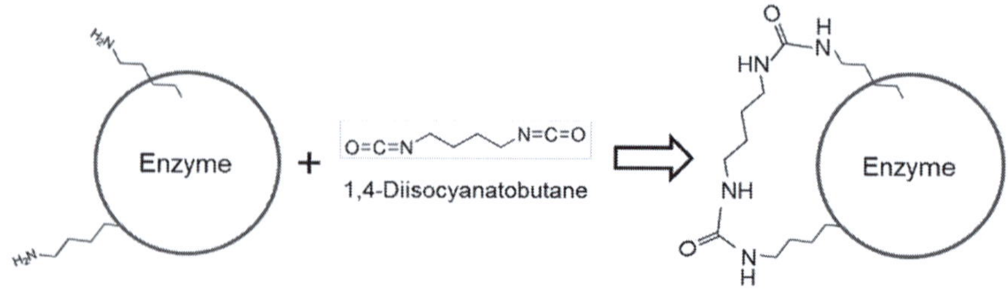

Figure 5.13 Reaction between the ε-amino groups of lysine residues on an enzyme and the 1,4-diisocyanatobutane.

5.14.2.1.4 Others

In addition to the crosslinkers mentioned above, numerous other crosslinkers are available. These include the homobifunctional crosslinkers such as diamines (Table 5.3) that react with the carboxylic acid groups on the enzyme. This condensation reaction releases a water molecule and an amide bond is formed between the enzyme and the crosslinker. A two-step reaction procedure may be used: in the first step the activating reagent *N*-hydroxysuccinimide (NHS) is added to a sufficiently dilute protein solution together with the carbodiimide reagent 1-ethyl-3-(3-dimethylaminopropyl)carbodiimide (EDC). The diamine is added to the reaction mixture in a second step.

The hydrocarbon backbone linking the two reactive functional groups of a bifunctional crosslinker is generally hydrophobic (Table 5.3), but can be modified in various ways to make it more hydrophilic. The reactions used in attaching various types of crosslinking and surface modifying molecules to an enzyme, have been thoroughly reviewed in the literature (Algar, 2017; Baslé et al., 2010; Glazer et al., 1976; Spicer and Davis, 2014; Wong, 1991).

5.15 Multimeric enzymes

Many industrial enzymes are multimeric (Fernandez-Lafuente, 2009). In a multimeric enzyme, unravelling of the multimeric arrangement, that is, the quaternary structure of the enzyme, as well as the changes to the tertiary structure of the individual monomers, result in denaturation. Often, the dissociation of monomers is the first step in denaturation of a multimeric enzyme (Fernandez-Lafuente, 2009). Dissociation is generally favored under conditions of a low enzyme concentration (Balcão and Vila, 2015; Fernandez-Lafuente, 2009). In general, multimeric enzymes are no more prone to thermal denaturation than the monomeric enzymes and may actually be more thermostable than many monomeric enzymes (Balcão and Vila, 2015; Iyer and Ananthanarayan, 2008). A higher percentage of multimeric enzymes has indeed been reported in hyperthermophiles compared to mesophiles (Fernandez-Lafuente, 2009). Stabilization of multimeric enzymes has been reviewed by Fernandez-Lafuente (2009).

Preventing dissociation of the multimers is the first objective in stabilizing multimeric enzymes (Fernandez-Lafuente, 2009). This can be achieved in many ways, including suitable formulation of the reaction medium, intramolecular chemical crosslinking, protein engineering, and enzyme immobilization (Fernandez-Lafuente, 2009). All the same methods used in stabilizing monomeric enzymes through immobilization on solid supports or within gel matrixes, could be used also with multimeric enzymes (Fernandez-Lafuente, 2009). Some of the same additives used in stabilizing monomeric enzymes (Section 5.9.2) also stabilize the multimeric enzymes. Stabilizing additives typically function by forcing the multimers into a tighter structure to minimize the surface area in contact with the surround medium.

Environmental factors such as pH and pressure may greatly influence interaction between the enzyme subunits (Fernandez-Lafuente, 2009). Metal ions such as Zn^{2+} may play a role in the stability of some multimeric enzymes, without directly contributing to maintaining the subunit association (Fernandez-Lafuente, 2009). Long-chain ionic polymers such as the cationic polyethylenimine (Fig. 5.6) may be used to coat the surface of a multimeric enzyme to physically prevent dissociation of the subunits (Fernandez-Lafuente, 2009). Polyethylenimine (PEI) has numerous ionized tertiary, secondary, and primary amino groups that interact electrostatically with the anionic groups on the enzyme. Adsorption of the polymer on the enzyme may be reversible by changing the pH. The polymer coating may enhance enzyme stability by preventing dissociation of the multimers, but does not stabilize through rigidification of the enzyme molecule. Steric hindrance of the enzyme active site by the adsorbed polymer may be a problem.

Protein engineering approaches for stabilizing multimeric enzymes include the introduction of intersubunit disulfide links and the substitution of native amino acids with others, to reinforce hydrophobic or electrostatic interaction between the subunits (Fernandez-Lafuente, 2009). Stability of many multimeric enzymes has been improved through protein engineering, as discussed by Fernandez-Lafuente (2009).

5.16 Enzymes in hydrophobic organic solvents

Enzyme stability is affected by the interactions of its surface with the solvent (Balcão and Vila, 2015). Water is the natural solvent of enzymes and a threshold level of water is necessary to fully activate the enzyme (Balcão and Vila, 2015). Activity is lost or greatly reduced in the presence of a water miscible organic solvent such as ethanol and acetone although stability may be improved. Increasing concentration of the organic solvent in an aqueous solution will reduce water activity around the enzyme, with a consequent loss in enzyme activity.

Hydrophobic substrates and products are generally most soluble in nonpolar hydrophobic solvents. Such solvents may be used for enzymatic reactions so long as the solvent contains a low level (< 3 g L^{-1}) of dissolved water. Enzymes cannot function in complete absence of water, but less water than the equivalent of a monolayer surrounding the enzyme molecule may be sufficient for catalytic action (Zaks and Klibanov, 1988). The minimum required amount of water is 0.4 g per g enzyme, or less than the amount needed to form a monolayer on the enzyme molecule (Balcão and Vila, 2015). Apparently, a few molecules of water need to be present in the vicinity of the active site for a reaction to occur. In hydrolytic reactions, water is involved directly as a substrate. In other reactions it may participate in the transition state. The low-water environment of a hydrophobic solvent shifts the thermodynamic equilibrium to favor synthesis over hydrolysis (Stepankova et al., 2013). Reactions that would not normally be catalyzed in an aqueous medium may be catalyzed in hydrophobic organic solvents containing a small amount of dissolved water (Klibanov, 1997). For example, in aqueous media, lipases (Sharma et al., 2001) hydrolyze lipids but may be used for esterification in hydrophobic organic solvents (Chuensangjun et al., 2012; Gumel et al., 2011, 2013b; Singh et al., 2008). Similarly, the protease chymotrypsin may be used for peptide synthesis in hydrophobic solvents. For some enzyme-catalyzed reactions, the rate in nonpolar organic solvents may be billions of fold greater than in an aqueous medium (Zaks and Klibanov, 1988).

Both thermal and storage stabilities of many enzymes are vastly greater in nonaqueous solvents compared to water (Zaks and Klibanov, 1988). The enhanced stability is a consequence of the surrounding organic solvent forcing the globular enzyme into a tighter structure to minimize the hydrophilic surface in contact with the solvent (Gianfreda and Scarfi, 1991; Zaks and Klibanov, 1988). This tightly folded molecule has an increased conformational rigidity. Inactivation is further minimized due to an absence of free water as free water molecules are required for nearly all inactivation processes (Gianfreda and Scarfi, 1991).

Enzymes in nonaqueous organic solvents exist in the form of undissolved dispersed particles. To minimize activity loss in nonaqueous organic media,

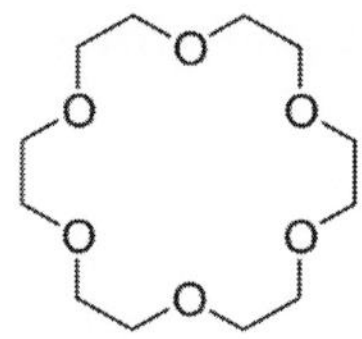

Figure 5.14 The crown ether 18-crown-6, or 1,4,7,10,13,16-hexaoxacyclooctadecane.

the enzymes must be suitably prepared for use. The enzyme is added to the solvent as a lyophilized powder (Stepankova et al., 2013). Lyophilization should be from an aqueous solution of a pH optimal for activity (Janeček, 1993) so that the functional groups at the active site are appropriately ionized. Another option is to use an enzyme powder prepared by dehydration with alcohols. In this method, the enzyme is precipitated from an aqueous buffer using alcohols. The precipitate is dried by rinsing with anhydrous *n*-propanol. The resulting powder is used in place of lyophilized enzyme in hydrophobic organic media. The activity is maximized by keeping the size of the suspended enzyme particles small and by vigorous mixing.

Many enzymes have been activated and stabilized for use in nonaqueous media by colyophilizing with the crown ether 18-crown-6 (Fig. 5.14). This activity enhancing effect is postulated to be a consequence of the interactions between the macrocyclic crown ether and the enzyme reducing the formation of salt bridges to enable better refolding of the enzyme to an active state (Stepankova et al., 2013).

Lyophilization with simple salts such as KCl has enhanced the activity of many enzymes in hydrophobic organic solvents (Stepankova et al., 2013). Through hydrogen bonding with water, salts increase the viscosity of fluid around an enzyme molecule (Illanes et al., 2012) and this apparently dampens the motion of the residual water molecules in the vicinity of the enzyme (Stepankova et al., 2013).

The molecular structure of the hydrophobic organic solvent can dramatically influence the enzyme activity, specificity and stability (Wang et al., 2016). Partition coefficient P of an organic solvent is a measure of its hydrophobicity. Partition coefficient is determined by measuring the equilibrium concentration of the solvent distributed between an organic phase (1-octanol is used as a standard) and an aqueous phase (water). Thus,

$$P = \frac{[\text{solvent}]_{\text{octanol phase}}}{[\text{solvent}]_{\text{water phase}}} \tag{5.24}$$

The above equation applies to nondissociating solvents. In practice, the log *P* value is used to characterize the solvent hydrophobicity. Enzyme activity in an organic solvent is affected by log *P* of the solvent, the solvent structure and functional groups (Wang et al., 2016).

Conformational changes are the most common cause of denaturation in hydrophobic organic solvents (Stepankova et al., 2013). Protein engineering has been successfully used to improve enzyme stability and activity in hydrophobic organic solvents (Singh et al., 2013; Stepankova et al., 2013). Enzyme use and stabilization for reactions in hydrophobic organic solvents are further discussed by others (Arnold, 1990; Doukyu and Ogino, 2010; Gupta, 1992; Klibanov, 1997; Lee and Dordick, 2002; Ogino and Ishikawa, 2001; Stepankova et al., 2013; Wang et al., 2016; Zaks and Klibanov, 1988). Many enzyme-mediated reactions in hydrophobic organic solvents have been reviewed by Stepankova et al. (2013). Enzyme engineering for improved catalytic performance in nonaqueous media has been reviewed by Arnold (1990) and Singh et al., 2016. Enzymes for non-natural chemistries may be obtained also through directed evolution (Hammer et al., 2017).

5.17 Enzymes in ionic liquids and supercritical fluids

Ionic liquids are emerging media for enzyme catalyzed reactions (Gumel et al., 2012, 2013b). Ionic liquids are good for dissolving polar substrates and their physicochemical properties may be tailored by changing the ionic components (Stepankova et al., 2013). In the context of enzyme use, an ionic liquid may be viewed as molten organic salt at room temperature. Enzymes coated with ionic liquids have been effectively used in hydrophobic organic solvents (Stepankova et al., 2013). Use and stabilization of enzymes in ionic liquids has been discussed in the literature (Moniruzzaman et al., 2010; Wang et al., 2016; Zhao, 2010).

Supercritical fluids, especially supercritical carbon dioxide, have been used to carry out many enzymatic reactions (Illanes et al., 2012; Wang et al., 2016; Wimmer and Zarevúcka, 2010). Carbon dioxide behaves as a supercritical fluid above its critical point (pressure = 7.39 MPa, temperature = 31.1°C). Supercritical carbon dioxide is able to solubilize many substrates, has a low viscosity and permits rapid diffusion of solutes. After the reaction, the product is easily recovered simply by releasing the pressure to convert the supercritical liquid to gas. Notwithstanding these advantages, the pressurization/depressurization required in processing with supercritical carbon dioxide results in inactivation of enzymes, preventing their reuse (Illanes et al., 2012). Enzyme activity and stability in supercritical media have been reviewed (Wang et al., 2016; Wimmer and Zarevúcka, 2010).

5.18 Cost versus benefit of enzyme stabilization

There is a cost associated with stabilizing an enzyme. For example, immobilization of an enzyme is estimated to increase its cost 4-fold compared to the cost of the enzyme used in the immobilization process, although on a per kilogram basis the cost of the final immobilized preparation would be less than that of the starting enzyme (Tufvesson et al., 2011). Estimates suggest that for production of bulk and specialty chemicals where the cost of the product is of the order of €1–5 per kg, an acceptable contribution of the enzyme to the final product cost is around 5%, or €0.05–0.25 per kg of product (Tufvesson et al., 2011). For higher value fine chemicals where the cost of the product is of the order of >€15/kg, the acceptable contribution of the enzyme to the final product cost may be as high as 10% (Tufvesson et al., 2011). Pharmaceutical products that typically cost >€100/kg, may use more expensive enzymes so long as the contribution of the enzyme to the final cost does not exceed around 10% (Tufvesson et al., 2011).

5.19 Conclusion

Enzymes are important biocatalysts. They are generally used in processes where conventional chemical catalysis cannot be used because: the required reaction conditions would be too severe for the reactants or products involved; the required specificity would not be achieved; or the contamination of the product with the catalyst would be unacceptable for reasons such as toxicity. Poor stability of enzymes, especially a low thermostability, limits their use. Enzymes with improved stability have been produced through isolation from extremophilic microorganisms, various types of protein engineering, and chemical modification of the less-stable native enzymes.

Chemical additives are widely used to improve storage stability of enzymes and the reaction conditions are often carefully selected to prolong operational stability. Immobilization on or within solid matrixes, is used primarily as a method of retaining and reusing enzymes in bioreactors, but also has a substantial stabilizing effect. Relatively stable enzymes originally isolated from extremophiles, usually thermophiles, are commonly produced in genetically modified mesophiles, for industrial use. Enzymes naturally immobilized within viable or inactivated microbial cells are also used in biotransformations, to eliminate the expense of recovering and purifying the enzyme.

Other than immobilization, chemical modification of the native enzyme for commercial use is not widely practiced at the moment, because of the additional expense of the stabilization treatment. Nonetheless, stabilization through hydrophilization of the surface functional groups by reacting with monofunctional modifying reagents, can be highly effective in enhancing enzyme stability. Similarly, crosslinking of surface functional groups of the enzyme can be quite effective. A priori prediction of the likely outcome of a stabilizing treatment is difficult, therefore, extensive experimental effort is required for developing suitable treatments that are consistently effective.

Processes being developed for enzymatic hydrolysis of lignocellulosic substrates, for example, continue to demand large amounts of cheap, highly thermostable, and soluble enzymes for single-use operations. As a general rule, thermostability of an enzyme is enhanced by increasing the rigidity of its folded active form, but increasing rigidity is a balancing act, as too rigid a structure may be highly stable, but less active specially at a low temperature that may be necessary because of poor thermostability of the products of the reaction, for example.

References

Aehle, W., 2004. Enzymes in Industry, Production, Applications, second ed. Wiley-VCH, Berlin, Germany.

Algar, W.R., 2017. A brief introduction to traditional bioconjugate chemistry. In: Algar, W.R., Dawson, P.E., Medintz, I.L. (Eds.), Chemoselective and Bioorthogonal Ligation Reactions: Concepts and Applications. Wiley-VCH, New York, N.Y., pp. 3–36.

Arnold, F.H., 1990. Engineering enzymes for non-aqueous solvents. Trends Biotechnol. 8, 244–249.

Balcão, V.M., Vila, M.M.D.C., 2015. Structural and functional stabilization of protein entities: State-of-the-art. Adv. Drug Delivery Rev. 93, 25–41.

Banerjee, U.C., Saxena, B., Chisti, Y., 1992. Biotransformations of rifamycins: Process possibilities. Biotechnol. Adv. 10, 577–595.

Banerjee, A., Chisti, Y., Banerjee, U.C., 2004. Streptokinase: A clinically useful thrombolytic agent. Biotechnol. Adv. 22, 287–307.

Baslé, E., Joubert, N., Pucheault, M., 2010. Protein chemical modification on endogenous amino acids. Chem. Biol. 17, 213–227.

Bennion, B.J., Daggett, V., 2003. The molecular basis for the chemical denaturation of proteins by urea. Proc. Nat. Acad. Sci. USA 100, 5142–5147.

Bernal, C., Rodríguez, K., Martínez, R., 2018. Integrating enzyme immobilization and protein engineering: An alternative path for the development of novel and improved industrial biocatalysts. Biotechnol. Adv. 36, 1470–1480.

Biesgen, C., Hillebrand, H., Herbers, K., 2002. Technical enzymes produced in transgenic plants. Phytochem. Rev. 1, 79–85.

Bommarius, A.S., Paye, M.F., 2013. Stabilizing biocatalysts. Chem. Soc. Rev. 42, 6534–6565.

Böttcher, D., Bornscheuer, U.T., 2010. Protein engineering of microbial enzymes. Curr. Opin. Microbiol. 13, 274–282.

Briggs, G.E., Haldane, J.B.S., 1925. A note on the kinetics of enzyme action. Biochem. J. 19, 338–339.

Brown, W.E., Green, A.H., Cedel, T.E., Cairns, J., 1987. Biochemistry of protein-isocyanate interactions: A comparison of the effects of aryl vs. alkyl isocyanates. Environ. Health Persp. 72, 5–11.

Brown, I., Dafforn, T.R., Fryer, P.J., Cox, P.W., 2013. Kinetic study of the thermal denaturation of a hyperthermostable extracellular α-amylase from *Pyrococcus furiosus*. Biochim. Biophy. Acta 1834, 2600–2605.

Buchholz, K., Kasche, V., Bornscheuer, U.T., 2012. Biocatalysts and Enzyme Technology, second ed. Wiley-Blackwell, Hoboken, NY.

Cao, E., Chen, Y., Cui, Z., Foster, P.R., 2003. Effect of freezing and thawing rates on denaturation of proteins in aqueous solutions. Biotechnol. Bioeng. 82, 684–690.

Cardi, T., Lenzi, P., Maliga, P., 2010. Chloroplasts as expression platforms for plant-produced vaccines. Expert Rev. Vaccine 9, 893–911.

Chen, R., 2001. Enzyme engineering: rational redesign versus directed evolution. Trends Biotechnol. 19, 13–14.

Chisti, Y., 2003. Sonobioreactors: Using ultrasound for enhanced microbial productivity. Trends Biotechnol. 21, 89–93.

Chisti, Y., 2007b. Principles of membrane separation processes. In: Subramanian, G. (Ed.), Bioseparation and Bioprocessing: A Handbook, vol. 1, second ed., Wiley-VCH, New York, N.Y., pp. 289–322.

Chisti, Y., 2007a. Strategies in downstream processing. In: Subramanian, G. (Ed.), Bioseparation, Bioprocessing: A Handbook, vol. 1, second ed. Wiley-VCH, New York, N.Y, pp. 29–62.

Chisti, Y., 2010a. Solid substrate fermentations, enzyme production, food enrichment. In: Flickinger, M.C. (Ed.), Encyclopedia of Industrial Biotechnology, Bioprocess, Bioseparation, and Cell Technology, vol. 7, Wiley, New York, N.Y, pp. 4516–4534.

Chisti, Y., 2010b. Fermentation technology. In: Soetaert, W., Vandamme, E.J. (Eds.), Industrial Biotechnology: Sustainable Growth and Economic Success. Wiley-VCH, New York, N.Y., pp. 149–171.

Chisti, Y., 2010c. Shear sensitivity. In: Flickinger, M.C. (Ed.), Encyclopedia of Industrial Biotechnology, Bioprocess, Bioseparation and Cell Technology, vol. 7, Wiley, New York, N.Y., pp. 4360–4398.

Chisti, Y., 2010d. Mass transfer. In: Flickinger, M.C. (Ed.), Encyclopedia of Industrial Biotechnology, Bioprocess, Bioseparation and Cell Technology, vol. 7, Wiley, New York, N.Y., pp. 3241–3277.

Chisti, Y., Moo-Young, M., 1986. Disruption of microbial cells for intracellular products. Enzyme Microb. Technol. 8, 194–204.

Chisti, Y., Moo-Young, M., 1999. Fermentation technology, bioprocessing, scale-up and manufacture. In: Moses, V., Cape, R.E., Springham, D.G. (Eds.), Biotechnology: The Science and the Business. second ed. Harwood Academic Publishers, New York, N.Y., pp. 177–222.

Chisti, Y., Moo-Young, M., 2002. Bioreactors. In: Meyers, R.A. (Ed.), Encyclopedia of Physical Science and Technology, Vol. 2, third ed. Academic Press, San Diego, CA, pp. 247–271.

Choi, J.-M., Han, S.-S., Kim, H.-S., 2015. Industrial applications of enzyme biocatalysis: Current status and future aspects. Biotechnol. Adv. 33, 1443–1454.

Chuensangjun, C., Pechyen, C., Chisti, Y., Sirisansaneeyakul, S., 2012. Lipase-catalyzed polymerization of lactic acid and the properties of the polymer. Adv. Materials Res. 506, 154–157.

Chutmanop, J., Chuichulcherm, S., Chisti, Y., Srinophakun, P., 2008. Protease production by *Aspergillus oryzae* in solid-state fermentation using agroindustrial substrates. J. Chem. Technol. Biotechnol. 83, 1012–1018.

Cole, M.F., Gaucher, E.A., 2011. Utilizing natural diversity to evolve protein function: Applications towards thermostability. Curr. Opin. Chemical Biol. 15, 399–406.
Combes, D., Yoovidhya, T., Girbal, E., Willemot, R.-M., Monsan, P., 1987. Mechanism of enzyme stabilization. Annals N.Y. Acad. Sci. 501, 59–62.
Cowan, D.A., Fernandez-Lafuente, R., 2011. Enhancing the functional properties of thermophilic enzymes by chemical modification and immobilization. Enzym. Microb. Technol. 49, 326–346.
Damborsky, J., Brezovsky, J., 2014. Computational tools for designing and engineering enzymes. Curr. Opin. Chem. Biol. 19, 8–16.
Daniel, R.M., Danson, M.J., 2013. Temperature and the catalytic activity of enzymes: A fresh understanding. FEBS Lett. 587, 2738–2743.
Daniell, H., Singh, N.D., Mason, H., Streatfield, S.J., 2009. Plant-made vaccine antigens and biopharmaceuticals. Trends Plant Sci. 14, 669–679.
Davies, M.J., 2003. Singlet oxygen-mediated damage to proteins and its consequences. Biochem. Biophys. Res. Commun. 305, 761–770.
Davis, B.G., 2003. Chemical modification of biocatalysts. Curr. Opin. Biotechnol. 14, 379–386.
DeSantis, G., Jones, J.B., 1999. Chemical modification of enzymes for enhanced functionality. Curr. Opin. Biotechnol. 10, 324–330.
Doukyu, N., Ogino, H., 2010. Organic solvent-tolerant enzymes. Biochem. Eng. J. 48, 270–282.
Dozier, J.K., Distefano, M.D., 2015. Site-specific PEGylation of therapeutic proteins. Int. J. Mol. Sci. 16, 25831–25864.
Eijsink, V.G.H., Bjørk, A., Gåseidnes, S., Sirevåg, R., Synstad, B., van den Burg, B., Vriend, G., 2004. Rational engineering of enzyme stability. J. Biotechnol. 113, 105–120.
Ellueche, S., Schröder, C., Sahm, K., Antranikian, G., 2014. Extremozymes: Biocatalysts with unique properties from extremophilic microorganisms. Curr. Opin. Biotechnol. 29, 116–123.
Eijsink, V.G.H., Gåseidnes, S., Borchert, T.V., van den Burg, B., 2005. Directed evolution of enzyme stability. Biomol. Eng. 22, 21–30.
Ellueche, S., Schäfers, C., Blank, S., Schröder, C., Antranikian, G., 2015. Exploration of extremophiles for high temperature biotechnological processes. Curr. Opin. Microbiol. 25, 113–119.
Erickson, H.P., 2009. Size and shape of protein molecules at the nanometer level determined by sedimentation, gel filtration, and electron microscopy. Biol. Proc. Online 11, 32–51.
Feller, G., 2010. Protein stability and enzyme activity at extreme biological temperatures. J. Physics: Condensed Matter 22, e323101.
Fernandes, P., 2010. Enzymes in food processing: A condensed overview on strategies for better biocatalysts. Enzym. Res. 2010, 862537.
Fernandez-Lafuente, R., 2009. Stabilization of multimeric enzymes: Strategies to prevent subunit dissociation. Enzyme Microb. Technol. 45, 405–418.
Fuhrmann, G., Leroux, J.-C., 2014. Improving the stability and activity of oral therapeutic enzymes—Recent advances and perspectives. Pharmaceut. Res. 31, 1099–1105.
Garrido, F., Banerjee, U.C., Chisti, Y., Moo-Young, M., 1994. Disruption of a recombinant yeast for the release of β-galactosidase. Biosep. 4, 319–328.
Gershenson, A., Arnold, F.H., 2000. Enzyme stabilization by directed evolution. In: Setlow, J.K. (Ed.), Genetic Engineering. Kluwer Academic Publishers, Dordrecht, The Netherlands, pp. 55–76.
Ghoshal, G., Banerjee, U.C., Chisti, Y., Shivhare, U.S., 2012. Optimization of xylanase production from *Penicillium citrinum* in solid-state fermentation. Chem. Biochem. Eng. Quarterly. 26, 61–69.

Gianfreda, L., Scarfi, M.R., 1991. Enzyme stabilization: State of the art. Mol. Cell. Biochem. 100, 97–128.
Glazer, A.N., DeLange, R.J., Sigman, D.S., 1976. Chemical Modification of Proteins. North-Holland Publishing Co., Amsterdam, The Netherlands.
Govardhan, C.P., 1999. Crosslinking of enzymes for improved stability and performance. Curr. Opin. Biotechnol. 10, 331–335.
Gumel, A.M., Annuar, M.S.M., Chisti, Y., 2013b. Ultrasound-assisted enzymatic synthesis of poly-ε-caprolactone: Kinetic behavior and reactor design. Int. J. Chem. Reactor Eng. 11, 609–617.
Gumel, A.M., Annuar, M.S.M., Chisti, Y., Heidelberg, T., 2012. Ultrasound assisted lipase catalyzed synthesis of poly-6-hydroxyhexanoate. Ultrason. Sonochem. 19, 659–667.
Gumel, A.M., Annuar, M.S.M., Chisti, Y., 2013a. Lipase catalyzed ultrasonic synthesis of poly-4-hydroxybutyrate-*co*-6-hydroxyhexanoate. Ultrason. Sonochem. 20, 937–947.
Gumel, A.M., Annuar, M.S.M., Heidelberg, T., Chisti, Y., 2011. Lipase mediated synthesis of sugar fatty acid esters. Process Biochem. 46, 2079–2090.
Gupta, M.N., 1992. Enzyme function in organic solvents. Eur. J. Biochem. 203, 25–32.
Haki, G.D., Rakshit, S.K., 2003. Developments in industrially important thermostable enzymes: A review. Bioresour. Technol. 89, 17–34.
Hammer, S.C., Knight, A.M., Arnold, F.H., 2017. Design and evolution of enzymes for non-natural chemistry. Curr. Opin. Green Sustainable Chem. 7, 23–30.
Hancock, R., 1957. Protein synthesis in a species of thermophilic *Bacillus*. J. Gen. Microbiol. 17, 480–490.
Hanefeld, U., Gardossi, L., Magner, E., 2009. Understanding enzyme immobilisation. Chem. Soc. Rev. 38, 453–468.
Headon, D.R., Walsh, G., 1994. The industrial production of enzymes. Biotechnol. Adv. 12, 635–646.
Hendrickx, M., Ludikhuyze, L., Van den Broeck, I., Weemaes, C., 1998. Effects of high pressure on enzymes related to food quality. Trends Food Sci. Technol. 9, 197–203.
Holmberg, K., 2018. Interactions between surfactants and hydrolytic enzymes. Colloids Surfaces B: Biointerfaces 168, 169–177.
Homaei, A.A., Sariri, R., Vianello, F., Stevanato, R., 2013. Enzyme immobilization: An update. J. Chem. Biol. 6, 185–205.
Hood, E.E., 2002. From green plants to industrial enzymes. Enzym. Microb. Technol. 30, 279–283.
Hwang, E.T., Gu, M.B., 2013. Enzyme stabilization by nano/microsized hybrid materials. Eng. Life Sci. 13, 49–61.
Illanes, A., Cauerhff, A., Wilson, L., Castro, G.R., 2012. Recent trends in biocatalysis engineering. Bioresour. Technol. 115, 48–57.
Imamura, K., Murai, K., Korehisa, T., Shimizu, N., Yamahira, R., Matsuura, T., Tada, H., Imanaka, H., Ishida, N., Nakanishi, K., 2014. Characteristics of sugar surfactants in stabilizing proteins during freeze-thawing and freeze-drying. J. Pharm. Sci. 103, 1628–1637.
Iyer, P.V., Ananthanarayan, L., 2008. Enzyme stability and stabilization: Aqueous and non-aqueous environment. Process Biochem. 43, 1019–1032.
Jain, D., Meena, V.S., Kaushik, S., Kamble, A., Chisti, Y., Banerjee, U.C., 2012. Production of nitrilase by a recombinant *Escherichia coli* in a laboratory scale bioreactor. Ferment. Technol. 1, 103.
Janeček, Š., 1993. Strategies for obtaining stable enzymes. Process Biochem. 28, 435–445.
Jegannathan, K.R., Nielsen, P.H., 2013. Environmental assessment of enzyme use in industrial production: A literature review. J. Cleaner Prod. 42, 228–240.
Jia, F., Narasimhan, B., Mallapragada, S., 2014. Materials-based strategies for multi-enzyme immobilization and co-localization: A review. Biotechnol Bioeng. 111, 209–222.

Kamerzell, T.J., Middaugh, C.R., 2008. The complex inter-relationships between protein flexibility and stability. J. Pharm. Sci. 97, 3494–3517.
Karimi, K., Chisti, Y., 2017. Bioethanol production and technologies. In: Abraham, M.A. (Ed.), Encyclopedia of Sustainable Technologies, vol. 3, Elsevier, London, U.K., pp. 273–284.
Kaspersen, J.D., Søndergaard, A., Madsen, D.J., Otzen, D.E., Pedersen, J.S., 2017. Refolding of SDS-unfolded proteins by nonionic surfactants. Biophys. J. 112, 1609–1620.
Kelly, S.M., Price, N.C., 2000. The use of circular dichroism in the investigation of protein structure and function. Curr. Prot. Pept. Sci. 1, 349–384.
Kerwin, B.A., Remmele, Jr., R.L., 2007. Protect from light: photodegradation and protein biologics. J. Pharm. Sci. 96, 1468–1479.
Khan, R.H., Rasheedi, S., Haq, S.K., 2003. Effect of pH, temperature and alcohols on the stability of glycosylated and deglycosylated stem bromelain. J. Biosci. 28, 709–714.
Kim, J., Grate, J.W., Wang, P., 2006. Nanostructures for enzyme stabilization. Chem. Eng. Sci. 61, 1017–1026.
Klibanov, A.M., 1979. Enzyme stabilization by immobilization. Anal. Biochem. 93, 1–25.
Klibanov, A.M., 1997. Why are enzymes less active in organic solvents than in water? Trends Biotechnol. 15, 97–101.
Lee, C.-K., Au-Duong, A.-N., 2018. Enzyme immobilization on nanoparticles: Recent applications. In: Chang, H.N. (Ed.), Emerging Areas in Bioengineering. Wiley-VCH, New York, N.Y., pp. 67–80.
Lee, M.-Y., Dordick, J.S., 2002. Enzyme activation for nonaqueous media. Curr. Opin. Biotechnol. 13, 376–384.
Lee, C.-W., Jang, S.-H., Chung, H.-S., 2017. Improving the stability of cold-adapted enzymes by immobilization. Catalysts. 7, 112.
Lutz, S., 2010. Beyond directed evolution: Semi-rational protein engineering and design. Curr. Opin. Biotechnol. 21, 734–743.
Madsen, J.K., Pihl, R., Møller, A.H., Madsen, A.T., Otzen, D.E., Andersen, K.K., 2015. The anionic biosurfactant rhamnolipid does not denature industrial enzymes. Front. Microbiol. 6, Article 292.
Manderson, D., Dempster, R., Chisti, Y., 2006a. A recombinant vaccine against hydatidosis: production of the antigen in *Escherichia coli*. J. Ind. Microbiol. Biotechnol. 33, 173–182.
Manderson, D., Dempster, R., Chisti, Y., 2006b. Production of an active recombinant Aspin antigen in *Escherichia coli* for identifying animals resistant to nematode infection. Enzym. Microb. Technol. 38, 591–598.
Martinek, K., Goldmacher, V.S., Klibanov, A.M., Berezin, I.V., 1975. Denaturing agents (urea, acrylamide) protect enzymes against irreversible thermoinactivation: A study with native and immobilized α-chymotrypsin and trypsin. FEBS Lett. 51, 152–155.
Mateo, C., Palomo, J.M., Fernandez-Lorente, G., Guisan, J.M., Fernandez-Lafuente, R., 2007. Improvement of enzyme activity, stability and selectivity via immobilization techniques. Enzym. Microb. Technol. 40, 1451–1463.
Mattanovich, D., Branduardi, P., Dato, L., Gasser, B., Sauer, M., Porro, D., 2012. Recombinant protein production in yeasts. Methods Mol. Biol. 824, 329–358.
Meldgaard, M., Svendsen, I., 1994. Different effects of *N*-glycosylation on the thermostability of highly homologous bacterial (1,3-1,4)-β-glucanases secreted from yeast. Microbiol. 140, 159–166.
Mevarech, M., Frolow, F., Gloss, L.M., 2000. Halophilic enzymes: proteins with a grain of salt. Biophys. Chem. 86, 155–164.
Michaelis, L., Menten, M.L., 1913. Die kinetik der invertinwirkung. Biochemische Zeitschrift 49, 333–369.
Moniruzzaman, M., Kamiya, N., Goto, M., 2010. Activation and stabilization of enzymes in ionic liquids. Org. Biomol. Chem. 8, 2887–2899.

Moore, E.J., Zorine, D., Hansen, W.A., Khare, S.D., Fasan, R., 2017. Enzyme stabilization via computationally guided protein stapling. Proc. Nat. Acad. Sci. USA 114, 12472–12477.

Mozhaev, V.V., Martinek, K., 1984. Structure-stability relationships in proteins: new approaches to stabilizing enzymes. Enzym. Microb. Technol. 6, 50–59.

Mozhaev, V.V., Melik-Nubarov, N.S., Šikšnis, V., Martinek, K., 1990. Strategy for stabilizing enzymes part two: Increasing enzyme stability by selective chemical modification. Biocatal. 3, 189–196.

Obembe, O.O., Popoola, J.O., Leelavathi, S., Reddy, S.V., 2011. Advances in plant molecular farming. Biotechnol. Adv. 29, 210–222.

Obón, J.M., Casanova, P., Manjón, A., Fernández, V.M., Iborra, J.L., 1997. Stabilization of glucose dehydrogenase with polyethyleneimine in an electrochemical reactor with $NAD(P)^+$ regeneration. Biotechnol. Prog. 13, 557–561.

Ó'Fágáin, C., 1995. Understanding and increasing protein stability. Biochim. Biophys. Acta 1252, 1–14.

Ó'Fágáin, C., 2003. Enzyme stabilization: Recent experimental progress. Enzym. Microb. Technol. 33, 137–149.

Ogino, H., Ishikawa, H., 2001. Enzymes which are stable in the presence of organic solvents. J. Biosci. Bioeng. 91, 109–116.

Ohmae, E., Miyashita, Y., Kato, C., *2013*. Thermodynamic and functional characteristics of deep-sea enzymes revealed by pressure effects. Extremophiles 17, 701–709.

Otzen, D., 2011. Protein-surfactant interactions: A tale of many states. Biochim. Biophys. Acta 1814, 562–591.

Otzen, D., 2015. Proteins in a brave new surfactant world. Curr. Opin. Colloid Interface Sci. 20, 161–169.

Overton, T.W., 2014. Recombinant protein production in bacterial hosts. Drug Disc. Today 19, 590–601.

Pai, O., Banoth, L., Ghosh, S., Chisti, Y., Banerjee, U.C., 2014. Biotransformation of 3-cyanopyridine to nicotinic acid by free and immobilized cells of recombinant *Escherichia coli*. Process Biochem. 49, 655–659.

Park, S.-H., Ong, R.G., Sticklen, M., 2016. Strategies for the production of cell wall-deconstructing enzymes in lignocellulosic biomass and their utilization for biofuel production. Plant Biotechnol. J. 14, 1329–1344.

Patel, R.N., 2000. Stereoselective Biocatalysis. Marcel Dekker, New York, N.Y.

Patel, A.K., Singhania, R.R., Pandey, A., 2016. Novel enzymatic processes applied to the food industry. Curr. Opin. Food Sci. 7, 64–72.

Patil, R., Banoth, L., Singh, A., Chisti, Y., Banerjee, U.C., 2013. Enantioselective bioreduction of cyclic alkanones by whole cells of *Candida* species. Biocatal. Biotransform. 31, 123–131.

Patil, M.D., Dev, M.J., Tangadpalliwar, S., Patel, G., Garg, P., Chisti, Y., Banerjee, U.C., 2017. Ultrasonic disruption of *Pseudomonas putida* for the release of arginine deiminase: Kinetics and predictive models. Bioresour. Technol. 233, 74–83.

Pawar, S.V., Meena, V.S., Kaushik, S., Kamble, A., Kumar, S., Chisti, Y., Banerjee, U.C., 2012. Stereo-selective conversion of mandelonitrile to (R)-(2)-mandelic acid using immobilized cells of recombinant *Escherichia coli*. 3 Biotech 2, 319–326.

Polaina, J., MacCabe, A.P., 2007. Industrial Enzymes: Structure, Function and Applications. Springer, New York, N.Y.

Prado, J.R., Segers, G., Voelker, T., Carson, D., Dobert, R., Phillips, J., Cook, K., Cornejo, C., Monken, J., Grapes, L., Reynolds, T., Martino-Catt, S., 2014. Genetically engineered crops: From idea to product. Ann. Rev. Plant Biol. 65, 769–790.

Qi, D., Tann, C.-M., Haring, D., Distefano, M.D., 2001. Generation of new enzymes via covalent modification of existing proteins. Chem. Rev. 101, 3081–3111.

Qureshi, A.S., Bhutto, M.A., Chisti, Y., Khushk, I., Dahot, M.U., Bano, S., 2012. Production of pectinase by *Bacillus subtilis* EFRL 01 in a date syrup medium. African J. Biotechnol. 11, 12563–12570.

Qureshi, A.S., Khushk, I., Ali, C.H., Chisti, Y., Ahmad, A., Majeed, H., 2016. Coproduction of protease and amylase by thermophilic *Bacillus* sp. BBXS-2 using open solid-state fermentation of lignocellulosic biomass. Biocatal. Agric. Biotechnol. 8, 146–151.

Ramos, A., Raven, N.D.H., Sharp, R.J., Bartolucci, S., Rossi, M., Cannio, R., Lebbink, J., Vand der Oost, J., de Vos, W.M., Santos, H., 1997. Stabilization of enzymes against thermal stress and freeze-drying by mannosylglycerate. Appl. Environ. Microbiol. 63, 4020–4025.

Raveendran, S., Parameswaran, B., Ummalyma, S.B., Abraham, A., Mathew, A.K., Madhavan, A., Rebello, S., Pandey, A., 2018. Applications of microbial enzymes in food industry. Food Technol. Biotechnol. 56, 16–30.

Reyes-de-Corcuera, J., Olstad, H.E., García-Torres, R., 2018. Stability and stabilization of enzyme biosensors: The key to successful application and commercialization. Ann. Rev. Food Sci. Technol. 9, 293–322.

Rosano, G.L., Ceccarelli, E.A., 2014. Recombinant protein expression in *Escherichia coli*: advances and challenges. Front. Microbiol. 5, 172.

Rouf, S.A., Moo-Young, M., Chisti, Y., 1996. Tissue-type plasminogen activator: Characteristics, applications and production technology. Biotechnol. Adv. 14, 239–266.

Sadana, A., 1988. Enzyme deactivation. Biotechnol. Adv. 6, 349–446.

Sadana, A., Henley, J.P., 1986. Influence of chemical modification on enzyme inactivation kinetics and stability. Biotechnol. Adv. 4, 27–74.

Salehmin, M.N.I., Annuar, M.S.M., Chisti, Y., 2013. High cell density fed-batch fermentations for lipase production: Feeding strategies and oxygen transfer. Bioprocess Biosys. Eng. 36, 1527–1543.

Salehmin, M.N.I., Annuar, M.S.M., Chisti, Y., 2014. High cell density fed-batch fermentation for the production of a microbial lipase. Biochem. Eng. J. 85, 8–14.

Schaertel, B.J., Firstenberg-Eden, R., 1988. Biosensors in the food industry: Present and future. J. Food Prot. 51, 811–820.

Schmid, R.D., 1979. Stabilized soluble enzymes. Adv. Biochem. Eng. 12, 41–118.

Schülke, N., Schmid, F.X., 1988. The stability of yeast invertase is not significantly influenced by glycosylation. J. Biol. Chem. 263, 8827–8831.

Sharma, R., Chisti, Y., Banerjee, U.C., 2001. Production, purification, characterization, and applications of lipases. Biotechnol. Adv. 19, 627–662.

Shaw, A., Bott, R., 1996. Engineering enzymes for stability. Curr. Opin. Struct. Biol. 6, 546–550.

Sheldon, R.A., 2007. Enzyme immobilization: The quest for optimum performance. Adv. Synth. Catal. **349**, 1289–1307.

Sheldon, R.A., 2011. Cross-linked enzyme aggregates as industrial biocatalysts. Org. Process Res. Develop. 15, 213–223.

Silva, J.L., Weber, G., 1993. Pressure stability of proteins. Ann. Rev. Phys. Chem. 44, 89–113.

Silva, C., Martins, M., Jing, S., Fu, J., Cavaco-Paulo, A., 2018. Practical insights on enzyme stabilization. Crit. Rev. Biotechnol. 38, 335–350.

Singh, R., Kumar, M., Mittal, A., Mehta, P.K., 2016. Microbial enzymes: industrial progress in 21st century. 3 Biotech 6, 174.

Singh, M., Singh, S., Singh, R.S., Chisti, Y., Banerjee, U.C., 2008. Transesterification of primary and secondary alcohols using *Pseudomonas aeruginosa* lipase. Bioresour. Technol. 99, 2116–2120.

Singh, A., Chisti, Y., Banerjee, U.C., 2011. Production of carbonyl reductase by *Metschnikowia koreensis*. Bioresour. Technol. 102, 10679–10685.

Singh, A., Chisti, Y., Banerjee, U.C., 2012. Stereoselective biocatalytic hydride transfer to substituted acetophenones by the yeast *Metschnikowia koreensis*. Process Biochem. 47, 2398–2404.

Singh, R.K., Tiwari, M.K., Singh, R., Lee, J.-K., 2013. From protein engineering to immobilization: Promising strategies for the upgrade of industrial enzymes. Int. J. Mol. Sci. 14, 1232–1277.

Sirisansaneeyakul, S., Worawuthiyanan, N., Vanichsriratana, W., Srinophakun, P., Chisti, Y., 2007. Production of fructose from inulin using mixed inulinases from *Aspergillus niger* and *Candida guilliermondii*. World J. Microbiol. Biotechnol. 23, 543–552.

Sirisha, V.L., Jain, A., Jain, A., 2016. Enzyme immobilization: An overview on methods, support material, and applications of immobilized enzymes. Adv. Food Nutr. Res. 79, 179–211.

Socha, R.D., Tokuriki, N., 2013. Modulating protein stability: Directed evolution strategies for improving protein function. FEBS J. 280, 5582–5595.

Solá, R.J., Griebenow, K., 2009. Effects of glycosylation on the stability of protein pharmaceuticals. J. Pharm. Sci. 98, 1223–1245.

Spicer, C.D., Davis, B.J., 2014. Selective chemical protein modification. Nature Commun. 5, 4740.

Stepankova, V., Bidmanova, S., Koudelakova, T., Prokop, Z., Chaloupkova, R., Damborsky, J., 2013. Strategies for stabilization of enzymes in organic solvents. ACS Catal. 3, 2823–2836.

Stoner, M.R., Dale, D.A., Gualfetti, P.J., Becker, T., Manning, M.C., Carpenter, J.F., Randolph, T.W., 2004. Protease autolysis in heavy-duty liquid detergent formulations: Effects of thermodynamic stabilizers and protease inhibitors. Enzym. Microb. Technol. 34, 114–125.

Sulaiman, A.Z., Ajit, A., Chisti, Y., 2013. Ultrasound mediated enzymatic hydrolysis of cellulose and carboxymethyl cellulose. Biotechnol. Prog. 29, 1448–1457.

Suplatov, D., Voevodin, V., Švedas, V., 2015. Robust enzyme design: Bioinformatic tools for improved protein stability. Biotechnol. J. 10, 344–355.

Tamer, I.M., Chisti, Y., 2001. Production and recovery of recombinant protease inhibitor α_1-antitrypsin. Enzym. Microb. Technol. 29, 611–620.

Tananchai, P., 2011. Stabilization of enzymes by chemical modifications. PhD thesis, Massey University, New Zealand.

Tananchai, P., Chisti, Y., 2010. Stabilization of invertase by molecular engineering. Biotechnol. Prog. 26, 111–117.

Tann, C.-M., Qi, D., Distefano, M.D., 2001. Enzyme design by chemical modification of protein scaffolds. Curr. Opin. Chem. Biol. 5, 696–704.

Tomazic, S.J., 1991. Protein stabilization. In: Dordick, J.D. (Ed.), Biocatalysis for Industry. Springer, New York, pp. 241–256.

Torchilin, V.P., Martinek, K., 1979. Enzyme stabilization without carriers. Enzym. Microb. Technol. 1, 74–82.

Tufvesson, P., Lima-Ramos, J., Nordblad, M., Woodley, J.M., 2011. Guidelines and cost analysis for catalyst production in biocatalytic processes. Org. Process Res. Develop. 15, 266–274.

Vaidya, A.M., Annapure, U.S., 2019. Enzymes in biosensors for food quality assessment. In: Kuddus, M. (Ed.), Enzymes in Food Biotechnology: Production, Applications, and Future Prospects. Academic Press, San Diego, CA, pp. 659–674.

van Beilen, J.B., Li, Z., 2002. Enzyme technology: An overview. Curr. Opin. Biotechnol. 13, 338–344.

Vellard, M., 2003. The enzyme as drug: Application of enzymes as pharmaceuticals. Curr. Opin. Biotechnol. 14, 444–450.

Vieille, C., Zeikus, G.J., 2001. Hyperthermophilic enzymes: Sources, uses, and molecular mechanisms for thermostability. Microb. Mol. Biol. Rev. 65, 1–43.

Wang, C., Eufemi, M., Turano, C., Giartosio, A., 1996. Influence of the carbohydrate moiety on the stability of glycoproteins. Biochemistry 35, 7299–7307.

Wang, S., Meng, X., Zhou, H., Liu, Y., Secundo, F., Liu, Y., 2016. Enzyme stability and activity in non-aqueous reaction systems: A mini review. Catalysts 6, 32.

Wee, L.L., Annuar, M.S.M., Ibrahim, S., Chisti, Y., 2011. Enzyme-mediated production of sugars from sago starch: Statistical process optimization. Chem. Eng. Commun. 198, 1339–1353.

Weijers, S.R., Van't Riet, K., 1992. Enzyme stability in downstream processing Part 1: Enzyme inactivation, stability and stabilization. Biotechnol. Adv. 10, 237–249.

Wijma, H.J., Floor, R.J., Janssen, D.B., 2013. Structure- and sequence-analysis inspired engineering of proteins for enhanced thermostability. Curr. Opin. Struct. Biol. 23, 588–594.

Wimmer, Z., Zarevúcka, M., 2010. A review on the effects of supercritical carbon dioxide on enzyme activity. Int. J. Mol. Sci. 11, 233–253.

Wong, S.S., 1991. Chemistry of Protein Conjugation and Cross-Linking. CRC Press, Boca Raton, FL.

Wong, S.S., Wong, L.-J.C., 1992. Chemical crosslinking and the stabilization of proteins and enzymes. Enzym. Microb. Technol. 14, 866–874.

Woodley, J.M., 2013. Protein engineering of enzymes for process applications. Curr. Opin. Chem. Biol. 17, 310–316.

Yabuki, S., 2017. How to lengthen the long-term stability of enzyme membranes: Trends and strategies. Catalysts 7, 36.

Yoshida, K., Matsui, T., Shinmyo, A., 2004. The plant vesicular transport engineering for production of useful recombinant proteins. J. Mol. Catal. B: Enzymatic 28, 167–171.

Yu, H., Huang, H., 2014. Engineering proteins for thermostability through rigidifying flexible sites. Biotechnol. Adv. 32, 308–315.

Yunus, G., 2019. Biosensors: An enzyme-based biophysical technique for the detection of foodborne pathogens. In: Kuddus, M. (Ed.), Enzymes in Food Biotechnology: Production, Applications and Future ProspectsAcademic Press, San Diego, CA, pp. 723–738.

Zaks, A., Klibanov, A.M., 1988. Enzymatic catalysis in nonaqueous solvents. J. Biol. Chem. 263, 3194–3201.

Zhao, H., 2010. Methods for stabilizing and activating enzymes in ionic liquids: A review. J. Chem. Technol. Biotechnol. 85, 891–907.

Zoldák, G., Zubrik, A., Musatov, A., Stupák, M., Sedlák, E., 2004. Irreversible thermal denaturation of glucose oxidase from Aspergillus niger is the transition to the denatured state with residual structure. *J. Biol. Chem.* 279, 47601–47609.

CHAPTER SIX

Immobilization of enzymes and their use in biotechnological applications

Selim Kermasha, Jagpreet K. Gill
McGill University, Montreal, Quebec, Canada

6.1 Introduction

Enzymes are natural catalysts that enhance the rate of chemical reactions in nearly all biological processes (Ding et al., 2015). As compared to conventional chemical catalysts, enzymes are known for their ability to catalyze reactions under milder conditions of temperature, pressure, and pH, as well as a relatively higher substrate specificity (Mohamad et al., 2015; Barbosa et al., 2015). The ease of enzyme production, their substrate specificity, and green chemistry makes biocatalysis an attractive choice for biotechnological applications (Datta et al., 2013). Nevertheless, the use of free enzymes is usually limited due to their sensitivity to the surrounding process conditions, including pH, temperature, substrate, and product concentrations. In addition, the recovery and reusability of the active enzyme, when used as a suspension is extremely difficult and may reduce the overall economic viability of the process (Brena et al., 2013).

Enzyme immobilization is a technique that is used to improve enzyme stability and reusability in biocatalytic processes, thereby promoting their use in various industrial applications (Ammann et al., 2014). Immobilization of enzymes involves the physical confinement or localization of the biocatalyst in a defined region of space, while retaining their catalytic activity (Khan and Alzohairy, 2010; Mohamad et al., 2015). As compared to free enzymatic suspensions, immobilized enzymes offer several advantages, including ease of process control, manipulation of the biocatalyst, enhanced enzyme stability and reusability, ease in recovery from the reaction components, limited protein contamination in the synthesized products, and reduced process costs (Mohamad et al., 2015). However, despite the various advantages of immobilized enzymes, the technique may result in changes to

Enzymes: Novel Biotechnological Approaches for the Food Industry.
http://dx.doi.org/10.1016/B978-0-12-800217-9.00006-X

their inherent catalytic properties that may result from the method used for immobilization and/or the surrounding microenvironment, which includes the support matrix and the synthesized end products. Hence, immobilization may improve or deteriorate the activity and stability of the enzyme depending on the process design as well as their interactions with carrier support (Mateo et al., 2007).

In general, immobilized enzymes involve interactions between mainly three components; the enzyme, the carrier support, and the mode of attachment. The methods used for immobilizing enzymes exploit the different functional groups that are present on the side chains of the protein's amino acids, where these groups may bind to the support matrix via various interactions (Mohamad et al., 2015). Therefore, the designed process for an immobilization must consider the surface properties of both, the enzyme as well as the support matrix. (Hanefeld et al., 2009). Moreover, the support matrix characteristics are important in determining the efficiency of the immobilized enzymes, where their choice may vary for different biocatalysts. An ideal support matrix must include characteristics such as stability, physical resistance, hydrophilicity, chemical inertness, biocompatibility, microbial resistance, and low costs (Brena et al., 2013; Mohamad et al., 2015). In general, the carrier support matrix may be organic (polysaccharides, polyacrylic, and polyvinylic) or inorganic (silica, titanium, and aluminum) in nature, where the former types are subdivided into natural and synthetic organic polymers (Zucca and Sanjust, 2014). Even though both types of carriers act as efficient matrices, inorganic polymers are preferred due to their relatively lower costs, range of porosities, ease of manipulation in terms of chemical modifications, mechanical stability, and microbial resistance (Yu et al., 2012. Zucca and Sanjust, 2014; Bilal and Asgher, 2015).

In addition to the choice of carrier supports, the type of immobilization method holds significant importance as it affects the catalytic activity of the immobilized biocatalyst. Typically, the immobilization methods are categorized into two types based on the type of interaction between the carrier support and the enzyme; namely, chemical and physical methods. The physical methods involve the interactions by weaker forces, including Van der Waals, hydrophobic interactions, hydrogen bonds, ionic and affinity binding, or the entrapment of the biocatalyst within the support matrix (Yi et al., 2007). On the other hand, chemical methods include stronger interactions by covalent binding via amide, ether, thio-ether or carbamate linkages between the enzyme and the carrier support (Hanefeld et al., 2009; Mohamad et al., 2015). There are four principal techniques for immobilization of enzymes, including adsorption or deposition, encapsulation or entrapment (Yi et al., 2006a), covalent binding, and self-immobilization (Gaffar et al., 1999;Yi et al., 2002; Brady and Jordaan, 2009; Hanefeld et al., 2009; Gill et al., 2018a,b)

6.2 Methods of immobilization

Regardless of the various immobilization methods, the choice of a certain method could differ for each enzyme and would depend on the biochemical properties of both the biocatalyst as well as the carrier support.

6.2.1 Adsorption

Adsorption (Fig. 6.1) is the simplest method of enzyme immobilization, where enzymes are physically adsorbed on to the surface of the carrier matrix. Adsorption may occur via various interactions, including ionic, hydrogen bonding, and other weak forces of attraction such as Van der Waals or hydrophobic interactions (Mohamad et al., 2015). In general, adsorption of enzymes on hydrophobic carriers occurs via Van der Waals forces that are accompanied by changes in the overall entropy. During this immobilization, a large number of water molecules are displaced by the enzyme from both its and the carrier support; this then results in a gain in the entropy and subsequently leads to hydrophobic interactions between them. In addition, these interactions may be further modified by varying parameters, such as pH, temperature and salt concentration (Hanefeld et al., 2009). Alternatively, adsorption via hydrogen bonding occurs between hydrophilic amino acid residues of the enzyme onto hydrophilic carriers supports like cellulose, silica gel, clay, porous glass, and celite (Khoshnevisan et al., 2011 & 2017). Adsorption via ionic interactions takes place using ion exchangers as carrier supports, where the immobilization depends on the pH of the surrounding solution and the isoelectric point of the enzyme. The pH and

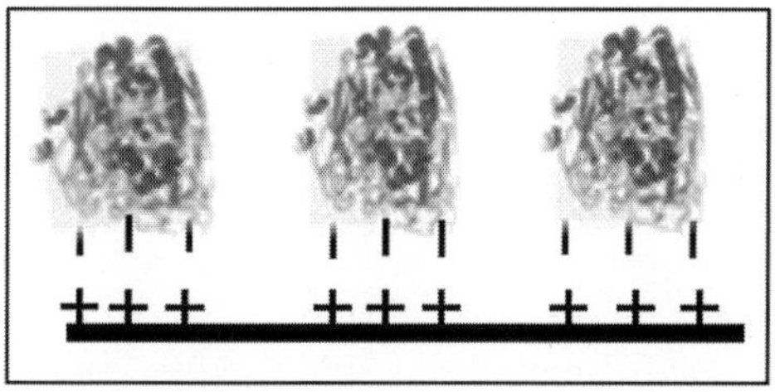

Figure 6.1 Enzymes immobilized via adsorption to support matrix.

the isoelectric point of the enzyme determine their surface charge distribution; hence any ion exchanger may act as a carrier and immobilize the enzyme via ionic or strong polar interactions. Nevertheless, the limitation of this method is its sensitivity to higher salt concentrations in the surrounding medium that may result ion exchange and a subsequent leakage of the enzyme from the support matrix (Gaffar et al., 1999; Hanefeld et al., 2009; Mohamad et al., 2015).

The general method for carrying out adsorption involves immersion of the carrier support into the enzyme solution for a certain period of time that could be the recovered by precipitation or evaporation from the aqueous medium. Alternatively, adsorption may be carried out by leaving the enzyme solution on the carrier support until dry, which is then washed to remove any unreacted enzyme.

Immobilization of enzymes by physical adsorption is an attractive commercial choice due to the simplicity of the process, low cost, relatively high enzymatic activity, and chemical-free enzyme-support interactions. On the other hand, this method has certain limitations, including enzyme leaching due to the weak hydrophobic interactions that are further enhanced during industrial applications due to the use of extreme conditions, such as high reactant and product concentrations, temperature as well as pH. In comparison to the other methods of immobilization, adsorption is more prone to be affected by changes in the external environmental conditions of pH, temperature, and buffer's ionic strength (Hanefeld et al., 2009; Mohamad et al., 2015).

6.2.2 Covalent binding

Covalent binding is the most widely used irreversible method of immobilization, where the enzyme is linked to the support matrix via covalent bonds (Fig. 6.2). In general, the interaction occurs between the activated functional groups of the carrier support and the nucleophilic residues of

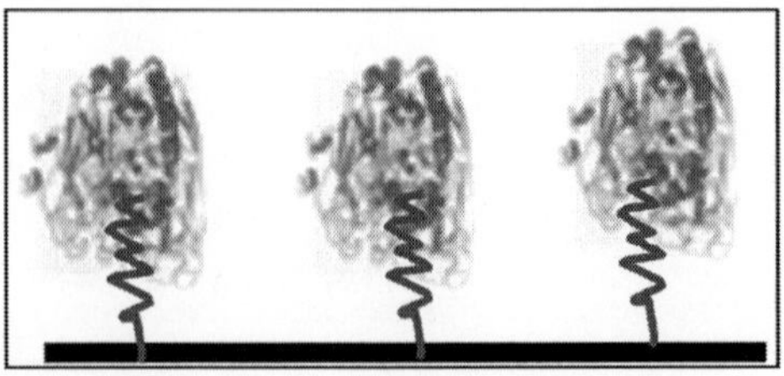

Figure 6.2 Immobilization of enzymes via covalent binding.

the enzymatic protein (Fernández-Fernández et al., 2013). The activation of the carrier groups is achieved using reagents that enhance their electrophilic nature and their ability to bind to the nucleophilic residues on the enzyme. The functional groups of the carrier usually bind via the side chains of the amino acid residues, including lysine, histidine, cysteine, arginine, and aspartic acid; however, the reactivity of the enzyme depends on the involved functional groups, such as amino, carboxylic, sulfhydryl, imidazole, and phenolic groups (Bickerstaff, 1997; Sirisha et al., 2016). Alternatively, multipoint covalent immobilization of enzymes may also be used as a method for. immobilization and is a variation of the original method. This method involves the surface modification of the enzymes by placing more than one amino acid residue on them and their subsequent covalent binding to highly preactivated carrier supports, using spacer arms which could result in a more rigid immobilized enzyme. Both long or short spacer arms may be used for this type of immobilization and are associated with their own respective advantages, where the use of the former allows for more conformational flexibility, the latter offers a higher thermal stability and rigidity (Mateo et al., 2007; Hanefeld et al., 2009).

In general, the catalytic activity of the covalently immobilized enzyme depends on the amino acid groups involved in binding with the support matrix; hence, the direction of the enzymatic binding is an important factor in predicting its stability. It has been reported that significantly higher enzymatic activities are obtained when the binding occurs between the amino acid residues that are not part of the active sites of the protein that could eventually lead to their inactivation. In order to avoid the involvement of such residues, the enzymes may be immobilized to modified support matrices with incorporated reactive functional groups or to modified support matrices that generate activated groups. Moreover, a relatively higher specific activity and stability is obtained using controlled protein orientation that involve peptide-modified support surfaces for enzyme immobilization (Mohamad et al., 2015; Sirisha et al., 2016). The immobilization of enzymes via covalent binding involves the use of strong binding forces and hence, increase the reusability of the enzyme in industrial processes as well as prevent their leakage into the reaction environment. In addition, this method has also been reported to enhance the thermal stability of the enzymes when linked to support matrices, such as mesoporous silica and chitosan (Mohamad et al., 2015). Nevertheless, the tight linkage of the enzyme to the support matrix also leads to their reduced conformational flexibility, while the involvement of the amino acid residues of the native protein could affect their inherent catalytic activity and function (Mateo et al., 2007; Hanefeld et al., 2009).

6.2.3 Self-immobilization

Self-immobilization or carrier-free immobilization is another irreversible method of enzyme immobilization that does not require the need for a carrier support matrix and hence, avoids the limitations associated with the presence of support matrices (Brady and Jordaan, 2009). This method involves the enzyme acting as its own carrier support, where the enzymes usually, cross-link via a bi- or multifunctional crosslinking agent (Brady and Jordaan, 2009). Glutaraldehyde is the most commonly used cross-linking agent due to its lower costs and availability. Cross-linking of enzymes (Fig. 6.3) may involve both, Schiff's base formation and Michael-type addition to unsaturated aldehyde moieties, where the specific mode of cross-linking could vary with the pH. In general, the use of carrier supports dilute the catalytic activity of the immobilized enzyme which may be resulted by noncatalytic ballast that can range from 90% to 99%, possibly leading to lower product yields and increased overall process costs (Mohamad et al., 2015).

Self-immobilization of enzymes may be carried out using different approaches, mainly including cross-linked enzyme crystals (CLECs) or their modified form, cross-linked enzyme aggregates (CLEAs). CLECs are prepared by cross-linking enzyme crystals using a suitable cross-linker; these immobilized enzymes are often associated with enhanced catalytic activity and stability (Brady and Jordaan, 2009). Similarly, CLEAs are prepared by a modification to the method by cross-linking enzyme aggregates that are obtained by their direct precipitation in precipitants, such as acetone, ammonium sulfate, and ethanol. The preparation of CLEAs combines the enzyme purification and immobilization into a single step and may be synthesized using three different approaches; the mixing of the prepolymers

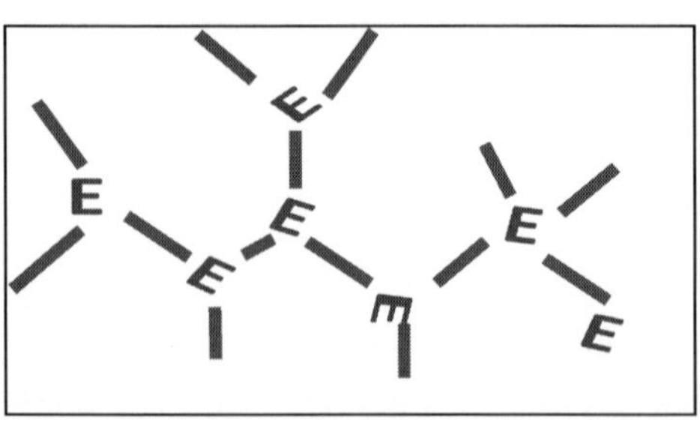

Figure 6.3 Enzyme immobilization via cross-linking.

with a photosensitizer (e.g., benzoin ethyl ether), mixing with the enzymatic solution and subsequent gelation using UV radiation and freezing of the enzyme-containing monomer solution as small beads. Following this, the polymerization is either initiated using gamma radiation or chemically by mixing the enzyme in a buffered solution containing the acrylamide monomer and cross-linker (Mohamad et al., 2015). Furthermore, another approach has been reported for self-immobilization of enzymes, named as sphere enzymes; these are synthesized by preparing water-in-oil emulsions of the enzymes in oil using a surfactant to produce spherical self-immobilized enzyme particles (Sheldon, 2007; Brady and Jordaan, 2009).

Recently, it has been reported that the immobilization of enzymes on nano-support matrices resulted in their enhanced residual catalytic activity that could be due to increased porosity and surface area (Sakai et al., 2010; Sirisha et al., 2016). One advantage of self-immobilization is that the structural and functional properties of the enzymes are maintained; however, this method imposes the need for high enzyme purity that could result in an increased operational cost.

6.2.4 Encapsulation/entrapment

Encapsulation is an irreversible method of enzyme immobilization that involves entrapment of enzymes in a support or a semi-permeable membrane that allow the free diffusion of the substrates and products, while retaining the enzymes inside the matrix (Sirisha et al., 2016). Alternatively, encapsulation may also be defined as the physical retention of the enzymes in a confined space or network. Encapsulation of enzyme is known to be one of the best approaches for immobilizing enzymes as it improves their stability, while limiting their leaching from the matrix. In addition, this method does not involve any chemical interaction of the enzyme with the support matrix and avoids any detrimental effects to the native enzymatic protein (Bickerstaff, 1997; Guisan, 2006; Mohamad et al., 2015). Furthermore, this method allows for modification to the encapsulating polymer material which could be created to be optimum for the specific biocatalyst to be encapsulated. For example, pH of the support matrix could be adjusted to the optimum value as that of the enzyme; this could be achieved with various types of polymers, including sol-gels, silica-organic polymer hybrids and other inorganic materials (Mohamad et al., 2015).

In general, the approaches for the encapsulation of enzymes involve their entrapment in an extensively cross-linked polymer matrix or by enzyme dissolution in a non-aqueous phase or separation of the enzyme from

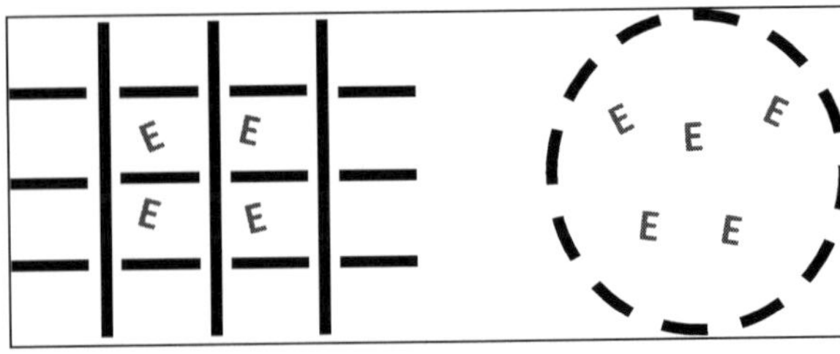

Figure 6.4 Enzyme immobilization by encapsulation or entrapment in a semi-permeable matrix.

the bulk solution by their encapsulation in a semi-permeable microcapsule (Sirisha et al., 2016). Although encapsulation allows for free diffusion of small biomolecules, it could also lead to membrane rupture and mass-transfer limitations resulted by the accumulation of products. Nevertheless, modification of the membrane in terms of porosity broadens their application to different types of biomolecules (Bickerstaff, 1997; Guisan, 2006). The mass-transfer limitations occurring in the encapsulated enzymes could be reduced by an alternative approach, microencapsulation. Microencapsulation is a method of encapsulating a bioactive agent in the core of micron-sized particles that are composed of semi-permeable membranes (Fernández-Fernández et al., 2012). In comparison to encapsulation, the release of microencapsulated bioactive compounds occurs at controlled rates and under specific conditions (Desai and Jin Park, 2005) (Fig. 6.4).

6.3 Choice of support matrix

The choice of an appropriate support matrix is one of the most crucial steps for the development of an efficient method of immobilization. An ideal support matrix must be easily available, biocompatible, inert, antimicrobial, robust, have a high-loading capacity, should offer regeneration ability, enhance enzyme specificity and activity, be cost-effective as well as ecofriendly (Es et al., 2015; Sirisha et al., 2016). In addition, the matrix preparation must be simple and should offer a potential for scale-up for production in bioreactors. The support material must also be approved for applications in food and pharmaceuticals (Es et al., 2015). Various support matrices may possess some of these ideal matrix characteristics but could differ for different biocatalysts to be immobilized.

The support matrices for immobilization are broadly classified into two categories based on their chemical composition, namely inorganic and organic supports, where organic matrices are further classified into natural and synthetic types (Sirisha et al., 2016). The inorganic support materials mainly include, silica gels, glass, and metal oxides; in comparison to organic supports, they are known to provide enhanced mechanical and thermal stability, rigidity, porosity as well as have antimicrobial properties. Alternatively, various natural polysaccharides, such as collagen, chitosan, alginate, cellulose, starch, agarose, carrageenan, etc., may be used as organic support materials for enzyme immobilization. Organic biopolymers are easily available, cost-efficient inert gels, biocompatible with a mild surrounding environment as well as high thermal and mechanical stability (Sirisha et al., 2016).

6.4 Whole cell immobilization and biotransformations

Over the recent years, the development of sustainable bio-based processes for the production of chemicals, pharma products, fuels, and other materials have been gaining increasing attention. In comparison to the conventional chemical synthesis, biocatalysis offers a more specific and environment-friendly route for the production of compounds with potential applications in the food, beverage, cosmetic, or pharmaceutical industries. (Lin and Tao, 2017). Whole cell biotransformations involve biocatalysis using the entire microorganism, where the reaction may require multiple enzymes or when the isolation of the enzyme from the cell is rather difficult (Kaushik et al., 2014). As compared to enzymatic catalysis, whole-cells exhibit significantly lower rates of reaction and increased mass-transfer limitations. In addition, whole cell biotransformations involve the use of expensive equipment, handling of large reaction volumes, reduced productivity, nonspecific side reactions, accumulation of toxic by-products, difficult product separation and recovery (Kaushik et al., 2014; Weatherley et al., 2018). On the other hand, whole-cell biotransformations offer several advantages, such as the use of low-cost feedstocks and raw materials, a large choice in natural or recombinant (designer cells) microorganisms, avoids the use or regeneration of cofactors, multiple transformations may be performed and reactions may be carried out at ambient conditions of temperature and pH (Weatherley et al., 2018).

In order to overcome the limitations of whole cells, an efficient immobilization method may be used. It is well known that the nature of prokaryotic cells is mobile that allows them to access essential nutrients required for their metabolism; however, at some stage of their life cycle they tend to attach themselves onto surfaces by adhesion. In industrial processes, flocculation, pellet formation and surface attachment may be considered as immobilized forms of whole cells. In general, the use of cell-immobilized systems for biotransformation in the industry are area specific and the techniques used for their immobilization may vary with different cells (Es et al., 2015). Immobilization of cells may be carried out by exploiting this nature of such microorganisms. As compared to whole cells, the immobilized cells exhibit relatively higher catalytic activity, long-term stability, regeneration ability, reusability, higher loading capacity, and high resistance to abrasion as well as microbial contamination (Es et al., 2015). Over the recent years, various whole cells have been immobilized for biotransformations and have been listed in Table 6.1. In general, whole-cell catalysis is classified into two categories based on its approach, including biotransformation or biocatalysis and fermentation bioprocesses.

Fermentation bioprocesses involve the biosynthesis of products from growth substrates by host cells as part of their natural metabolism; the products synthesized during fermentation bioprocesses are released into the broth which also contains their metabolic intermediate and hence, make the downstream processing more difficult. Alternatively, whole-cell biotransformation involves the separation of the enzyme manufacturing cell-growth phase and the production phase, where the substrates are converted into products by resting cells that result in reduced downstream processing (Ellaiah et al., 2004; Lin and Tao, 2017).

6.4.1 Advantages of whole-cell biotransformation

There are numerous advantages using whole cells in biotechnological applications, for the bioconversion of raw and inexpensive materials into added-value products. Other advantages may also include the processing in a multistep reaction, the lower cost of separation and purification of a given product and the potential of the reuse of cells. In addition, the immobilization of the whole cells could protect and enhance the stability of the enzymes. The overall advantages of using whole-cell biotransformation, particularly when immobilized, are as follows.

6.4.1.1 Biocatalyst cost

Natural or genetically modified whole-cell biocatalysts are the most low-cost catalysts that may be used in biocatalytic reactions; where, as compared to enzymes, these biocatalysts have been reported to be 10 times less expensive (Wachtmeister and Rother, 2016). Moreover, the use of whole-cell biocatalysts reduces the cost associated with enzyme processes, such as cell lysis and enzyme purification. In addition, whole-cell biocatalysts are relatively

Table 6.1 Recent applications of immobilized whole-cells of selected microorganisms in biocatalysis and biotransformations.

Microorganism	Support matrix	Substrate	Product	References
Rhizopus oryzae ATCC #34612	Polyethylene	Cotton seed oil	Fatty acid methyl ester	Athalye et al. (2013)
Gluconobacter oxydans WSH-003	Calcium alginate	D-sorbitol	L-sorbose	Wang et al. (2013)
Bacillus subtilis	Calcium alginate	L-glutamic acid	α-ketoglutaric acid	Hossain et al. (2014)
Rhizopus oryzae NBRC 4697	Polyurethane foam	Soyabean oil/ Methanol	FAME biodiesel	Kyeong & Yeom (2014)
Echerichia coli $_{SpNox}$ & *E. coli*$_{HjLAD}$	Calcium alginate	L-arabinitol	L-xylulose	Gao et al. (2015)
Echerichia coli BL21(DE3)	Polyvinyl alcohol	3-azabicyclo [3,3,0] octane	Telprevir	Zajkoska et al. (2015)
Methylocella tundrae	Calcium alginate	Methane	Methanol	Mardina et al. (2016)
Aspergillus oryzae	Polyurethane foam	Soyabean oil/ Methanol	FAME biodiesel	Amoah et al. (2016)
Pseudomonas mendocina	Fe_3O_4-chitosan	Soyabean oil/ Methanol	FAME biodiesel	Chen et al. (2016)
Pseudomonas putida KT2440	Calcium alginate	Arginine	Citrulline	Patil et al. (2017)
Meyerozyma guilliermondii SC1103	Calcium alginate	HMF	BHMF	Xu et al. (2018)
Methylocella tundrae	Silica gel	Methane	Methanol	Patel et al. (2018)
Methylomonas methanica	Silica gel	Methane	Methanol	Patel et al. (2018)
Aspergillus niger EXF 4321	Lyophilized	Caffeic acid (CA)	CA phenethyl ester	Rajapriya et al. (2018)
Aspergillus oryzae	Activated carbon	Soyabean oil/ Methanol	FAME biodiesel	Quayson et al. (2019)
Meyerozyma guilliermondii SC1103	Calcium alginate	Furfural	Furfuryl alcohol	Zhang et al. (2019)

easy to prepare as they involve low-cost of fermentation and have a higher reusability (Lin and Tao, 2017).

6.4.1.2 Higher efficiency and use of cofactors

In general, whole-cell biocatalysts involve a cascade of enzymatic pathways that allow for multiple reactions with a renewable supply of cofactors that are required for such biotransformations (Lin and Tao, 2017). Hence, the use of whole-cells as biocatalysts reduces the need for adding cofactors as they are self-generated by the host cells; this feature is especially advantageous for cofactor-dependent reactions, where the alternative use of whole-cells reduces the associated costs. Whole-cell biocatalysts are often a preferred choice for reactions catalyzed by oxidoreductases that depend on nicotinamide cofactors, where the cofactors are naturally regenerated by conversion of a co-substrate or by an enzyme-coupled approach in which more than one enzyme are used in a combination (Wachtmeister and Rother, 2016). In addition, the presence of protective cellular envelopes on the whole cells result in an enhancement of the enzyme stability and may allow for biocatalysis under relatively extreme reaction conditions. Overall, the close proximity of the substrates and biocatalysts, as well as the presence of inherent cofactors, results in an improved biocatalytic efficiency with reduced process costs (Lin and Tao, 2017).

6.4.1.3 Reduced downstream processing

The synthesis of bioproducts via whole cells comprises mainly of two phases; the growth of the biocatalyst and the conversion of the substrate to products. After the first phase, the process involves the harvesting of the cells, where they are isolated, washed using a buffer solution, and subsequently used for biocatalysis (Lin and Tao, 2017). Due to this process, components such as unused substrates, nutrients and intermediate metabolites are removed from the fermentation broth, thereby resulting in a better productivity and reduced downstream processing. In addition, the removal of nutrients leads to an arrest in cell growth, where resting cells have been reported to produce significantly higher yields; this may be due to the use of carbon source solely for product synthesis and not for biomass production (de Carvalho, 2016).

6.4.1.4 Increased applicability in nonconventional media

In general, biocatalytic reactions are often limited by low substrate or product loads which may be due to the use of conventional aqueous medium as the surrounding environment (Wachtmeister and Rother, 2016). It has been reported that various industrial processes involve the use of hydrophobic substrates and products; hence, the reactions catalyzed in aqueous environments result in a relatively lower productivity. This limitation may be overcome by the use of organic solvents or nonconventional solvent systems, such as ionic liquids, microaqueous and solvent-free conditions. However, the use of nonconventional solvents is known to be associated with reduced enzymatic activity that may be detrimental to the enzyme's native protein structure. The use of solvent-tolerant microorganism whole-cells provide an effective solution to this problem, where they provide a more protective natural environment to the enzymes and prevent the loss of their catalytic activity that may occur due to the change in the structural conformation of the native protein (de Carvalho, 2016).

6.4.2 Limitations of whole-cell biotransformations

Although the use of whole cell in biotransformations has many advantages, there are certain disadvantages that may include the need for additional costly equipment and materials, the potential of altering the enzymatic activity and the risk of cell contamination. Indeed, despite the various advantages of whole-cell catalyzed biotransformations, there are some limitations involved, mainly lower stability and increased mass-transfer limitations. As mentioned earlier, to facilitate the recovery, stability and recyclability of the whole-cells, immobilization may be carried out. For example, Lentikat is a well-known technology that was developed for the immobilization of whole cells in polyvinyl alcohol matrix, where an advancement in this technology has reported the high strength of this matrix and is particularly suitable for stirred tank bioreactors (Wachtmeister and Rother, 2016). In addition, *E. coli* whole-cells immobilized in Lentikats have been shown to be stable for 15 months in terms of the reaction rates and catalytic activity, as compared to their free forms (Cárdenas-Fernández et al., 2012). Another immobilization matrix for *E. coli* whole-cells, that are grown on ion-exchange resins is silicone coating, where both hydrophobic and hydrophilized coatings have been reported to maintain their catalytic activity as compared to the uncoated whole-cells (Findeisen et al., 2014).

Increased mass-transfer limitations, due to the presence of cell membranes is known to be one of the major disadvantages of whole-cell biotransformation. Wachtmeister and Rother (2016) suggested that this limitation may be overcome by deliberate permeabilization of cell membranes using surfactants, chelating agents or organic solvents. Alternatively, co-expression of membrane transporters could also be carried out to reduce mass-transfer limitations, where a significant enhancement in product yields has been reported, as a consequence of an increased uptake of substrate (Wachtmeister and Rother, 2016).

6.5 Mass transfer effects

6.5.1 Generality

Immobilized enzymes exhibit a relevant case of heterogenous biocatalysis, where the reaction takes place, both at the surface and inside the biocatalyst; however, as compared to the bulk reaction medium, where the reaction is continuously monitored, the conditions in the two environments of the enzyme are quite different. In an isotropic system, the changes in enzyme kinetic behavior may be caused by two main factors. The first factor is the change in the enzyme or its immediate surroundings due to immobilization; this could lead to conformational changes and/or restriction in accessibility to their active as well as allosteric sites. The second factor is a heterogenous local environment, which is formed if the concentration of product, substrate and effectors is different than that from the bulk reaction medium (Illanes et al., 2008).

Overall, the immobilization of enzymes could result in conformational and micro-environmental effects that subsequently affect the enzyme kinetics during biocatalytic reactions (Illanes et al., 2008). The conformational changes refer to structural changes to the enzymatic protein's native three-dimensional structure and such changes could exhibit a different kinetic behavior as compared to the free one. Furthermore, changes in the enzyme micro-environment lead to partition and mass transfer limitations, including partition of substrates or products as well as their mass transport to and from the bulk medium (Illanes et al., 2008). In order to understand these two effects in detail there are certain terms that need to be considered, including the micro/macroenvironment of immobilized enzymes, intrinsic kinetics, inherent kinetics, and the effective rate (apparent) kinetics.

6.5.1.1 Micro/macroenvironment of immobilized enzymes

Enzymatic reactions that are carried out in well-stirred homogenous systems result in a uniform concentration for reactants throughout the medium; however, for immobilized enzymes the interactions between the carrier matrix and substrates with diffusional limitations could result in non-uniformities (Wingard et al., 1976). Microenvironment is the area within immediate surrounding of the immobilized enzyme, whereas the bulk medium is the macroenvironment. Immobilized enzyme kinetics is determined by the local concentration of the substrates or products in the microenvironment; hence, their kinetic parameters cannot be determined by only monitoring the bulk medium reaction (Illanes et al., 2008; Kobayashi and Laidler 1973).

6.5.1.2 Intrinsic Kinetics

The intrinsic kinetics of immobilized enzymes represent their true kinetic nature and are determined in the absence of partition and mass transfer limitations, where the concentrations of the substrate, product and effectors are same in both, the micro and the macroenvironment (Chaplin and Bucke, 1990). On the other hand, the intrinsic kinetics may not be the same as that of the free enzyme as their immobilized forms undergo conformational changes (Illanes et al., 2008; Kobayashi and Laidler 1973).

6.5.1.3 Inherent kinetics

The inherent rate of reaction may be defined as the rate that is observed in the absence of diffusional or mass transfer limitations; hence, the rate of transfer of substrates and products could be infinitely fast. In addition, the inherent rate of reaction may differ from the intrinsic rate when partition effects prevail, as the interactions between matrix and various soluble species could result in a concentration difference between the micro as well as the macroenvironment (Illanes et al., 2008; Kobayashi and Laidler 1973).

6.5.1.4 Effective rate (apparent) kinetics

The effective kinetic parameters for immobilized enzymes are observed when diffusional or mass-transfer limitations occur in the presence or the absence of the partition effects. In general, the first step for analyzing the kinetics of immobilized enzymes involves determination of the effective reaction rate and kinetic parameters. In order to obtain a deeper insight, the diffusional limitations are separated to determine the inherent kinetic parameters, which are subsequently removed by removing the partition effects; hence, leading to the intrinsic reaction parameters. Following this, a comparison may be made between the free and immobilized enzymes along with the evaluation of the effects of immobilization on their structure as well as the significance of enzyme-matrix interactions (Illanes et al., 2008; Kobayashi and Laidler 1973).

6.5.2 Diffusional limitations

An enzyme-catalyzed reaction involves the diffusion of the substrate molecules from the bulk medium to the enzyme active sites and diffusion of the product molecules from the matrix into the bulk medium. This system involves both molecular and convective diffusions, where concentration gradients are obtained in the surrounding medium; these gradients may be due to diffusional or partitional limitations and generally occur together (Illanes et al., 2008). The depletion of substrate molecules or accumulation of the product in the microenvironment usually lead to concentration gradients in the reaction system. In addition, the extent of substrate depletion or product accumulation is a function of their molecular size, where larger size molecules exhibit lower diffusivity and higher diffusional limitations.; these limitations may occur both externally or internally for the immobilized enzymes. In general, external diffusional limitations (EDR) occur for immobilized enzymes with impervious carrier matrices, whereas internal ones (IDR) may be result with gels or microporous solid matrices (Illanes et al., 2008; Kobayashi and Laidler 1973). It has been reported that IDR is usually more severe than EDR as the substrate and products need to be diffused through a medium, with relatively lower mass transfer rates than those observed for liquid supports.

6.5.2.1 Effects of external diffusional limitations on kinetic parameters

The effects of external diffusional limitations (EDR) *on Kinetic Parameters* may be observed for enzymes that are immobilized on porous supports. In assuming that the enzyme is homogenously distributed throughout an impervious support matrix, the conversion of the substrate occurs in three main steps; transfer of substrate from the bulk medium to the enzymatic surface, enzyme-catalyzed conversion of substrate to product and transport of product from surface to the bulk medium. The enzymatic reaction is located at its surface, with lower substrate and higher product concentration at the surface than that in the bulk medium. The magnitude of the substrate depletion or product accumulation may vary with the enzyme catalytic activity and the transport phenomena (Illanes et al., 2008; Kobayashi and Laidler 1973).

6.5.2.2 Effects of internal diffusional limitations on kinetic parameters

For enzymes immobilized in porous membranes, the mass transfer at both external and internal porous membranes needs to be considered. To develop the simplest model for determination of these diffusional limitations, consider the enzyme uniformly distributed on the membrane, where the membrane is exposed to the substrate solution at one end, while sealed at the other one. On the other hand, the membrane may be exposed at both ends to the same substrate solution. The heterogenous system may be characterized by two limiting cases. At sufficiently low enzyme activity, the concentration of the substrates and products are same in the external and internal membrane; hence, the reaction is kinetically controlled and is independent of any transport phenomena. Contrarily, at higher enzyme activity all substrate molecules are converted to products at the surface of the particle and the reaction rate is equals the maximum rate of mass transport in the bulk solution.

6.5.3 Partition effects

The immobilized enzymes bound to a charged membrane may influence the bulk environment by both charge and hydrophobicity of their surface. The surface charge produced by ion-exchange matrices repel molecules of same charge and attract opposite charges, where the nature of attraction is significant until a certain distance and decreases with the square of the distance from the surface. This decrease in attraction results in a partitioning of the charged particles between the bulk and the microenvironment, where equally charged species are expelled into the bulk environment and that of the opposite charge are partitioned into the microenvironment.

6.6 Effect of immobilization on the stability of enzymes

Immobilization can affect greatly the stability of a given enzyme, when such procedure would result in an alteration in the conformation of the enzymatic protein molecule that could lead to its denaturation. Nevertheless, immobilization should enhance the enzyme's stability since it limits the movement of the enzyme at higher temperatures preventing its thermal denaturation. Indeed, selective alteration of chemical and physical properties of a given enzyme may result by its stability as compared to the free one. The kinetics of an immobilized enzyme could differ from that of the free one. The success in obtaining a more efficient immobilized enzyme is pending on the nature of the bonding between the enzyme and the chemical used for its immobilization. This type of bonding should have the flexibility of forming reversal unstrained covalent or noncovalent interactions that would stabilize the resultant immobilized enzyme. In addition, immobilization of a given enzyme in a bioprocess may also protect it from the effects of other enzymes mainly the proteolytic ones and microbiological attacks. However, its important to ensure that the surfaces of the enzyme and support be complementary with the formation of non-covalent bonds. Nevertheless, other factors are also considered in the immobilization process, such as the cost, the need for a specific support material and ensuring that that the substrate is not sterically hindered from diffusing to the active site of the immobilized enzyme.

6.7 Selected of biotechnological applications of immobilized enzymes

The increasing demand in the use of enzymes in a wide range of biotechnological applications resulted in the development of approaches for the enhancement of their activity and stability. Despite the many advantages of using enzymes in biotechnological applications, there are limits to their uses (Sheldon, 2011). This is due to the lack of their stability as well as for their cost. One of the major approaches to achieve a successful use of enzymes is their immobilization (Datta et al., 2013). Using novel approaches in the entrapment and immobilization of enzymes for selective biotechnological applications is crucial for their industrial success. The scope of potential use of immobilized enzymes in food and pharmaceutical applications are itemized as follow:

6.7.1 Selected immobilized enzymes in bioprocesses

There is a spectacular increase in the applications of enzymes in various bioprocesses to achieve desired conversions, under controlled conditions, for obtaining well-targeted products. These bioconversions can be achieved by a single enzyme extract, an enzyme complex or a whole cell. Although the sources of biocatalysts can be of plant and animal origin, the microbial one is the most predominant and used in biotechnological applications. Enzymes are used in a wide range of applications in food processing in bioconversion aspects as well as the production of natural food ingredient and nutraceuticals. In addition, enzymes are used as tools to facilitate the downstream food processing.

Although a biocatalyst could be as a free enzyme extract (in vitro) or within a whole cell (in vivo) as well as immobilized one, the immobilization has great consequence not only on the biocatalytic activity but also on the downstream process in terms of substrate supply and mass transfer (Gibbs et al., 1999; Kermasha et al., 2017). In addition, the immobilization of whole cells could influence the physiological conditions of their growth.

The list of enzymes used in biotechnological applications in food industry is long. It would be of interest to review briefly the immobilization of selected enzymes of industrial importance. These enzymes would include esterases, lipases, proteases, amylases, cellulases, pectinases, laccases, lipoxygenases, hydroperoxide lyases, and chlorophyllase.

Esterases are a group of hydrolases that split in chemical reactions, in the presence of water, esters into acids and alcohols. They are involved in a wide range of catalytic activities, such as the development of flavors (Kermasha et al., 1999, 2000; Goulet et al., 2012). Their immobilization is well-reported in literature (Shaw et al., 2006; Del Monte-Martinez et al., 2019; Doraiswamy et al., 2019; Luu Le et al., 2020).

Although they are considered as subclass of esterases, lipases are enzymes that catalyze the hydrolysis of triacylglycerols of lipids fats (lipids) into free fatty acids and glycerol molecule. In addition to their presence in most if not all living organisms, lipases are an important group of hydrolytic enzymes involved in selected bioconversion processes, such as the production of natural flavors (Aziz et al., 2015), fatty acid transesterification (Kermasha and Bisakowski, 1998; Khodadadi and Kermasha, 2014), lipid synthesis (Tweddell et al., 1999; Bai et al., 2012) and structured lipids (Bai et al., 2012; Khodadadi et al., 2013, 2014) as well as the production of novel nutraceutical biomolecules (Aziz et al., 2014). The immobilization of lipases (Ellaiah et al., 2004; Mendes et al., 2013; Manoel et al., 2015 as well as their encapsulation in vitro (Buisson et al., 2001; Zheng et al., 2013; Rodrigues et al., 2019; Nadar and Rathod, 2018) and in vivo (Su et al., 2016) are well reported in the literature.

Proteases, peptidases or proteinases are a group of enzymes that catalyze the breakdown of protein macromolecules into smaller polypeptides or in single amino acids. The enzymatic action of proteases proceeds with the cleavage of the peptide bonds within the protein molecules. A review of their mechanism ands applications were recently reported (Razzaq et al., 2019; Urano et al., 2020). Although there is a wide range of techniques used for the immobilization of proteases, the selection of each one is pending on their nature as well on their applications (Silva et al. 2006; Abida Anwar et al., 2009; Elchinger et al., 2015; Küchler et al., 2015; Bavaro et al., 2016; Abdel Mehdi et al., 2018).

Among the first group of enzymes described in literature are amylases, which is a class that catalyze the hydrolysis of starch macromolecules into smaller units, such as α-glucose and maltose. Amylases are used extensively in the conversion of starch into liquid glucose syrup. Their immobilization is also reported (Sing and Kumar, 2011; Swarnalatha et al., 2015; Al-Najada et al., 2019).

Cellulases are a group of enzymes that break down the cellulose molecule into single monosaccharides, β-glucose or shorter chain polysaccharides, such as oligosaccharides. Because of the strong binding β-glyosidic bond, the hydrolysis of cellulose is more difficult as compared that of starch in the case of α-glyosidic bond. However, cellulases are used especially in the textile industry (Kuhad et al., 2011; Jayasekara and Ratnayake, 2019). Their immobilization is also reported (Wu et al., 2005; Khoshnevisan et al., 2011, 2017).

Pectinases are a group of enzymes that catalyze pectic substance degradation through depolymerization (hydrolases and lyases) and deesterification (esterases) reactions (Pedrolli et al., 2009). Pectinases are paricular y useful in the fruit juice industry (Nighojkar et al., 2019). The industrial microbial production of pectinases (Alazi et al., 2018) and their immobilization (Li et al. 2007; Rajdeo et al., 2016; Sojitra et al., 2017) are also reported.

Alcohol dehydrogenases are a group of dehydrogenase enzymes that catalyze the interconversion between alcohols and aldehydes or ketones. They are often used in biotransformation applications mainly for the synthesis of enantiomerically pure stereoisomers of chiral alcohols. Often, high chemo- and enantioselectivity can be achieved. In addition, in vivo the alcohol dehydrogenases can be used in the production of ethanol (Aziz et al., 2016; Zheng et al., 2017). The immobilization or alcohol dehdrogenases in vitro (Shinde et al., 2018; Zeuner et al., 2018) as well as in vivo (Moreno-Garcia et al., 2018) is reported.

Glucose oxidase is an oxidoreductase that catalyses the oxidation of glucose to hydrogen peroxide and D-glucono-δ-lactone. Glucose oxidase This enzyme is produced by certain species of fungi and insects and displays antibacterial activity when oxygen and glucose are present. Glucose oxidase is commonly used to stabilize liquid egg white by converting glucose to D-gluconic acid, thereby limiting the potential for Maillard browning reactions. Its immobilization is also reported in literature (Balisteri et al., 2016; Özdem et al., 2014; Phiri et al., 2019).

Laccases are multicopper oxidases found in plants, fungi, and bacteria. Laccases oxidize a variety of phenolic substrates, performing one-electron oxidations, leading to crosslinking. Laccases have versatile industrial applications (Brijwani et al., 2010; Moreno et al., 2019). These applications may include the detoxification of industrial waste materials, mostly from the paper and pulp, textile, and petrochemical industries. In addition, Laccases are also used cross-linking proteins for the development novel food ingredients (Sato et al., 2015). The immobilization of laccases (Lu et al., 2007; Liu et al., 2012) and their encapsulation (Gill et al., 2018a, b) are also reported

Lipoxygenases (LOXs) are dioxygenases that catalyze the formation of corresponding hydroperoxides from polyunsaturated fatty acids, such as linoleic acid and arachidonic acid. In addition to their role in the physiological in cells (Riley et al., 1996; Mashima and Okuvama, 2015), they have major impact on the biogeneration of flavor precursors (Shen et al., 2015). The immobilization of lipoxygenase was the subject of a series of research work aimed at its application in the production of fatty acid hydropeoxides, considered to be flavor precusrsors (Battu et al., 1994; Hsu et al., 1997a, b; Kermasha et al., 2002; Vega et al., 2005; Karout et al., 2007; Hall et al., 2008; Aziz et al., 2014).

Hydroperoxide lyases are enzymes that catalyze the cleavage of C-C bonds in the hydroperoxides of fatty acids. They belong to the cytochrome P450 enzyme family (Noordermeer et al., 2001). Hydroperoxde lyases not only involved in the ripening of fruits (Riley et al., 1996), but also in the development of volatile flavor compounds in vivo (Salas and Sanchez, 1999; Kuroda et al., 2005). More recent research work in molecular biology is aimed at the development of production of selected flavors (Stolterfoht et al., 2019). The immobilization of hydroperoxide lyases (Hall et al., 2008; Nunez et al., 2010; Liu et al., 2013a, b) and their biocatalysis (Kuldamrong et al., 2013) are reported.

Chlorophyllase is the key enzyme in chlorophyll metabolism and although it catalyzes the hydrolysis of chlorophyll into chlorphyllide and phytol, it can carry out the reverse reaction of the esterification of chlorophyllide (Khalyfa et al., 1995). Chlorophyllase has the potential to be used in certain potential biotechnological applications for the removal of green pigments from edible oils (Ward et al., 1994; Khamessan and Kermasha, 1996) and plant protein concentrates (Fang et al., 1997). The immobilization of chlorophyllase (Gaffar et al., 1999; Kermasha et al., 2000; Karboune et al., 2005; Yang et al., 2014) as well as its encapsulation (Yi et al., 2002; Yi et al., 2006a, b; Yi et al., 2008;) are reported.

6.7.2 Encapsulation of enzymatic bioconversion products

As a result of the involvement of enzymes in biocatalysis processes, there is an increase interest in the encapsulation of enzymatic end products, in particular food additives and nutraceuticals. Indeed, the encapsulation is aimed at the protection of the product during the various processes and storage. In other words, this interest is due to the impact of the encapsulation on these products in terms of preservation of nutraceutical and organoleptic qualities, control of their release, facilitate their transport and better management of their marketing (Madene et al., 2006). The encapsulation of additives, such as flavors (Given, 2009; Renu and Zahra, 2015; Soottitantawat et al., 2015), and nutraceuticals, such krill oil (Aziz et al., 2014) and its esterified phenolic lipid product (Kermasha et al., 2017) are good examples of the usefulness of the entrapment processes.

6.7.3 Food and beverages

Immobilized enzymes are important in the food and beverage industry, where they have been used in cheese making, starch processing, food preservation, beer stabilization, improvement in food function, flavor, and quality, lipid hydrolysis as well as food preservation. In addition, immobilization of enzymes allows the use of enzymes at large-scale commercial level as they result in improved production yields, enzyme reusability, and recovery as well as increased economic viability (Es et al., 2015).

Xylanases are naturally synthesized by microorganisms, including bacteria, fungi, and actinomycetes, to cleave xylan which is a major component of hemicellulose. Xylanases have been reported to enhance the specific bread volume as the breakdown of hemicellulose results in an increased binding of water in the dough; hence, the xylanse treated dough is relatively softer and results in a delayed crumb formation in the bread (Raveendran et al., 2018). Carvalho et al. (2017) reported the use of thermostable xylanases, obtained from *Trichoderma stromaticum,* in bread making, where the results showed an increased softness and volume of both, wheat and whole-grain bread. Elegbede and Lateef (2018) investigated the effect of xylanases that were obtained from eight fungal strains; each fungal strain resulted in an improved dough rising of bread by 1.87 to 2.2-fold as well as the clarification of orange juice by 58.12%–74.22%. Recently, xylanases in combination with other enzymes, have found applications in the extraction, clarification, and stabilization of fruit juices and have been reported to lead to an increased yield, vitamins, mineral salts, pigments, etc. They have also been used in beer industry for the hydrolysis of the cell wall of barley that result in release of arabinoxylans and lower oligosaccharides hence, reducing the beer haziness as well as its viscosity (Raveendran et al., 2018). Kumar et al. (2014) investigated the effect of *Bacillus pumilus* VLK-1 xylanase immobilized on glutaraldehyde activated aluminium oxide pellets, in juice enrichment from grape and orange pulps, where the results showed an enhanced juice yield, clarity, reducing sugars, acidity, and filterability as well as a decrease in the turbidity and juice viscosity. Similarly, Shahrestani et al. (2016) reported that the use of xylanase immobilized on 1,3,5-triazine-functionalized-silica magnetic microcapsules had a significant effect on apple, pineapple, and orange juice clarification as well as their enrichment. Furthermore, xylanase hydrolysed xylan leads to the formation of xylo-oligosaccharides that have been reported to have prebiotic effects and may be used as functional foods. Xylo-oligosaccharides promote the growth of healthy microorganisms and limit the growth of harmful ones; this property allows their use in various beverage products such as tea, coffee, and soymilk. Additionally, xylanases have also been reported to be used in the treatment of animal feeds as they facilitate the release of nutrients from macromolecules by improving their bioavailability as well as the nutritional value of the diet (Kumar et al., 2018).

Lactase or β-galactosidase catalyzes one of the major reactions involved in food processing, the hydrolysis of lactose. β-galactosidases find their main application in the milk and dairy products, where their use has been associated with an enhancement in creaminess as well as a reduction in requirement of additional sugars. In addition, they find potential application in hydrolysis of lactose in whey and help reduce environmental issues such as high biological and chemical oxygen demands (BOD & COD). β-Galactosidase may also be used in the synthesis of galactooligosaccharides, which are functional food ingredients (Raveendran et al., 2018). Lactozym is a commercially available β-galactosidase preparation, which is immobilized on cellulose beads and was used for hydrolysis of lactose; the results showed a 90% and 60% conversion for whey and milk lactose, respectively (Panesar et al., 2010). α-Amylases (EC 3.2.1.1) are starch-degrading enzymes that hydrolyze the α-1,4 glycosidic bonds in polysaccharides to form dextrins. They have wide applications in the food industry, including baking, brewing and starch liquefaction. In baking they are associated with improvement of bread quality, color, and flavor. They are also used as glazing agents in rice cakes and dry foods. Gupta et al. (2013) investigated the saccharification of starch using α-amylase immobilized on Dowex MAC-3 via adsorption; the results confirmed the potential use of the enzyme in starch hydrolysis. Furthermore, He et al. (2017) used α-amylase, immobilized on magnetic nanoparticles, to investigate their applicability in bread quality; the authors suggested an increase in specific volume and cohesiveness as well as a decrease in gumminess as compared to the control.

Laccases (EC 1.10.3.2) are multi-copper blue oxidases and have a wide application in food and beverages. They are known for their ability to enhance food color, remove undesirable phenolic compounds that may cause browning, haze and turbidity formation in juices, wines or beers (Fernández-Fernández et al., 2013). Laccases have also found an application in baking sector due to their cross-linking property. Lettera et al. (2016) reported the use of laccase, immobilized in poly-(methacrylate beads), for clarification of fruit juice, where the results indicated a significant reduction in phenol content and an improved sensory profile.

6.7.4 Wastewater treatment

Phenolic compounds are the major components in industrial wastewater with some severely being toxic to plant, animal, and human health. As a result, a variety of physical, biological, and chemical methods have been developed for the removal or degradation of these compounds. Enzymatic wastewater treatment has been reported as an efficient technology for the degradation of phenol and phenolic compounds with a relatively higher specificity and minimum environmental impact. Such enzymes including polyphenol oxidases, laccases, horseradish peroxidase (HRP), and other oxygenases have mainly been used for the biological treatment of industrial wastewaters. Laccase, a multi-copper blue oxidase, has been extensively reported for treatment of wastewater due to its relatively wide substrate specificity, with the ability to catalyze a broad range of phenolic as well as nonphenolic substrates in the presence of a mediator.

Wang et al. (2012) reported the use of laccase immobilized on magnetic mesoporous silica nanoparticles as an efficient method for phenol degradation, where the degradation rate was 2-fold higher and the enzyme retained 71% activity after 10 successive reaction cycles. This degradation was significantly enhanced in continuous treatment process and resulted in 91% degradation of phenol after a period of 40 hours. Another study reported the use of tyrosinase immobilized as cross-linked enzyme aggregates for the treatment of phenolic wastewater; the method efficiently eliminated phenol, *p*-cresol, *p*-chlorophenol and bisphenol A within 0.5–3 hours, which was confirmed by the *Hydra sinesis* test (Xu and Yang, 2013).

HRP immobilized on magnetic Fe_3O_4 nanoparticles, deposited on graphene oxide sheets, were used for the removal of 2-chlorophenol, 4-chlorophenol, and 2,4-dichlorophenol; the results indicated that immobilized HRP was an efficient biocatalyst for the degradation of phenolic compounds and retained 66% catalytic activity after use in four reaction cycles (Chang et al., 2015). Bilal et al. (2016) reported the immobilization of manganese peroxidase onto glutaraldehyde activated chitosan beads and its subsequent application in detoxification of dyes in textile wastewater; the results showed reduction in COD, TOC, and BOD by 82%, 78%, and 92%, respectively, with a 97% color removal from the water. In addition to this, the cytotoxicity and mutagenicity of the water was significantly reduced, while the enzyme retained 60% of its activity after 10 continuous decolorization batches (Bilal et al., 2016).

6.7.5 Biosensors

The need for selective, rapid, cost-effective, and portable screening methods for environmental monitoring, detection of biomolecules, and clinical analyses have boosted the development of biosensors that may be used as complementary tools to the conventional analytical methods. However, as compared to the traditional analytical methods, enzyme-based analytical devices have various advantages such as, higher sensitivity, specificity, portability, economic viability and ability for mass production (Ispas et al., 2012). Recent advances in nanotechnology has widened the interest in enzyme immobilization due to their potential applications in biotechnology, biosensing, and biomedicals, where the technology allows ease of handling as well as separation and reusability of the enzymes in reactions (Ansari and Husain, 2012). Enzyme immobilization is a key factor in development of efficient biosensors as they result in good operational stability, storage stability, high sensitivity and selectivity, short response times as well as high reproducibility; however, the immobilization method used must retain the biocatalyst's native structure, activity and biological function (Sassolas et al., 2012).

Phenolic compounds are ubiquitous in nature and form an important component of human and animal diets. Phenolic compounds are also broadly present in various industrial processes, including coal-conversion, refining of petroleum, pharmaceuticals, dyes, pesticides, surfactants, resins and plastics; hence, these compounds are largely released into ground and surface water. Many phenolic compounds show harmful effects on plants, animals and humans as a result of their toxicity. Various conventional methods have been used for the determination of phenolic compounds; however, they are time consuming and expensive. Oxidative enzymes such as laccase, tyrosinase horseradish peroxidase (HRP) have been used as electrochemical biosensors for the detection of phenolic compounds (Nazari et al., 2015). Laccase immobilized polyaniline via covalent bonding were electrodeposited on carbon electrode using glutaraldehyde for the development of an electrochemical biosensor to detect catechol. The results showed that the laccase modified electrode exhibited a direct electron transfer reaction between the enzyme and the electrode, where the biosensor demonstrated a linear range, sensitivity and detection limit of 3.2×10^{-6} to 19.6×10^{-6} M, 706.7 mALmol^{-1} and 2.07×10^{-6} M, respectively (Nazari et al., 2015).

Similarly, Camargo et al. (2018) reported the development of an immobilization matrix for tyrosinase that involved the combining of carbon nanodiamonds with potato starch which were then deposited on a glassy carbon electrode as a homogenous rough film. The authors suggested that the resulting tyrosinase-nanodiamond-potato starch deposited on glassy carbon electrode biosensor was suitable for the detection of catechol and resulted in a detection limit of 3.9×10^{-7} mol L^{-1} in the range between 5.0×10^{-6} and 7.4×10^{-4} mol L^{-1}. As compared to other reported biosensors, this biosensor displayed a higher sensitivity which may be attributed to the small nanodiamonds that imparted a large surface area (Camargo et al., 2018).

Lactate is an important biomarker as it is excessively produced by the body during anaerobic metabolism; existing electrochemical lactate detection methods require an external power source for the development of a positive potential to the working electrode. A novel self-powered amperometric lactate biosensor was developed by using a dimethylferrocene-modified linear poly(ethylenimine) hydrogel that immobilized and mediated electron transfer from lactate oxidase, simultaneously, at the anode as well as the earlier mentioned cathode. The author reported a detection range between 0 mM and 5 mM lactate with a sensitivity of 45 ± 6 μA cm^{-2} mM for the biosensor coupled with the enzymatic bicathode (Hickey et al., 2016). More recently, Bäcker et al. (2017) reported the development of a glucose oxidase biosensor for the detection of glucose, where the enzyme was immobilized using tobacco mosaic virus (TMV) nanotubes or coat proteins (CP); the results indicated that the presence of TMV nanotubes on the surface of the sensor allowed more precise enzyme positioning for binding without causing a significant effect on their catalytic activity. In addition, the sensor exhibited a superior performance as compared to glucose biosensors with streptavidin conjugated immobilized glucose oxidase and hence, confirmed their potential for future biosensor applications (Bäcker et al., 2017).

6.7.6 Pharmaceuticals

The use of immobilized enzymes in the pharmaceutical industry is growing and involves various enzymes, including asparaginase, cysteine proteases, urokinase, hyaluronidase and glucocerebrosidase. One of the major applications of immobilized enzymes in pharmaceuticals is immobilization of penicillin amidase (Es et al., 2015). Penicillin is known to be the most used antibiotic due to their enhanced antibacterial activity and lower toxicity. Penicillin G acylase (PGA) catalyzes the hydrolysis of penicillin G to synthesize 6-aminopenicillanic acid (6-APA) that may be used to produce semisynthetic penicillin. PGA was immobilized on metal affinity membrane and showed an enhanced pH, temperature and storage stability as well as an increased reusability (Es et al., 2015).

Osteoporosis is a well-known condition that has been reported to affect around 200 million people world-wide and is responsible for most of the hip fractures in the United States and Europe. Cathepsin K inhibitors are important antiresorptive agents that prevent bone loss and assist in the bone formation process. Odanacatib is a potent, orally bioavailable cathepsin K inhibitor and its synthesis depends on an enzyme-mediated production of a key chiral intermediate, fluoroleucine. Earlier studies involved the use of Novozyme 435 for the synthesis of fluroleucine ethyl ester; however, the deactivation of the immobilized enzyme was a major issue (Truppo and Hughes, 2011). Truppo and Hughes (2011) developed an immobilized form of *Candida antartica* lipase, on Sepabeads and compared it to the available commercial enzyme preparations. The results showed that the immobilized lipase demonstrated an enhanced enzyme activity, stability and reusability, while showing a significant reduction in the process cost as well as 3-fold decrease in the E-factor for the manufacture of fluoroleucine, as compared to that for the commercial one.

Bang et al. (2011) reported the immobilization of lysosomal enzymes on titanium (IV) oxide particles and their use as effective antimicrobial agents; the results showed complete immobilization efficiency with an increased antimicrobial activity as compared to that of the free one. Furthermore, pharmaceutical compounds such as aminodeacetoxycephalosporanic acid (7-ADCA) and aminocephalosporanic acid are also of considerable importance as they may be used for synthesis of semi-synthetic antibiotics, including penicillins and cephalosporins. Ma et al. (2015) suggested the one-pot biosynthesis of D-7-ACA that was carried out using cephalosporin C acylase (CPCA) and deacetylase (CAH); both enzymes were immobilized on epoxy carriers by covalent binding, where CPCA was initially used for the hydrolysis of CPC to 7-ACA, subsequent to which, the immobilized CAH carried out the deacetylation of 7-ACA to D-7-ACA. The authors suggested an increased product yield with higher enzyme reusability and stability; hence, the industrial application and feasibility of the process was confirmed.

L-Asparaginase is also an important enzyme that has been reported to have potential applications for the treatment of lymphoblastic leukemia and lymphosarcoma. Ghosh et al. (2012) investigated the immobilization of l-asparaginase on polyaniline nanofibers, while Bahreini et al. (2014) carried out their microencapsulation in chitosan triphosphate nanoparticles; both studies claimed an increased enzyme activity and reusability.

6.7.7 Textiles

The use of enzymes in textiles is expanding and gaining interest due to low toxicity and ecological properties. One of the main enzymes use in textiles is cellulase which is used during denim finishing and cotton softening. Other enzymes include amylase for de-sizing, pectate lyase for scouring, laccase for bleaching, peroxidases for dye removal and catalase for termination of bleach (Es et al., 2015).

Yu et al. (2012) investigated the covalent immobilization of cellulase on Eudragit S-100 via carbodiimide coupling and evaluated its effect on treatment of cotton fibres; the results showed that the immobilized cellulase demonstrated lower weight loss, higher tensile strength and resulted in a softer whiter cotton fabric as compared to those treated by the free enzyme. The method was reported to be an efficient method for improving cotton properties that could be used to overcome cotton-finishing challenges (Yu et al., 2012). Daâssi et al. (2014) investigated the immobilization of laccase in calcium alginate beads for decolorization of various synthetic dyes; the immobilized laccase demonstrated a higher pH, temperature stability and was able to decolorize the anthraquinoe dye, Remazol Brilliant Blue R with high efficiacy. On the other hand, the enzyme showed lower efficiency for decolorizing diazo dyes, Reactive Back 5 and Bismark Brown R as well as metal textile dye Lanaset Grey G; however, the decolorizing yield was improved upon addition of a mediator, 1-hydroxybenzotriazole. Similarly, immobilized manganese peroxidase, encapsulated in calcium alginate beads were used to investigate their decolorization efficiency of textile dyes, where the enzyme decolorized Sandal-fix Black CKF, Sandal-fix Turq Blue GWF, Sandal-fix Foron Blue, Sandal-fix Black CKF and Sandal-fix Golden Yellow CRL dyes by 88, 82, 89, 96 and 83 %, respectively (Bilal and Asgher, 2015).

6.8 Conclusion

An increasing demand for the development of sustainable and environmentally friendly bioprocesses have led to the use of enzymes in a wide range of industrial applications. The immobilization of enzymes and whole microbial cells would be an efficient approach method for the enhancement of their stability for the preservation of their catalytic activity. In addition, the immobilization of enzyme and whole microbial cells would promote their separation from the reaction media and allow for their re-usability downstream the bioprocessing. In addition to an overview of the technical terms used in immobilization, a summarized review includes the use of immobilized enzymes and whole microbial cells in selected biotechnological applications, such as selected enzymes used in biocatalysis and bioprocesses for the development of natural food additives and nutraceuticals as well as their encapsulation, food and beverages, wastewater treatment, and in biosensors. Pharmaceuticals and textiles are also reported. Hence, future studies should be directed towards the development of efficient immobilization strategies that can expand the use of immobilized enzymes extracts (in vitro) and whole microbial cells (in vivo) in a wide range of industrial applications.

References

Abdel Mehdi, W., Mehde, A.A., Özacar, M., Özacar, Z., 2018. Characterization and immobilization of protease and lipase on chitin-starch material as a novel matrix. Int. J. Biol. Macromol. 117, 947–958.

Abida Anwar, A., Ul Qader, A.A., Raiz, A., Iqbal, S., Abid Azhar, A., 2009. Calcium alginate: A support material for immobilization of proteases from newly isolated strain of *Bacillus subtilis* KIBGE-HAS. World Appl. Sci. J. 7, 1281–1286.

Alazi, E., Knetsch, T., Di falo, M., Reid, I.D., Arentshorst, Visser, J., Tsang, A., Ram, A.F.J., 2018. Inducer-independent production of pectinases in *Aspergillus niger* by overexpression of the D-galacturonic acid-responsive transcription factor *gaaR*. Appl. Microbiol. Biotechnol. 102, 2723–2736.

Al-Najada, A.R., Almulaiky, Y.Q, Aldhahri, M., El-Shishtawi, R.M., Mohamed, S.A., Baeshen, M., Al-Faraj, M., 2019. Immobilisation of α-amylase on activated amidrazone acrylic fabric: A new approach for the enhancement of enzyme stability and reusability. Sci. Reports 9, 12672.

Ammann, E.M., Gasser, C.A., Hommes, G., Corvini, P.F.X., 2014. Immobilization of defined laccase combinations for enhanced oxidation of phenolic contaminants. Appl. Microbiol. Biotechnol. 98, 1397–1406.

Amoah, J., Ho, S.H., Hama, S., Yoshida, A., Nakanishi, A., Hasunuma, T., Ogino, C., Kondo, A., 2016. Converting oils high in phospholipids to biodiesel using immobilized *Aspergillus oryzae* whole-cell biocatalysts expressing *Fusarium heterosporum* lipase. Biochem. Eng. J. 105, 10–15.

Ansari, S.A., Husain, Q., 2012. Potential applications of enzymes immobilized on/in nano materials: A review. Biotechnol. Adv. 30, 512–523.

Athalye, S., Sharma-Shivappa, R., Peretti, S., Kolar, P., Davis, J.P., 2013. Producing biodiesel from cottonseed oil using *Rhizopus oryzae* ATCC# 34612 whole cell biocatalysts: culture media and cultivation period optimization. Energy Sustain. Dev. 17, 331–336.

Aziz, M., Ben Akacha, N., Husson, F., Kermasha, S., 2014. Synthesis of linoleic acid hydroperoxides as flavor precursors, using selected substrate sources. J. Am. Oil Chem. Soc. 91, 1867–1876.

Aziz, M., Husson, F., Kermasha, S., 2015. Optimization of the hydrolysis of safflower oil for the production of linoleic acid, used as flavor precursor. International J. Food Sci., (Doi: org/10.115/2015/594238).

Aziz, M., St-Louis, R., Husson, F., Kermasha, S., 2016. Selected dehydrogenases in Yarrowia lipolytica JMY 861: Their role in the synthesis of flavor compounds. Biosci. Biotechnol. Biochem. 80, 2184–2191.

Aziz, S., Gill, J., Dutilleul, P., Neufeld, R., Kermasha, S., 2014. Microencapsulation of krill oil using complex coacervation. J. Microencapsul. 31, 774–784.

Bäcker, M., Koch, C., Eiben, S., Geiger, F., Eber, F., Gliemann, H., Schöning, M.J., 2017. Tobacco mosaic virus as enzyme nanocarrier for electrochemical biosensors. Sensor. Actuators B: Chem. 238, 716–722.

Bahreini, E., Aghaiypour, K., Abbasalipourkabir, R., Mokarram, A.R., Goodarzi, M.T., Saidijam, M., 2014. Preparation and nanoencapsulation of L-asparaginase II in chitosan-tripolyphosphate nanoparticles and in vitro release study. Nanoscale Res. Lett. 9, 340.

Bai, S., Aziz, S., Khodadadi, Bou-Metri, C., St-Louis, R., Kermasha, S., 2012. Lipase-catalyzed synthesis of mediu-long-medium type structured lipids using tricaprylin and trilinolenin as substrate models. J. Am. Oil Chem. Soc. 90, 377–389.

Balisteri, N., Gaboriau, D., Jolivalt, C., Launav, F., 2016. Covalent immobilization of glucose oxidase on mesocellular silica foams: Characterization and stability towards temperature and organic solvents. J. Mol. Catal. B: Enzym. 127, 26–33.

Bang, S.H., Jang, A., Yoon, J., Kim, P., Kim, J.S., Kim, Y.H., Min, J., 2011. Evaluation of whole lysosomal enzymes directly immobilized on titanium (IV) oxide used in the development of antimicrobial agents. Enzyme Microb. Technol. 49, 260–265.

Barbosa, O., Ortiz, C., Berenguer-Murcia, Á., Torres, R., Rodrigues, R.C., Fernandez-Lafuente, R., 2015. Strategies for the one-step immobilization-purification of enzymes as industrial biocatalysts. Biotechnol. Adv. 33, 435–456.

Battu, S., Rabinovitch-Chable, H., Neneytout, J.-L., 1994. Effectiveness of talc as adsorbent for purification and immobilization of plant lipoxygenases. J. Agric. Food Chem. 42, 2115–2119.

Bavaro, T., Cattaneo, G., Serra, I., Benucci, I., Pregnolato, M., Terreni, M., 2016. Immobilization of neutral protease from Bacillus subtilis for regioselective hydrolysis of acetylated nucleosides: Application to capecitabine synthesis. Molecules 21, 10621.

Bickerstaff, G.F., 1997. Immobilization of enzymes and cells: Some practical considerations. In: Bickerstaff, G.F. (Ed.), Methods in Biotechnology I: Immobilization of Enzymes and Cells. Humana Press, Totowa, N.J, pp. 1–11.

Bilal, M., Asgher, M., 2015. Dye decolorization and detoxification potential of Ca-alginate beads immobilized manganese peroxidase. BMC Biotechnol. 15, 111.

Bilal, M., Asgher, M., Iqbal, M., Hu, H., Zhang, X., 2016. Chitosan beads immobilized manganese peroxidase catalytic potential for detoxification and decolorization of textile effluent. Int. J. Biol. Macromol. 89, 181–189.

Brady, D., Jordaan, J., 2009. Advances in enzyme immobilisation. Biotechnol. Lett. 31, 1639–1650.

Brena, B., González-Pombo, P., Batista-Viera, F., 2013. Immobilization of enzymes: A literature survey. Methods Mol. Biol. 1051, 15–31.

Brijwani, K., Rigdon, A. and Vadlan, P.V. (2010). Fungal laccases: Production, function, and applications in food processing. In *Enzymes as Additives or Processing Aids in Food Biotechnol.* (Doi: org/10.4061/2010/149748).

Buisson, P., Hernandez, C., Pierre, M., Pierre, A.C., 2001. Encapsulation of lipases in aerogels. J. Non-Crystal. Solids 285, 295–302.

Camargo, J.R., Baccarin, M., Raymundo-Pereira, P.A., Campos, A.M., Oliveira, G.G., Fatibello-Filho, O., Janegitz, B.C., 2018. Electrochemical biosensor made with tyrosinase immobilized in a matrix of nanodiamonds and potato starch for detecting phenolic compounds. Anal. Chim. Acta 1034, 137–143.

Cárdenas-Fernández, M., Neto, W., López, C., Álvaro, G., Tufvesson, P., Woodley, J.M., 2012. Immobilization of *Escherichia coli* containing ω-transaminase activity in LentiKats®. Biotechnol. Prog. 28, 693–698.

Carvalho, E.A., dos Santos Góes, L.M., Uetanabaro, A.P.T., da Silva, E.G.P., Rodrigues, L.B., Pirovani, C.P., da Costa, A.M., 2017. Thermoresistant xylanases from *Trichoderma stromaticum*: Application in bread making and manufacturing xylo-oligosaccharides. Food Chem. 221, 1499–1506.

Chang, Q., Jiang, G., Tang, H., Li, N., Huang, J., Wu, L., 2015. Enzymatic removal of chlorophenols using horseradish peroxidase immobilized on superparamagnetic Fe_3O_4/graphene oxide nanocomposite. Chinese J. Catal. 36, 961–968.

Chaplin, F.M., Bucke, C., 1990. The preparation and kinetics of immobilized enzymes. In: *Enzyme Technology,* Press Syndicate, Cambridge, U.K., pp. 80–136.

Chen, G., Liu, J., Qi, Y., Yao, J., Yan, B., 2016. Biodiesel production using magnetic whole-cell biocatalysts by immobilization of *Pseudomonas mendocina* on Fe_3O_4-chitosan microspheres. Biochem. Eng. J. 113, 86–92.

Daâssi, D., Rodríguez-Couto, S., Nasri, M., Mechichi, T., 2014. Biodegradation of textile dyes by immobilized laccase from *Coriolopsis gallica* into Ca-alginate beads. Intern. Biodeterior. Biodegrad. 90, 71–78.

Datta, S., Christena, L.R., Rajaram, Y.R.S., 2013. Enzyme immobilization: An overview on techniques and support materials. Biotech 3, 1–9.

de Carvalho, C.C., 2016. Whole cell biocatalysts: Essential workers from nature to the industry. Microb. Biotechnol. 10, 250–263.

Del Monte-Martinez, A., González-Bacerio, J., Varela, C.M., Vega-Villasante, F., Lalana-Rueda, R., Nolasco, H., Diaz, J., Guisán, J.M., 2019. Screening and immobilization of interfacial esterases from marine invertebrates as promising biocatalyst derivatives. Appl. Biochem. Biotechnol. 189, 903–918.

Desai, K.G.H., Jin Park, H., 2005. Recent developments in microencapsulation of food ingredients. Drying Technol. 23, 1361–1394.

Ding, S., Cargill, A.A., Medintz, I.L., Claussen, J.C., 2015. Increasing the activity of immobilized enzymes with nanoparticle conjugation. Curr. Opin. Biotechnol. 34, 242–250.

Doraiswamy, N., Sarathi, M., Pennathur, G., 2019. Cross-linked esterase aggregates (CLEAs) using nanoparticles as immobilization matrix. Prep. Biochem. Biotechnol. 49, 270–278.

Elchinger, P.-H., Delattre, C., Faure, S., Roy, O., Badel, S., Bernardi, T., Taillefumier, C., Michaud, P., 2015. Immobilization of proteases on chitosan for the development of films with anti-biofilm properties. Int. J. Biol. Macromol. 72, 1063–1068.

Elegbede, J.A., Lateef, A., 2018. Valorization of corncob by fungal isolates for production of xylanase in submerged and solid-state fermentation media and potential biotechnological applications. Waste Biomass Valori. 9, 1273–1287.

Ellaiah, P., Prabhakar, T., Ramakrishna, B., Thaer Taleb, A., Adinarayana, K., 2004. Production of lipase by immobilized cells of *Aspergillus niger*. Proc. Biochem. 39, 525–528.

Es, I., Vieira, J.D.G., Amaral, A.C., 2015. Principles, techniques and applications of biocatalyst immobilization for industrial application. Appl. Microbiol. Biotechnol. 99, 2065–2082.

Fang, Z., Bouwkamp, J.C., Salomos, T., 1997. Chlorophyllase ativities and chlorophyll degradationduring leaf senescence in non-yellowing mutant and wild type of *Phaseolus vulgaris*. J. Exper. Botany 49, 503–510.

Fernández-Fernández, M., Sanromán, M.Á., Moldes, D., 2013. Recent developments and applications of immobilized laccase. Biotechnol. Adv. 31, 1808–1825.

Findeisen, A., Thum, O., Ansorge-Schumacher, M.B., 2014. Biocatalytically active silCoat-composites entrapping viable *Escherichia coli*. Appl. Microbiol. Biotechnol. 98, 1557–1566.

Gaffar, R., Kermasha, S., Bisakowski, B., 1999. Biocatalysis of immobilized chlorophyllase in a ternary micellar system. J. Biotechnol. 74, 45–55.

Gao, H., Kim, I.W., Choi, J.H., Khera, E., Wen, F., Lee, J.K., 2015. Repeated production of L-xylulose by an immobilized whole-cell biocatalyst harboring L-arabinitol dehydrogenase coupled with an NAD^+ regeneration system. Biochem. Engineer. J. 96, 23–28.

Ghosh, S., Chaganti, S.R., Prakasham, R.S., 2012. Polyaniline nanofiber as a novel immobilization matrix for the anti-leukemia enzyme l-asparaginase. J. Mol. Catal. B: Enzym. 74, 132–137.

Gibbs, B., Kermasha, S., Alli, I., Mulligan, C.N., 1999. Encapsulation in the food industry: A review. Int. J. Food Sci. Nutr. 50, 213–224.

Gill, J., Orsat, V., Kermasha, S., 2018a. Optimization of encapsulation of a microbial laccase enzymatic extract using selected matrices. Process Biochem. 65, 55–61.

Gill, J.K., Orsat, V., Kermasha, S., 2018b. Screening trials for the encapsulation of laccase enzymatic extract in silica sol-gel. J. Sol-Gel Sci. Technol. 85, 653–657.

Given, Jr., P.S., 2009. Encapsulation of flavors in emulsions for beverages. Current Opin. Colloid Interface Sci. 14, 43–47.

Goulet, C., Mageroy, M.H., Lam, N.B., Floystad, A., Tieman, D.M., Klee, H.J., 2012. Role of an esterase in flavor volatile variation within the tomato clade. PNAS 109, 19009–19014.

Guisan, J.M., 2006. Immobilization of enzymes as the 21st century begins. In: Guisan, J.M. (Ed.), Immobilization of Enzymes and Cells. 2nd edition. Humana Press, Totowa, N.J, pp. 1–13.

Gupta, K., Jana, A.K., Kumar, S., Maiti, M., 2013. Immobilization of α-amylase and amyloglucosidase onto ion-exchange resin beads and hydrolysis of natural starch at high concentration. Bioproc. Biosyst. Engineer. 36, 1715–1724.

Hall, C.E., Karboune, S., Husson, F., Kermasha, S., 2008. Immobilization of an enzymatic extract from *Penicillium camemberti* containing lipoxygenase and hydroperoxide lyase activities. J. Mol. Catal. B: Enzym. 52-53, 88–95.

Hanefeld, U., Gardossi, L., Magner, E., 2009. Understanding enzyme immobilization. Chem. Soc. Rev. 38, 453–468.

He, L., Mao, Y., Zhang, L., Wang, H., Alias, S.A., Gao, B., Wei, D., 2017. Functional expression of a novel α-amylase from Antarctic psychrotolerant fungus for baking industry and its magnetic immobilization. BMC Biotechnol. 17, 22.

Hickey, D.P., Reid, R.C., Milton, R.D., Minteer, S.D., 2016. A self-powered amperometric lactate biosensor based on lactate oxidase immobilized in dimethylferrocene-modified LPEI. Biosens. Bioelectron. 77, 26–31.

Hossain, G.S., Li, J., Shin, H.D., Liu, L., Wang, M., Du, G., Chen, J., 2014. Improved production of α-ketoglutaric acid (α-KG) by a *Bacillus subtilis* whole-cell biocatalyst via engineering of L-amino acid deaminase and deletion of the α-KG utilization pathway. J. Biotechnol. 187, 71–77.

Hsu, A.-F., Foglia, T.A., Piazza, G.J., 1997a. Immobilization of lipoxygenase in an alginate-silicate solgel matrix: formation of fatty acid hydroperoxides. Biotehnol. Lett. 19, 71–74.

Hsu, A.F., Wu, E., Shen, S., Foglia, T.A., Jones, K., 1997b. Immobilized lipoxygenase in a packed-bed column bioreactor: Continuous oxygenation of linoleic acid. Biotechnol. Appl. Biochem. 30, 245–250.

Illanes, A., Fernandez-Lafuente, R., Guisan, J.M., Wilson, L., 2008. Heterogeneous enzyme kinetics. In: Illanes, A. (Ed.), Enzyme Biocatalysis: Principles and Applications. Springer, New York, N.Y, pp. 155–197.

Ispas, C.R., Crivat, G., Andreescu, S., 2012. Recent developments in enzyme-based biosensors for biomedical analysis. Anal. Lett. 45, 168–186.

Jayasekara, S., Ratnayake, R., 2019. Microbial cellulases: An overview and applications. Cellulase. IntechOpen Publisher. doi: 10.5772/intechopen.84531.

Karboune, S., Neufeld, R.J., Kermasha, S., 2005. Immobilization and biocatalysis of chlorophyllase in selected organic solvent systems. J. Biotechnol. 120, 273–283.

Karout, A., Chpard, C., Pierre, A.C., 2007. Immobilization of a lipoxygenase in silica gels for application in aqueous media. J. Mol. Catal. B: Enzym. 44, 117–127.

Kaushik, N., Biswas, S., Singh, J., 2014. Biocatalysis and biotransformation processes-an insight. J. Scitech 8, 15–22.

Kermasha, S., Bisakowski, B., 1998. The effect of immobilization on the hydrolytic/interesterification activity of lipase from *Rhizopus niveus*. J. Am. Oil Chem. Soc. 75, 1791–1799.

Kermasha, S., Aziz, S., Gill, J., Neufeld, R., 2017. Microencapsulation of esterified krill oil, using complex coacervation. J. Microencapsul. 35, 36–48.

Kermasha, S., Bisakowski, B., Morin, A., Ismail, S., 1999. Biogeneration of short-chain fatty acid esters by *Pseudomonas fragi* CRDA 037. Biocatal. Biotransform. 17, 269–282.

Kermasha, S., Bisakowski, B., Morin, A., Ismail, S., 2000. Effect of homogenization process on the release of esterase activity from cell debris of *Pseudomonas fragi* CRDA 037. Food Res. Int. 33, 747–767.

Kermasha, S., Dioum, N., Bisakowski, B., Vega, M., 2002. Biocatalysis of immobilized lipoxygenase in a ternary micellar system. J. Mol. Catal. B: Enzym. 19-20, 305–317.

Kermasha, S., Gaffar, R., Bisakowski, B., 2000. Biocatalysis of silica gel-immobilized chlorophyllase in a ternary micellar system. Process Biochem. 35, 1103–1109.

Khalyfa, A., Kermasha, S., Marsot, P., Goetghebeur, M., 1995. Purification and characterization of chlorophyllase from alga *Phaeodactylum tricornutum* by preparative native electrophoresesis. Appl. Biochem. Biotechnol. 53, 11–27.

Khamessan, A., Kermasha, S., 1996. Biocatalysis of chlorophyllase in canola oil using organic solvent systems. J. Food Biochem. 20, 311–328.

Khan, A.A., Alzohairy, M.A., 2010. Recent advances and applications of immobilized enzyme technologies: A review. Res. J. Biol. Sci. 5, 565–575.

Khodadadi, M., Kermasha, S., 2014. Modeling lipase-catalyzed interesterification of flaxseed oil and tricaprylin for the synthesis of structured lipids. J. Mol. Catal. B: Enzym. 102, 33–40.

Khodadadi, M., Aziz, S., St-Louis, R., Kermasha, S., 2013. Lipase-catalyzed synthesis and characterization of flaxseed oil-based structured lipids. J. Functional Foods 5, 424–433.

Khoshnevisan, K., Bordbar, a.-K., Zar, D., Davoodi, D., Noruzi, M., Barkhi, M., Tabatabaei, M., 2011. Immobilization of cellulose enzyme on supermagnetic nanoparticles and determination of its activity and stability. Chem. Engineer. J. 171, 669–673.

Khoshnevisan, K., Vakhshiteh, F., Barkhi, M., Boharifacr, H., Poor-Akbar, E., Zari, N., Stametis, H., Barkhi, M., 2017. Immobilization of cellulose enzyme onto supermagnetic nanoparticles: Applications and recent advances. Molecular Catal. 442, 66–73.

Kobayashi, T., Laidler, K., 1973. Kinetic analysis for solid-supported enzymes. Biochim. Biophys. Acta 302, 1–12.

Küchler, A., Bleich, J.N., Sebastian, B., Dittrich, P.S., Walde, P., 2015. Stable and simple immobilization of proteinase K inside glass tubes and microfluidic channels. ACS Appl. Mater. Interfaces 46, 25970–25980.

Kuhad, R.C., Gupta, R., Singh, A., 2011. Microbial cellulase and their industrial applications. Enzyme Res, e280969 (2011) (Doi: 10.4061/2011/280696).

Kuldamrong, W., Husson, F., Kermasha, S., 2013. Biocatalysis with hydroperoxide lyase in extracts from *Penicillium camaberti* in neat organic solvent media. Biocatal. Biotransform. 31, 94–99.

Kumar, L., Kumar, D., Nagar, S., Gupta, R., Garg, N., Kuhad, R.C., Gupta, V.K., 2014. Modulation of xylanase production from alkaliphilic *Bacillus pumilus* VLK-1 through process optimization and temperature shift operation. Biotech 4, 345–356.

Kumar, V., Dangi, A.K., Shukla, P., 2018. Engineering thermostable microbial xylanases toward its industrial applications. Mol. Biotechnol. 60, 226–235.

Kuroda, H., Kojima, H., Kaneda, H., Takashio, M., 2005. Characterization of 9-fatty acid hydroperoxide lyase-like activity in germinating barley seeds that transforms 9(S)-hydroperoxy-10(E),12(Z)-octadecadienoic acid into 2(E)-nonenal. Biosci. Biotechnol. Biochem. 69, 1661–1668.

Kyeong, J.S., Yeom, S.H., 2014. Preparation of immobilized whole cell biocatalyst and biodiesel production using a packed-bed bioreactor. Bioproc. Biosyst. Engineer. 37, 2189–2198.

Lettera, V., Pezzella, C., Cicatiello, P., Piscitelli, A., Giacobelli, V.G., Galano, E., Sannia, G., 2016. Efficient immobilization of a fungal laccase and its exploitation in fruit juice clarification. Food Chem. 196, 1272–1278.

Li, T., Wang, N., Zhao, Q., Guo, M., Zhang, C., 2007. Optimization of covalent immobilization of pectinase on sodium alginate support. Biotechnol. Lett. 29, 1413–1416.

Lin, B., Tao, Y., 2017. Whole-cell biocatalysts by design. Microb. Cell Factories 16, 106.

Liu, Q., Hua, Y., Kong, X., Zhang, C., Chen, Y., 2013a. Covalent immobilization of hydroperoxide lyase on chitosan hybrid hydrogels and production of C6 aldehydes by immobilized enzyme. J. Mol. Catal. B: Enzym. 95, 89–98.

Liu, Q., Kong, X., Zhang, C., Chen, Y., Hua, Y., 2013b. Immobilisation of a hydroperoxide lyase and comparative enzymological studies of the immobilised enzyme with membrane-bound enzyme. J. Sci. Food Agric. 93, 1953–1959.

Liu, Y., Zeng, Z., Zen, G., Tang, L., Pang, P., Li, Z., Liu, C., Lei, X., Wu, M., Ren, P., Liu, Z., Chen, M., Xie, G., 2012. Immobilization of laccase on magnetic bimodal mesoporous carbon and the application in the removal of phenolic compounds. Bioresource Technol. 115, 21–26.

Lu, J., Zhang, M., Wang, Y., 2007. Immobilization of laccase by alginate–Chitosan microcapsules and its use in dye decolorization. World J. Microbiol. Biotechnol. 23, 159–166.

Luu Le, L.T.H., Yoo, W., Jeon, S., Kim, K.K., Kim, T.D., 2020. Screening and immobilization of interfacial esterases from marine invertebrates as promising biocatalyst derivatives. Int. J. Molecular Sci. 21. doi: 10.3390/ijms21010091.

Ma, X., Deng, S., Su, E., Wei, D., 2015. One-pot enzymatic production of deacetyl-7-aminocephalosporanic acid from cephalosporin C via immobilized cephalosporin C acylase and deacetylase. Biochem. Engineer. J. 95, 1–8.

Madene, A., Scher, J., Jacquot, M., Desobry, S., 2006. Flavour encapsulation and controlled release: A review. Int. J. Food Sci. Technol. 41, 1–21.

Manoel, E.A., dos Santos, J.C.S., Freire, D.M.G., Nazzoly, R., Fernandez-Lafuente, R., 2015. Immobilization of lipases on hydrophobic supports involves the open form of the enzyme. Enzyme Microb. Technol. 71, 53–57.

Mardina, P., Li, J., Patel, S.K., Kim, I.W., Lee, J.K., Selvaraj, C., 2016. Potential of immobilized whole-cell *Methylocella tundrae* as a biocatalyst for methanol production from methane. J. Microbiol. Biotechnol. 26, 1234–1241.

Mashima, R., Okuvama, T., 2015. The role of lipoxygenases in pathophysiology; New insights and future perspectives. Redox Biol. 6, 297–310.

Mateo, C., Palomo, J.M., Fernandez-Lorente, G., Guisan, J.M., Fernandez-Lafuente, R., 2007. Improvement of enzyme activity, stability and selectivity via immobilization techniques. Enzyme Microb. Technol. 40, 1451–1463.

Mendes, A.A., de Castro, H.F., Andeade, G.S.S., Tardioli, P.W., Giordano, R., de, L.C., 2013. Preparation and application of epoxy–chitosan/alginate support in the immobilization of microbial lipases by covalent attachment. Reactive Functional Polymers 73, 160–167.

Mohamad, N.R., Marzuki, N.H.C., Buang, N.A., Huyop, F., Wahab, R.A., 2015. An overview of technologies for immobilization of enzymes and surface analysis techniques for immobilized enzymes. Biotechnol. Biotechnol. Equip. 29, 205–220.

Moreno, A.D., Ibarra, D., Tomás-Pejó, E., 2019. Laccases as versatile enzymes: from industrial uses to novel applications. J. Chem. Technol. Biotechnol. 95, 481–494.

Moreno-Garcia, J., Garcia-Martinez, T., Mauricio, J.C., Moreno, J., 2018. Yeast immobilization systems for alcoholic wine fermentations: Actual trends and future perspectives. Front. Microbiol. 9, 241. doi: 10.3389/fmieb.2018.00241.

Nadar, S.S., Rathod, V.K., 2018. Encapsulation of lipase within metal-organic framework (MOF) with enhanced activity intensified under ultrasound. Enzyme Microb. Technol. 108, 11–20.

Nazari, M., Kashanian, S., Rafipour, R., 2015. Laccase immobilization on the electrode surface to design a biosensor for the detection of phenolic compound such as catechol. Spectrochim. Acta A: Mol. Biomol. Spectrosc. 145, 130–138.

Nighojkar, A., Patidar, M.K., Nighojkar, S., 2019. Pectinases: production and applications for juice beverages. In: *Process and Sustainability of Beverages*, vol. 2, Woodhead Publisher-Elsevier, Amsterdam, The Netherlands, pp.235-273.

Noordermeer, M.A., Veldink, G.A., Vliegenthart, J.F.G., 2001. Fatty acid hydroperoxide lyase: A plant cytochrome P450 enzyme involved in wound healing and pest resistance. Chem. Eur. 2, 494–504.

Nunez, A., St-Armand, G., Foglia, T.A., Piazza, G., 2010. Immobilization of hydroperoxide lyase from *Clorella*. Biotechnol. Appl. Biochem. 25, 75–80.

Özdem, N., Özkan, E.H., San, N., Arslan, F., Tümtürk, H., 2014. Immobilization of glucose oxidase attached to new nanospheres including azomethine. Macromol. Res. 22, 1282–1287.

Panesar, P.S., Kumari, S., Panesar, R., 2010. Potential applications of immobilized β-galactosidase in food processing industries. Enzyme Res. 2010. doi: 10.4061/2010/473137.

Patel, S.K., Kumar, V., Mardina, P., Li, J., Lestari, R., Kalia, V.C., Lee, J.K., 2018. Methanol production from simulated biogas mixtures by co-immobilized *Methylomonas methanica* and *Methylocella tundrae*. Bioresour. Technol. 263, 25–32.

Patil, M.D., Dev, M.J., Tangadpalliwar, S., Patel, G., Garg, P., Chisti, Y., Banerjee, U.C., 2017. Ultrasonic disruption of *Pseudomonas putida* for the release of arginine deiminase: Kinetics and predictive models. Bioresour. Technol. 233, 74–83.

Pedrolli, D.B., Monteiro, A.C., Gomes, E., Carmona, W.C., 2009. Pectinases are a group of enzymes that catalyze pectic substance degradation through depolymerization (hydrolases and lyases) and deesterification (esterases) reaction. Open Biotechnol. J. 3, 9–18.

Phiri, M.M., Mulder, D.W., Mason, S., Vorster, B.C., 2019. Facile immobilization of glucose oxidase onto gold nanostars enhanced binding affinity. Royal Soc. Open Sci. 6, e190205. doi: 10.1098/rsos.190205.

Quayson, E., Amoah, J., Hama, S., Yoshida, A., Morita, K., Kondo, A., Ogino, C., 2019. Valorization of activated carbon as a reusable matrix for the immobilization of *Aspergillus oryzae* whole cell expressing *Fusarium heterosporum* lipase towards biodiesel synthesis. ACS Sustainable Chem. Engineer. 7, 5010–5017.

Rajapriya, G., Morya, V.K., Mai, N.L., Koo, Y.M., 2018. *Aspergillus niger* whole-cell catalyzed synthesis of caffeic acid phenethyl ester in ionic liquids. Enzyme Microb. Techol. 111, 67–73.

Rajdeo, K., Harini, T., Lavanva, K., Fadnavis, N.W., 2016. Immobilization of pectinase on reusable polymer support for clarification of apple juice. Food Bioproducts Process. 99, 12–19.

Raveendran, S., Parameswaran, B., Ummalyma, S.B., Abraham, A., Mathew, A.K., Madhavan, A., Pandey, A., 2018. Applications of microbial enzymes in food industry. Food Technol. Biotechnol. 56, 16–30.

Razzaq, A., Shamsi, S., Ali, A., Ali, Q., Sajjad, M., Mlik, A., Ashraf, M., 2019. Microbial proteases applications. Front. Bioengineer. Biotechnol. 7. doi: 10.3389/fbioe.2019.00110.

Renu, R., Zahra, F., 2015. Micrencapsulation of flavours. Int. J. Basic Appl. Biol. 2, 333–338.

Riley, J.C.M., Willemot, C., Thompson, J.E., 1996. Lipoxygenase and hydroperoxide lyase activities in ripening tomato fruit. Postharvest Biol. Technol. 7, 97–107.

Rodrigues, R.C., Virgen-Ortiz, J.J., dos Santos, J.C.S., Berenguer-Murcia, A., Alcantara, A.R., Barbosa, Q., Ortiz, C., Fernandez-Lafuente, R., 2019. Immobilization of lipases on hydrophobic supports: Immobilization mechanism, advantages, problems, and solutions. Adv. Biotechnol. 37, 746–770.

Sakai, S., Liu, Y., Yamaguchi, T., Watanabe, R., Kawabe, M., Kawakami, K., 2010. Immobilization of *Pseudomonas cepacia* lipase onto electrospun polyacrylonitrile fibers through physical adsorption and application to transesterification in nonaqueous solvent. Biotechnol. Lett. 32, 1059–1062.

Salas, J.J., Sanchez, J., 1999. The decrease of virgin olive oil flavor produced by high malaxation temperature is due to inactivation of hydroperoxide lyase. J. Agric. Food Chem. 47, 809–812.

Sassolas, A., Blum, L.J., Leca-Bouvier, B.D., 2012. Immobilization strategies to develop enzymatic biosensors. Biotechnol. Adv. 30, 489–511.

Sato, A.C.K., Perrechil, F.A., Costa, A.A.S., Santana, R.C., Cunha, R.L., 2015. Cross-linking proteins by laccase: Effects on the droplet size and rheology of emulsions stabilized by sodium caseinate. Food Res. Int. 75, 244–251.

Shahrestani, H., Taheri-Kafrani, A., Soozanipour, A., Tavakoli, O., 2016. Enzymatic clarification of fruit juices using xylanase immobilized on 1,3,5-triazine-functionalized silica-encapsulated magnetic nanoparticles. Biochem. Engineer. J. 109, 51–58.

Shaw, S.-Y., Chen, Y.-J., Ou, J.-J., Ho, L., 2006. Preparation and characterization of *Pseudomonas putida* esterase immobilized on magnetic nanoparticles. Enzyme Microb. Technol. 39, 1089–1095.

Sheldon, R.A., 2007. Enzyme immobilization: The quest for optimum performance. Adv. Synth. Catal. 349, 1289–1307.

Sheldon, R.A., 2011. Characteristic features and biotechnological applications of cross-linked enzyme aggregates (CLEAs). Appl. Microbiol. Biotechnol. 92, 467–477.

Shen, J., Tieman, D., Jones, J.B., Taylor, M.G., Schmelz, E., Huffaker, A., Bies, D., Chen, K., Klee, H.J., 2015. A 13-lipoxygenase, TomloxC, is essential for synthesis of C5 flavour volatiles in tomato. J. Exper. Botany 65, 419–428.

Shinde, P., Musameh, M., Gao, Y., Robinson, A.J., Kyartizis, I.L., 2018. Immobilization and stabilization of alcohol dehydrogenase on polyvinyl alcohol fibre. Biotechnol. Rep 19, e00260.

Silva, J.S.M., Zhang, Q., Shen, J., Cavaco-Paulo, A., 2006. Immobilization of proteases with a water soluble–insoluble reversible polymer for treatment of wool Carla. Enz. Microb. Technol. 39, 634–640.

Sing, V., Kumar, P., 2011. 'Carboxymethyl tamarind gum–silica nanohybrids for effective immobilization of amylase. J. Mol. Catal. B: Enzym. 70, 67–73.

Sirisha, V.L., Jain, A., Jain, A., 2016. Enzyme immobilization: An overview on methods, support material, and applications of immobilized enzymes. Adv. Food Nutr. Res. 79, 179–211.

Sojitra, U.V., Nadar, S.S., Rathod, V.K., 2017. Immobilization of pectinase onto chitosan magnetic nanoparticles by macromolecular cross-linker. Carbohydrates Polymers 157, 677–685.

Soottitantawat, A., Partanen, R., Neoh, T.L., Yoshii, H., 2015. Encapsulation of hydrophilic and hydrophobic flavors by spray drying. Japan J. Food Sci. 16, 37–52.

Stolterfoht, H., Rinnofner, C., Winkler, M., Pichler, H., 2019. Recombinant lipoxygenases and hydroperoxidelLyases for the synthesis of green leaf volatiles. J. Agric. Food Chem. 67 (49), 13367–13392.

Su, F., Li, G., Fan, Y., Yan, Y., 2016. Enhanced performance of lipase via microcapsulation and its application in biodiesel preparation. Sci. Rep. 6, 29670.

Swarnalatha, V., Esther, R.A., Dhamodharan, R., 2015. Immobilization of α-amylase on gum acacia stabilized magnetite nanoparticles, an easily recoverable and reusable support. J. Mol. Catal. B: Enzym. 96, 6–13.

Truppo, M.D., Hughes, G., 2011. Development of an improved immobilized CAL-B for the enzymatic resolution of a key intermediate to odanacatib. Org. Process Res. Dev. 15, 1033–1035.

Tweddell, R.J., Kermasha, S., Combes, D., Marty, A., 1999. Immobilization of lipase from *Rhizopus niveus*: A way to enhance its synthetic activity in organic solvent. Biocatal. Biotransform. 16, 411–426.

Urano, Y., Kamiya, M., Tachibana, R., 2020. Molecular design of flurogenuc probes based on quantum chemical prediction on intramolecular spirocyclization. Commun. Chem. 3, 82.

Vega, M., Karboune, S., Kermasha, S., 2005. Stability of immobilized soybean lipoxygenase in selected organic solvent media. Appl. Biochem. Biotechnol. 127, 29–42.

Wachtmeister, J., Rother, D., 2016. Recent advances in whole cell biocatalysis techniques bridging from investigative to industrial scale. Curr. Opin. Biotechnol. 42, 169–177.

Wang, F., Hu, Y., Guo, C., Huang, W., Liu, C.Z., 2012. Enhanced phenol degradation in coking wastewater by immobilized laccase on magnetic mesoporous silica nanoparticles in a magnetically stabilized fluidized bed. Bioresour. Technol. 110, 120–124.

Wang, X., Liu, J., Du, G., Zhou, J., Chen, J., 2013. Efficient production of L-sorbose from D-sorbitol by whole cell immobilization of *Gluconobacter oxydans* WSH-003. Biochem. Engineer. J. 77, 171–176.

Ward, K., Scarth, R., Daun, I.K., Thorsteinsen, C.T., 1994. Characterization of chlorophyll pigments in ripening canola seed (*Brassica napus*). J. Am. Oil Chem. Soc. 7, 1327–1331.

Weatherley, L.R., Gangu, A.S., Scurto, A.M., Petera, J., 2018. Process intensification of enzymatic biotransformation processes. In: Górak, A., Stankiewicz, A. (Eds.), Intensification of Biobased Processes. Royal Society of Chemistry, London, U.K., pp. 268–288.

Wu, L., Yuan, X., Sheng, J., 2005. Immobilization of cellulase in nanofibers PVP membranes by electrospinning. J. Membrane Sci. 250, 167–173.

Xu, D.Y., Yang, Z., 2013. Cross-linked tyrosinase aggregates for elimination of phenolic compounds from wastewater. Chemosphere 92, 391–398.

Xu, Z.H., Cheng, A.D., Xing, X.P., Zong, M.H., Bai, Y.P., Li, N., 2018. Improved synthesis of 2,5-bis (hydroxymethyl) furan from 5-hydroxymethylfurfural using acclimatized whole cells entrapped in calcium alginate. Bioresour. Technol. 262, 177–183.

Yang, C.H., Yen, C.C., Yen Jheng, J.J., Wang, C.Y., Chen, S.S., Huang, P.Y., Huang, K.S., Shaw, J.F., 2014. Immobilization of *Brassica oleracea* chlorophyllase 1 (BoCLH1) and *Candida rugosa* lipase (CRL) in magnetic alginate beads: An enzymatic evaluation in the corresponding proteins. Molecules 19, 11800–11815.

Yi, Y., Kermasha, S., Neufeld, R.J., 2006a. Optimization of tetramethoxysilane-derived sol gel entrapment protocol stabilizes highly active chlorophyllase. J. Sol-Gel Sci. Technol. 38, 251–259.

Yi, Y., Kermasha, S., Neufeld, R.J., 2006b. Characterization of sol-gel entrapped chlorophyllase. Biotechnol. Bioengineer. 95, 840–849.

Yi, Y., Kermasha, S., Neufeld, R.J., 2008. Nanoporous sol-gel enzymatic hydrolysis of chlorophyll in organic media. In: Kim, J., Kim, S.H., Wang, P. (Eds.), Nano-Scale Science and Technology in Biomolecular Catalysis. ACS Symposium Series 986, Washington, D.C, pp. 199–213.

Yi, Y., Kermasha, S., L'Hocine, L., Neufeld, R., 2002. Encapsulation of chlorophyllase in hydrophobically modified hydrogel. J. Mol. Catal. B: Enzym. 19-20, 319–325.

Yi, Y., Neufeld, R.J., Kermasha, S., 2007. Controlling sol-gel properties enhancing entrapped membrane protein activity through dopin additives. J. Sol-Gel Sci. Technol. 43, 161–170.

Yu, Y., Yuan, J., Wang, Q., Fan, X., Wang, P., 2012. Covalent immobilization of cellulases onto a water-soluble-insoluble reversible polymer. Appl. Biochem. Biotechnol. 166, 1433–1441.

Zajkoska, P., Rosenberg, M., Heath, R., Malone, K.J., Stloukal, R., Turner, N.J., Rebroš, M., 2015. Immobilised whole-cell recombinant monoamine oxidase biocatalysis. Appl. Microbiol. Biotechnol. 99, 1229–1236.

Zeuner, B., Ma, N., Berendt, K., Meyer, Andric, P., Hoffmann Jorgensen, Pinelot, M., 2018. Immobilization of alcohol dehydrogenase on ceramic silicon carbide membranes for enzymatic CH_3OH production. J. Chem. Technol. Biotechnol. 93, 2952–2961.

Zhang, X.Y., Xu, Z.H., Zong, M.H., Wang, C.F., Li, N., 2019. Selective synthesis of furfuryl alcohol from biomass-derived furfural using immobilized yeast cells. Catalysts 9, 70.

Zheng, Y.-G., Yin, H.-H., Yu, D.-F., Chen, X., Tang, X.-L., Zhang, X.-J., Xue, Y.-P., Wang, Y.-J., Liu, Z.-Q., 2017. Recent advances in biotechnological applications of alcohol dehydrogenases. Appl. Mirobiol. Biotecnnol. 101, 987–1001.

Zheng, Z.Y., Liu, J., Hahn, M., Qiao, S., Middelberg, A.P.J., Lizhong He, L., 2013. Encapsulation of lipase in mesoporous silica yolk shell spheres with enhanced enzyme stability. RSC Adv. 3, 22008–220013.

Zucca, P., Sanjust, E., 2014. Inorganic materials as supports for covalent enzyme immobilization: methods and mechanisms. Molecules 19, 14139–14194.

Wingard, B.L., Katzir-Katchalski, E., Goldstein, L., 1976. Diffusion and kinetics with immobilized enzymes. In: Wingard, B.L., Katzir-Katchalski, E., Goldstein, L. (Eds.), Immobilized Enzyme Principles. Academic Press, New York, N.Y, pp. 128–216.

CHAPTER SEVEN

Biocatalysis of enzymes in nonconventional media

Selim Kermasha, Jagpreet K. Gill
McGill University, Montreal, Quebec, Canada

7.1 Introduction

The use of enzymes in various industrial applications such as baking, cheese making, winemaking, etc., is well known. More than 60% of the total enzyme market finds its use in the food, beverage, animal feed, and the detergent industry; while, the remaining percentage is involved in starch processing, leather as well as the pulp and paper industry (Sheldon and Pereira, 2017). Biocatalysis may be defined as the use of enzymes, or enzyme containing cells that are involved in chemical transformations. In general, there are two types of biotransformations, one which involves growing microbial cells while the other involves resting cells. The former biotransformation which involves the growth of microbial cells are fermentations, that if grown in vivo lead to the production of more microbial cells and if in vitro result in the synthesis of chemical products (Sheldon and Pereira, 2017). Over the last few decades, biocatalysis has been mainly applied in mainstream organic synthesis, specifically in the pharmaceutical industry and may be attributed to the various environmental as well as economic benefits. The environmental benefits are linked to the use of biological catalysts (enzymes) that are derived from renewable resources, and are biocompatible, biodegradable, and nonhazardous or nontoxic in nature, as compared to the chemical ones. Moreover, enzymatic reactions occur under milder conditions of pH, temperature, and pressure, hence, are more energy efficient.

Overall, the use of enzymes for biotransformations results in significantly higher substrate specificities, relatively reduced reaction times as well as waste generation, as compared to the conventional methods that make use of chemical catalysts (Sheldon, 2016). The continuous efforts made to improve the activity and stability of enzymes for potential industrial applications (Klibanov, 2001; Kaul and Asano, 2008; Robinson, 2015), resulted in

Enzymes: Novel Biotechnological Approaches for the Food Industry.
http://dx.doi.org/10.1016/B978-0-12-800217-9.00007-1

the development of novel approaches for their biocatalysis in nonconventional media.

Even though enzymes find a wide applicability in several industries, they are often associated with limitations such as their increased sensitivity to their surrounding environment, including, pH, temperature, substrate concentration, and media type. In order to overcome such limitations, enzymes may be immobilized using various methods, including covalent binding, adsorption, encapsulation, microencapsulation, self-immobilization, etc. Enzyme immobilization has broadened the scope of their applications by increasing their pH and temperature stability, reusability, enhanced catalytic activity, while allowing them to be used in various conventional aqueous as well as nonconventional reaction systems (Illanes et al., 2012).

7.2 Biocatalysis in nonconventional media

Although biocatalysis by enzymes are carried out in aqueous medium (Lamer et al., 1996; Krishna, 2002; Chao et al., 2010), their reactions in nonaqueous media broadens their applications in several industrial sectors (Madamwar and Soni, 2003; Sheldon, 2005; Sabally et al., 2005a,b; Tong et al., 2005; Sabally et al., 2007; Gupta and Khare, 2009). It has been reported that in comparison to the conventional aqueous systems, biocatalysis carried out in nonaqueous media offer unique advantages including, increased enantioselectivity, possibility of altering enzyme regio- and/or enantio-selectivity, reverse thermodynamic equilibrium toward synthesis, decreased water-dependent side chain reactions and reduced bacterial contamination (Wang et al., 2016). In addition, an interest in nonaqueous biocatalysts has increased after the recognition that enzymes, cell organelles, and whole cells remain active in all kinds of reaction media including organic solvents, gases, ionic liquids as well as supercritical fluids. In general, the application of biocatalysts in such reaction systems is very attractive for organic synthesis routes that require biocatalysts with high-substrate specificity and that involve substrates insoluble in water. In nonconventional reaction systems, if water is one of the products after synthesis, the reaction equilibrium may be shifted in the favor of synthesis by lowering the water activity or by varying the partitioning characteristics of substrates and products, between relevant phases of the reaction (Priyadarshini and Pandey, 2018). The use of nonconventional media is particularly known in the pharmaceutical industry, for production of flavor compounds, oleochemicals, and drug intermediates. Organic solvents have been used for bioconversion of steroids, due to low substrate and products solubility in water (Priyadarshini and Pandey, 2018).

The advantages of carrying out the biocatalysis of enzymes in nonconventional media over the conventional aqueous medium are numerous, including chemo-, regio-, and enantioselectivity of enzymes, which are significant to produce fine chemicals, flavors, fragrances, agrochemicals, and pharmaceuticals (Ghaffari-Moghaddam et al., 2015). Indeed, the great interest in enzyme biocatalysis in nonconventional aqueous medium is due to many kinetic parameters, including good enantioselectivity, reverse thermodynamic equilibrium, and absence of water-dependent side reactions (Wang et al., 2016). It has been demonstrated that an enzyme has high activity and stability in nonaqueous media, and the variation of enzyme activity is attributed to its conformational modifications (Stepankova et al., 2013). Hence, it is of importance to indicate the relationship between the microenvironment surrounding the enzyme surface and its stability as well as its activity (Adlercreutz, 2013). In addition, the biocatalysis of enzymes is not limited to their conditions in vitro, but also for the whole cell in vivo (Nicolova and Ward, 1993; Schmidt et al., 2001). In addition, the insolubility of many substrates in aqueous medium would lead to an unfavorable thermodynamic equilibrium of many bioprocesses and the difficulties of product recovery. Hence, there was a need for the development of enzyme catalysis in nonaqueous media (Krishna, 2002; Klibanov, 2003; Konieczny et al., 2012). The development of nonconventional media for enzyme biocatalysis has resulted in biotechnological processes, including synthesis of chirals, modified fats and oils, specialty pharmaceuticals (Krishna, 2002; Xu et al., 2011), food additives, flavor esters, biopolymers, peptides, and proteins (Krishna, 2002).

The list of nonconventional media may include not only the use of green and safe organic solvents media for the biocatalysis of enzymes, but also other approaches, including solvent-free media (SFM), ionic liquids, supercritical fluids, and solid-gas.

7.2.1 Biocatalysis in organic solvent media

7.2.1.1 Generality

Enzymes are excellent biocatalysts that are increasingly in demand for their use in industry. This approach is aimed mainly at the synthesis of selected natural food ingredients and nutraceuticals (Gupta et al., 2013) as well as for their involvement in the bioprocesses in food industry (Gupta, 1992; Gupta and Roy, 2004).

Although the aqueous media are generally the traditional solvents used for biocatalysis, their use limits enzyme applications for large scale industrial processes. Compared to the conventional aqueous systems organic solvents are characterized by a relatively higher solubility for several organic compounds, greater stability of biocatalysts, allows for an ease in product recovery, facilitates the reuse of enzymes due to their insolubility in organic media and are more economically viable for use in industrial bioprocesses (Zaks and Klibanov, 1985).

Despite the various advantages associated with the use of organic solvents for biocatalysis, such properties are not applicable for all biocatalysts as some organic solvent systems may denature enzymes and subsequently result in mass transfer limitations (Adlercreutz, 2000; Doukyu and Ogino, 2010). It is also important to establish certain thermodynamic predictions for biocatalysis in nonconventional media (Halling, 1994). Moreover, due to the hazardous nature of organic solvents, they are a concern to the environment, thereby pressing the need for the development of relatively greener and sustainable technologies that can be used in biotransformations (Sheldon, 2005).

The main reason for using organic media is that in addition to the hydrophobic nature of organic media to preserve the conformation of the enzymatic protein, the recovery of the poorly aqueous-soluble products could be easily recovered from the reaction medium (Adlercreutz, 1996).

Nevertheless, organic solvents used in enzyme biocatalysis should exhibit in addition to their chemical properties, low volatility, miscibility, and biodegradability. They should also have the characteristics of green solvents, including a high degree of environmental health and safety (Capello et al., 2007).

Although organic solvent media contribute to the development of novel bioprocesses with enhancement of enzyme specificity (Mine et al., 2003; Konieczny et al., 2012), they show certain drawbacks in terms of activity and stability (Mb et al., 2005; Doukyu and Ogino, 2010; Goswami et al., 2012; Milasinovic et al., 2012).

7.2.1.2 Types of organic solvent media

There are different types of organic solvent systems that have been used for the biocatalytic reactions, including neat organic solvent systems, monophasic organic solvent systems, and multiphasic systems.

7.2.1.2.1 Neat organic solvent systems

Aqueous media with less than 20% organic solvents do not fall under the category of organic media for enzyme biocatalysis (Adlercreutz, 2008;

Kuldamrong et al., 2013). Common solvents that may be used for solutions containing > 20% organic solvents may include, dimethyl sulfoxide (DMSO), *N*-*N*-dimethylformamide (DMF), acetonitrile, methanol, ethanol, and 1, 4-dioxane. In such media, the biocatalyst is considered hydrated as the water content is usually higher than the water-stripping capacity of the organic solvents present in the medium. The addition of solvent causes an increase in the pH due to the decrease in the dielectric constant, however, Tris maintains the deviation to < 1.5 units throughout 0%–90% volume of the solvent (Sellek and Chaudhuri, 1999).

The effect of such solvent systems on the enzyme may be assessed under three categories, denaturation, stabilization, and inhibition. Denaturation involves conformational changes to the native enzyme structure that results in protein unfolding and is irreversible. The disruption of the enzyme's active sites leads to a reduction in catalytic activity, which is followed by irreversible aggregation. The stabilization also results in a reduced enzyme activity due to their rigidification; however, this is reversible. The addition of soluble organic solvent to the aqueous media results in a reduction of water activity, thereby contributing to rigidifying effects. Furthermore, inhibition may also occur in such reaction media, where solvent molecules may bind with the active sites of the enzyme. It has been reported that thermolysin (metalloproteinases) was inhibited by alcohols as result of the solvent molecules binding to this enzyme's active sites (Sellek and Chaudhuri, 1999). The prediction of whether the enzyme is denatured, stabilized or inhibited solely relies on the protein-solvent interactions, however, the final considerations also involve interactions of solvents with substrates and/or products as well as the effects of bulk medium on solvation process or desolvation one, which is a thermodynamically driven, self-assembly process for polymeric materials to prepare nanoparticles (Sellek and Chaudhuri, 1999).

7.2.1.2.2 Monophasic organic solvent systems

The perception that biocatalysis of enzymes in conventional aqueous medium is the only environment for reactions is to occur as enzymes are naturally active in water as organic solvents could result in their denaturation and inactivation. However, enzymatic catalysis in nonaqueous media significantly extends conventional aqueous-based biocatalysis (Dordick, 1989).

In a monophasic organic solvent system, the medium is either a nonpolar solvent, such as tertamyl alcohol (2-methylbutan-2-ol), hexane and toluene, or a supercritical fluid, usually carbon dioxide. Supercritical fluids will be discussed separately in later sections of this chapter (Sellek and

Chaudhuri, 1999). In general, native enzymes are insoluble in organic systems, where an increased solubility may be attained by processes such as enzyme modification, lyophilization, or immobilization. Due to the low-water content of the medium, a high rigidity to the biocatalyst occurs resulting in lower enzyme mobility. Alternatively, this system has found applications in synthesis and transesterification of esters and peptides as well as transformation of different organic compounds (Doukyu and Ogino, 2010). The biocatalysis of lipases (Karam et al., 2009; Karboune et al., 2005b; Lue et al., 2005), tyrosinase (Kermasha et al., 2001a, 2002a; Kermasha and Tse, 2000; Karboune et al., 2005a), lipxygenase (Kermasha et al., 2001b, 2002b), proteases (Wangikar et al., 1996; Bordusa, 2002), chlorophyllase (Karboune et al., 2005b; Khamessan et al., 1995a,b; Khamessan and Kermasha, 1996b; Arriagada-Strodthoff et al., 2007), and laccase (Ma et al., 2009).

7.2.1.2.3 Multiphasic-organic solvent systems

Most of organic solvent media used in enzyme biocatalysis consist of more than one phase. One of these phases, is most likely a nonpolar liquid that makes up the major portion of the reaction media and is the continuous phase. Biphasic organic solvent systems include long chain alcohols and aliphatic/branched hydrocarbons such as reversed micellar systems. In such media, the immiscibility of water with the nonpolar solvents forms a biphasic system and is exploited for carrying out biocatalysis (Khamessan et al., 1993, 1994, 1995; Bitar et al., 2004). The solubility of nonpolar substrate in the organic phase of the medium allows the use of relatively higher substrate concentrations, while avoiding the substrate inhibition effects. In addition, the reactions yield in such media can be increased when the products are simultaneously extracted from the organic phase (Sellek and Chaudhuri, 1999). The aqueous portion of this media contains the hydrophilic part of the enzyme and hence the biocatalysis occurs in this phase of the media. The organic phase of the solvent system is for the extraction of the synthesized product during biocatalysis. An enhanced enzyme stability is observed in this reaction media resulting from minimum contact of the enzyme with the hydrophobic organic phase of the reaction media. On the other hand, the biphasic nature of the reaction system also allows for an easy product recovery as well as an increased enzyme reusability. Hence, biocatalysis in such a reaction system offers applicability in continuous industrial processes (Khmelnitsky et al., 1988; Ogino and Ishikawa, 2001; Doukyu and Ogino, 2010). Contrarily, a disadvantage of this reaction media is that the process conditions for increasing the interphase mass-transfer effects results in a simultaneous increase in the inactivation of the biocatalyst. In such cases, immobilization of the enzyme may be necessary for an effective biocatalysis. In addition, although the solvent molecules are poorly soluble in the aqueous phase, the dissolved solvent molecules may lead to a disruption of the enzyme activity (Sellek and Chaudhuri, 1999; Llerena-Suster et al., 2012).

The surfactant-free ternary systems consist of mixture of certain organic solvents, such as n-hexane, short-chain alcohols or surfactants and water used to maximize the catalytic performance of enzymes (Tzialla et al., 2009). The results indicate that such ternary systems can be efficiently employed as reaction media leading to highly enhanced stability and increase CPO potential as practical catalyst for oxidation of large amounts of hydrophobic compounds in a single-step biocatalytic process.

The presence of surfactants, which are amphiphilic in nature has many advantages in organic solvent media where reversed micelles are formed (Samaha and Kermasha, 1997). Due to their safety use and stronger solubility power (Ismail et al., 1970; Attwood and Florence, 1983; Myers, 1988). The overall literature indicates that the ternary systems can be efficiently employed as reaction media leading to highly enhanced stability of enzymes (Khamessan and Kermasha, 1996a,b; Gaffar et al., 1999; Kermasha et al., 2000; Zoumpanion et al., 2006, 2010).

7.2.1.3 Parameters affecting enzymatic synthesis in organic media

There are various parameters affecting biocatalytic synthesis of enzymes in organic media, including effect of type of organic solvent systems, water activity, substrate concentration, and enzyme concentration.

7.2.1.3.1 Substrate concentration

Various solvents show different abilities to solubilize substrates, and this influences the catalytic activity of the enzymes in biocatalysis (Wehtje and Adlercreutz, 1997). Organic solvent media usually result in limiting substrate solubility and diffusion, hence the kinetic falls into two categories with respect to the diffusion susceptibility. In case of poorly active enzymes, the enzyme kinetics depend more on catalytic activity and not by the substrate diffusion rate. On the other hand, in case of enzymes with relatively higher catalytic activity, substrate diffusion is the primary limiting factor. The substrate diffusion limitation in organic solvent media can be severe enough to make kinetic analysis nearly impossible. Nevertheless, various methods may be used to decrease such diffusional limitations, where in case of external diffusional limitations, an increased agitation speed, increased interfacial surface, or alteration of enzyme solubility in organic solvent media may result in overcoming such limitations (Klibanov, 2001; Illanes et al., 2008). The synthesis of (z)-3-hexen-l-y1 acetate using immobilized lipase from *Candida antarctica* was reported in n-hexane, where a 94% of ester yield was obtained in less than 27 h; moreover, the lipase carried out the reaction efficiently in the absent of solvent while using 9% (w/w) of reactants (Bourg-Garros et al., 1998a,b).

7.2.1.3.2 Water activity

One of the most fundamental factors affecting enzymes in nonconventional solvent systems is the optimal level of hydration in the enzyme's surrounding environment. Water is an absolute necessity for all biocatalytic reactions and is involved in all noncovalent interactions that helps maintain the active enzymatic conformation. A complete dehydration of enzymes is impossible as the residual water is tightly bound to the enzyme with limited mobility and a reduced ability to form hydrogen bonds in hydrophobic environments (Gorman and Dordick, 1992). Zaks and Klibanov (1988) reported the effect of organic solvents on the catalytic activity of enzymes, where the results showed that enzyme's activity was found to be a function of the amount of water bound to the enzyme and not the amount present in the surrounding solvent system. The effect of water on enzyme activity is generally determined by the thermodynamic water activity (a_w) of the biocatalyst and can be measured by the vapor phase above the reaction mixture (Valivety et al., 1992).

7.2.1.3.3 Temperature

Temperature affects most biochemical reactions, where usually, the increase in temperature increases the rate of reaction; however, significantly higher temperatures may result in reduced reaction rates due to denaturation of the enzyme (Stamatis et al., 2001). Although significantly higher temperatures are associated with enzyme inactivation, a reduced water activity such in the case of organic solvent systems has been reported with enhanced thermostability of enzymes. In addition, an increased amount of water activity may result in increased enzyme inactivation rate, while enzyme molecules with small amounts of associated water have been reported to be more rigid thereby reducing the rate of their inactivation (Halling et al., 1995).

7.2.1.3.4 Effect of pH

In organic solvent systems, the surrounding pH has comparatively less significance on the enzyme activity. It has been reported that enzymes present a "pH memory" in organic solvents, where the enzyme's catalytic activity reflects the pH of the last aqueous solution to which they were exposed. This phenomenon results due the fact that the ionized group of the enzymes retain their last ionization state after dehydration and their subsequent exposure in an organic solvent. Alternatively, in systems of minimal water content the addition of appropriate buffering system enhances their catalytic activity (Klibanov, 2001). In case of reverse micellar systems, the surfactant head group's charge may play an important role on the enzyme's degree of pH dependency (Biasutti et al., 2008).

7.2.1.3.5 Molecular sieves

The immobilization of enzymes via adsorption onto solid surfaces has gained increasing attention as it broadens the applications of enzymes in various bioprocesses. However, in case of medical and food industries, it is essential to remove adsorbed proteins as any remaining adsorbed protein may result in adverse biological conditions, such as blood clotting and heart disease. The discovery of mesoporous silicate molecular sieves has open new possibilities in various sectors of chemistry and material sciences. Molecular sieves are associated with relatively high surface areas, pore volumes, and well-ordered structures with uniform mesopores with diameters ranging from 1.5 to 30 nm. As compared to microporous zeolites, mesoporous silicates are often preferred for biotransformation of large organic molecules (Hartmann, 2005).

In general, zeolites and porous silicas are considered the most important porous materials with respect to their applications in separation and catalysis. Zeolites are derived from the crystalline family aluminosilicates. The term molecular sieve was obtained from McBain in 1932 after he found that the mineral chabazite had a property of selective adsorption of molecules < 5 Å in diameter. Hence, molecular sieves retain the particles that fit within their channels and let the larger ones pass through; hence, they are associated with selective sorption properties. Ordered mesoporous materials that are based on Mobile Crystalline Material (MCM-41) are silicates that are obtained by hydrothermal synthesis and liquid templating mechanism. These materials offer several advantages such as well-defined pore sizes and uniform shapes with arrays of nonintersecting hexagonal channels. Additionally, these materials offer high surface areas of about 1000 m^2/g and show excellent thermal, hydrothermal and hydrolytic stabilities. The channel diameters may be controlled by changing the length of the template molecule. The functionalization of these materials with organic or inorganic functional groups may lead to new physical and chemical properties and can be used in catalysis, adsorption as well as chromatographic separation (Zaharri et al., 2012). Zeolite membranes have also been used for molecular sieving-based separations in pervaporation (PV) separation of aqueous and organic mixtures. As compared to distillation, PV is more environmentally friendly since there is no third component such as benzene added in the process. In addition, for liquid feed streams, the total coverage area on the side of the feed is in proximity with saturation, such that the transport occurs via membranes (Bhat et al., 2006).

In organic medium biocatalysis molecular sieves are used for controlling the water content by drying the organic solvents (Hinzmann et al., 2019). Solvent engineering using molecular sieves has been reported to optimize the ionization state of the biocatalyst in an organic solvent medium. Hudson et al. (2005) elaborated that enzyme immobilization on novel polymeric supports as well as the use of zeolite molecular sieves can result in an increased solvent tolerance, enhanced catalytic activity, and improved enantioselectivity.

7.2.2 Biocatalysis in solvent-free media

7.2.2.1 Generality

The limitations observed using solvents may be overcome by avoiding the use of solvents completely. In general, solvent free systems are two types, the first when one of the reagents is in liquid or in gaseous form, whereas the second type includes both reagents in solid form and hence is termed as solid to solid biocatalysis. The latter type of biocatalysis is often characterized by mass transfer limitations with low reaction rates and reduced product yields. A major concern with solvent-free biocatalysis is the effect of substrates on the stability and activity of the enzymes. In such reactions systems, the substrates used also act as surrounding solvents for the biocatalyst, which can often have detrimental effects on the catalytic activity of the enzyme (Aziz et al., 2012; Sorour et al., 2012a,b; Gill, 2018).

Such limitations associated with solvent-free biocatalysis may be overcome using immobilized enzymes (Hobbs and Thomas, 2007). Several studies have been reported involving transesterification reactions that make use of immobilized enzymes in solvent-free conditions; however, these reaction systems mostly make use of liquid lipid substrates (Weber and Mukherjee, 2004; Petersson et al., 2005; Aziz et al., 2012). Such type of reactions is similar to the ones carried out in organic solvents; however, they may result in higher bioconversions and initial rates of the reaction.

In addition to numerous technical and cost advantages, the solvent-free media (SFM) is considered the most environmentally friendly condition for the biocatalysis of enzymes in nonconventional media (Ghaffari-Moghaddam et al., 2015). Consequently, the biocatalysis in SFS has gained considerable consideration, particularly for its application in the food industry (Santos et al., 2007; Sorour et al., 2012b). Nevertheless, there is a drawback in such media mainly due to the limitation of contact between enzyme and substrate. The enhancement of enzyme-substrate interaction and consequently the increase in productivity could be achieved throughout with the use of immobilized enzymes (Jin et al., 2003; Chaibakhsh et al., 2009; Kuperkar et al., 2014).

7.2.2.2 Type solvent-free media

There are two main approaches for solvent-free biocatalysis, including heterogeneous eutectic reaction systems and solid to solid reaction systems (Hobbs and Thomas, 2007; Bódalo et al., 2009).

7.2.2.2.1 Heterogeneous eutectic reaction systems

Heterogeneous eutectic reactions systems require low melting point eutectic substrates, generally with melting points <60°C, which are either self-maintained or may be maintained by addition of a small amount of adjuvant which is compatible with the thermal stabilities of the biocatalyst used.

7.2.2.2.2 Solid-solid reaction systems

Solid-solid reaction systems for biocatalysis were first demonstrated by Kuhl et al. (1995), where a micro-aqueous system was used for the synthesis of peptides, using sonication and fluidization technologies. The results showed that a yield of up to 85% was obtained using sonication, while 10%–40% was obtained in case of fluidized-bed reactors. In addition to relatively higher reaction yields, solid to solid reaction systems have another advantage as compared to the conventional aqueous or organic solvent systems, where a low-water content in the medium results in fewer side chain products and is directed towards condensation or synthesis.

The water content required in solid-solid reaction systems may be incorporated by a direct addition via syringe, as salt hydrates and/or using a wet air stream (Kuhl et al., 1995). The thermodynamics and energy efficiency of solid-solid reaction systems has been elaborated by Ulijn and Halling (2004), where the authors have investigated the overall enthalpy change as compared to the conventional chemical methods for catalysis.

The authors suggested a better atom efficiency and one-third less heat production, using an enzyme catalyzed reaction system in solid to solid reaction systems. Contrarily, one of the major concerns associated with this system is associated with the difficulty in scale-up of reactions, which results in limiting their use in industrial applications. Few studies have reported the scale up of such biocatalysis using stirred tank reactors (Erbeldinger et al., 2001). This area of biocatalysis is yet to be explored more which may also be due to the mass-transfer limitations associated with such reaction media.

7.2.2.3 Parameters affecting enzymatic synthesis in solvent-free media (SFM)

The different operational parameters would not only affect the yield and rate of enzymatic reaction in SFM, but also the stability of the enzyme (Sun et al., 2012a).

7.2.2.3.1 Enzyme concentration

Under solvent-free conditions, an increase in the concentration of the biocatalyst results in a simultaneous enhancement in the substrate conversion yield; however, after a certain concentration any further increase in the enzyme concentration leads to a decrease in the yield. It has been reported that this phenomenon resulted in the formation of enzyme aggregates at high enzyme loadings that leads to a decrease in the enzyme surface area and a subsequent decline in the biocatalytic activity of the enzyme. In addition, the mass-transfer limitations that occur in such cases decrease the substrate availability for the enzyme particles that are present inside the aggregates (Gawas et al., 2016). Similar findings were reported by Sun et al., 2012b, where the authors suggested that although relatively higher enzyme loadings may result in higher bioconversions yields, an excess of enzyme particles in the reaction mixture lead to a decrease in the mass-transfer efficiency, especially under solvent-free conditions. In general, the viscosity in a solvent-free reaction mixture is relatively higher than other reaction media; hence, increased amounts of enzyme loading results in further increase in the viscosity that results in less effective transfer of substrates to the active sites of the excess enzyme molecules that are present in the bulk of the enzyme particles. Gill (2018) carried out the biocatalysis of phenolic substrates using encapsulated laccase in a completely solvent-free media, where the author suggested that a higher catalytic activity was obtained at relatively lower concentrations of the enzyme. The author explained that this could be due to the saturation of the enzyme active sites by the substrate molecules and any further increase in the protein content would not affect the enzyme activity. In addition, the author suggested that the absence of substrate molecules in such conditions would result in unoccupied active sites and a simultaneous decrease in the catalytic activity of the enzyme.

7.2.2.3.2 Reaction temperature

In most cases, an increase in the reaction temperature results in an enhancement in the catalytic activity of the enzyme in solvent-free media. Reaction temperature is an important parameter that affects both enzyme activity and stability in biocatalysis. However, beyond a certain reaction temperature, a reduction in the catalytic activity may be obtained due to enzyme inactivation (Sun et al., 2012b). In general, higher reaction temperatures lead to high reaction velocity resulted by a reduced viscosity and greater collision frequency of reaction molecules. On the contrary, a significant increase in the reaction temperature may result in damage to the native structure of the enzyme causing denaturation and inactivation (Yadav and Thorat, 2012; Gawas et al., 2016). Gómez et al. (2011) reported the effect of reaction temperature in SFM, where they suggested that this parameter affected the catalytic reaction rate, reaction viscosity and the evaporation of water from the reaction system. However, the authors also suggested that beyond a reaction temperature of 50°C, adverse effects were observed on the enzyme activity that could be due to the loss of the native structural enzyme conformation. Gill (2018) reported the effect of reaction temperature on encapsulated laccase under solvent-free conditions, where the results showed that the enzyme activity increased with an increase in temperature; however, a decrease in the catalytic activity was observed after a reaction temperature of 60°C and could be attributed to enzyme deactivation beyond this temperature.

7.2.2.3.3 Effect of agitation speed

Under solvent-free conditions, an effective agitation speed is important for efficient mixing of the reaction components and overcoming mass-transfer limitations. On the other hand, a significant increase in the agitation speed may result in a decrease in the enzyme's catalytic activity; immensely high agitation speeds can result in mechanical damage to the enzyme molecules due to high shear forces (Gawas et al., 2016). Similar findings have been reported in other studies, where the authors have suggested that an effective agitation speed, up to a certain point could result in an increase in the enzymatic activity in SFM due to enhanced mass-transfer to the surface of the enzyme particles; however, beyond a certain speed, the absence in mass-transfer limitations can either result in a constant reaction rate or lead to a decline in the catalytic activity (Sun et al., 2012b; Yadav and Thorat, 2012). More recently, Gill (2018) reported that the specific activity of encapsulated laccase in SFM increased with an agitation speed up to 150 rpm; however, beyond this speed, a decline in the activity was obtained which could be attributed to the fact that the mass-transfer limitation was reduced to a minimum level.

7.2.2.3.4 Effect of water activity (a_w)

The effect of water activity varies with different enzymes as their optimal activity is determined by different amounts of water content in their surrounding environment (Zaks and Klibanov, 1985). Although a minimum amount of water is required for maintaining the enzyme's catalytic activity, an excess of water may result in undesired hydrolysis (Elfman-Borjesson and Harrod, 1999). It has been reported that the catalytic properties of enzymes are more influenced by the amount of bound water, rather than by water's total concentration (Petersson et al., 2007). Lipases have been reported to tolerate relative lower a_w environment and hence need only a small amount to maintain their native conformation that is required for catalytic activity. Hence, numerous reports of solvent-free reactions involve the use of lipases. On the contrary, recent studies have shown that the use of other enzymes if immobilized, can be used in SFM biocatalysis (Gill, 2018).

7.2.2.3.5 Molecular sieves

In general, an esterification reaction generates water that may induce hydrolysis, that generally results in a lower bioconversion yield. As mentioned in the previous section, molecular sieves have the capacity to remove water molecules and hence shift the reaction equilibrium towards esterification. Molecular sieves have been previously used for esterification, where reports have suggested an increase in the substrate conversion yield with the addition of molecular sieves. It has been explained that molecular sieves promote the lipase-catalyzed synthesis reactions by removing water; however, an excess of these sieves results in removing if the necessary water that is required for maintaining the catalytic activity of the enzymes (Li et al., 2007). Karboune et al. (2005b) reported that an addition of 10 mg/mL molecular sieves to the transesterification reaction mixture of cinnamic acid and triolein ratios of 1:3 and 1:4.5 resulted in 28% and 35% decrease in the bioconversion yield of cinnamoylated lipids, respectively.

7.2.3 Biocatalysis in ionic liquids

7.2.3.1 Properties of ionic liquids (ILs)

Ionic liquids solution (ILs) are organic salts with very low-melting points which are composed purely of ion pairs that are liquid at room temperature (Naushad et al., 2012); Ghanem, 2006; Kragl et al., 2002; Thomas et al., 2011). Since the physicochemical properties of ionic liquids are determined by the cation and/or anion pair, it is therefore possible to modify them in favor of targeting synthesis rather than hydrolysis (Naushad et al., 2012; Ghanem, 2006; Kragl et al., 2002). In addition to their thermostability, ionic liquid solutions have the capacity to dissolve many polar and nonpolar compounds and their miscibility with conventional solvents to form multiple phase systems of great interest as media for enzyme biocatalysis. These characteristics made ILs an appropriate environment for the enzymes in terms of activity, stability and enantioselectivity (Naushad et al., 2012; Ghanem, 2006; Kragl et al., 2002; Thomas et al., 2011; Milasinovic et al., 2012; Vossenberg et al., 2012; Fernández-Fernández et al., 2014).

Indeed, ionic liquids are compounds consisting of ions that remain liquids at room temperatures, usually lower than 100°C (von Rantwijk and Sheldon, 2007; Zhao, 2010). Room temperature ionic liquids are mostly used as solvents for biocatalytic reactions and are recognized for their low vapor pressures, low flammability, wider liquid range, high thermal and ionic conductivity, high thermal and chemical stabilities and a relatively wider substrate range as compared to other reaction media. Most commonly used ionic liquids cations are alkylammonium, *N, N'*-dialkylimidazolium, pyrrolidinium (which are nitrogen containing) and alkylphosphonium (phosphorus containing). On the other hand, the anions include those of halides such as BF^{4-}, PF^{6}, NO^{3-}, CH_3CO^{2-}, and CF_3CO^{2-}.

In general, the selection of cations or anions plays a very important role in design of ionic liquids, where they can be made water miscible, partially miscible or immiscible; ionic liquids can also be designed to have various viscosities that are required for specific applications (Zhao, 2010). A review by Moniruzzaman et al. (2010) describes the advances of biocatalysis that are carried out in ionic liquids such as lipase catalyzed and non-lipase catalyzed reactions. In most studies, it was found that biocatalytic reactions carried out in ionic liquids showed significantly superior results, as compared to organic solvents.

Ionic liquids are associated with several advantages, where the major one includes the ability to be able to design and produce them as per the required applications, using different cations and anions. This advantage remains exclusive in case of ionic liquids and is not possible for other reaction systems such as aqueous or organic solvents. In addition, ionic liquids allow for biocatalysis of both hydrophilic and hydrophobic substrates, including carbohydrates and amino acids which are generally insoluble in hydrophobic media. Hence, the range of substrate that can be used for biotransformation in ionic liquids are significantly wider, as compared to other reaction systems. Moreover, it has been reported that ionic liquids show enhanced enzyme operational and thermal stabilities, as compared to organic media (Moniruzzaman et al., 2010). van Rantwijk and Sheldon (2007) investigated the stability of enzymes in ionic liquids of different viscosities; the results showed that enhanced catalytic activity was obtained in relatively higher viscosity ionic liquids due to a slower change in their native structural conformation, as compared to the ionic liquids with a low viscosity. Ionic liquids allowed for easy reusability and product recovery as they are immiscible in organic solvents such as hexane and diethyl ether; due to this property a one step product recovery can be attained. For product recovery in ionic liquids, hexane, diethyl ether or even supercritical carbon dioxide may be used, where the enzyme remains in the ionic liquid phase and can be re-used for subsequent catalytic cycles (van Rantwijk and Sheldon, 2007; Moniruzzaman et al., 2010).

Despite several advantages that are associated with the use of ionic liquids, a major problem encountered in such medium is the insolubility of free enzymes. Enzymes are noticed to show lower activities, where they are generally insoluble and exist as powdered forms. In addition, some enzymes may exhibit low solubility via weak hydrogen bonding, which in turn affects their native structural conformation. To overcome such limitations, enzyme microemulsions (Kermasha et al., 1995; Safari et al., 1994a,b), immobilized enzymes or whole cells may be used, where the enzymes show improved catalytic activity in ionic liquids (Moniruazzaman et al., 2010).

7.2.3.2 Biocatalysis of selected enzymes in ionic liquids

Lipases have been shown to be active in ILs, they have been termed as the "work horses" of biocatalysis. It is well known that polar organic solvents such as methanol and DMF lead to inactivation of lipases; even though the polarity of ILs is similar to these solvents, ionic liquids, including [$C_n$mim][BF_4] and [$C_n$mim][PF_6], do not deactivate lipases. Hence, ionic liquids have shown to close the polarity gap between water and organic solvents with a log $P > 2$, that allows the use of lipases for carrying out reactions in polar media (Cantone et al., 2007). It has been reported that the acetylation of 1-phenylethanol catalyzed by lipase was as efficient and enantioselective in ILs as in toluene, whereas the acetylation of glucose using CaL-B was observed to be more regioselective in ILs. The significantly enhanced solubility of glucose in ILs favours the synthesis of monoacetylated products instead of mono- and diacetylated product mixtures, that are obtained in organic solvents. Other reactions if lipases that are carried out in ILs include, direct esterification, transesterifications, enantioselective transesterifications, enantioselective hydrolyses, enantioselective acylation of amines and ammoniolysis (Cantone et al., 2007).

Although most of enzymes reported to be active in ILS are lipases, other biocatalysts such as proteases, dehydrogenases, peroxidases, and glucose oxidases have been also investigated. Many proteases have demonstrated catalytic activity in ILs; Z-aspartame was synthesized from Z-aspartate and L-phenylalanine methyl ester hydrochloride using thermolysin in [bmim][PF_6], where the enzyme exhibited an enhanced stability and reaction rates, as compared to that in acetonitrile (Cantone et al., 2007). Similarly, the catalytic activity *a*-chymotrypsin in ILs was reported for transesterification of *N*-acetyl-L-phenylalanine ethyl ester with 1-propanol in in [bmim][PF_6] and [omim][PF_6], where the enzyme showed similar activity in other ILs and organic solvents. The biocatalysis of papain was reported in a diluted aqueous solution of the IL, [bmim][BF_4] for hydrolysis of D,L, *p*-hydroxyphenylglycine methyl ester, where the enzyme exhibited improved stability and catalytic activity as compared to the aqueous mixtures of polar solvents acetonitrile or 2-propanol. Furthermore, the hydrolysis of epoxides in ILs [bmim][PF_6], [bmim][Tf_2N] and [bmim][BF_4] was investigated using epoxide hydrolase, where the bioconversion rate was similar to that obtained in buffer solutions; however, the enantioselectivity of the enzyme was significantly higher in [bmim][PF_6] as compared to the buffer solutions where the spontaneous hydrolysis was suppressed (Cantone et al., 2007).

The catalytic activity of dehydrogenases has also been reported in biphasic ILs, where the use of enzymes involves cofactors such as NADPH or $NADP^+$. In such systems, the enzyme and cofactor are in the aqueous phase, hence are physically separated from the substrates and products that are present in the IL phase. An advantage of such a reaction medium is the ease of enzyme reusability of the cofactors that results in an enhanced catalytic activity and enzyme stability as compared to the organic solvent media (Cantone et al., 2007).

Peroxidases have been reported to demonstrate activity in ILs [bmim][PF_6], [omim][PF_6] and [bmim][Tf_2N], where the authors have suggested that the hydrophobic, noncoordinated environment of ILs is less disruptive than conventional solvents of similar polarity. In addition, the nature of ILs accelerates the peroxidase reaction by stabilizing the highly charged intermediates that are generated during the rate-determining step (Cantone et al., 2007). The multi-copper oxidase, laccase has also been reported to show catalytic activity in ILs [bmp][BF_4] and [bmim][PF_6], where the enzyme was observed to show moderate concentration of the IL; however, with increasing concentration of the IL, laccase demonstrated reduce catalytic activity that subsequently resulted in enzyme precipitation at an IL concentration of 50/50 v/v. Peroxidase and glucose oxidase were reported to catalyze the oxidation of thioanisoles in [bmim][PF_6], with a stereoselectivity similar to that in water. Furthermore, ß-galactosidase demonstrated catalytic activity in ILs [mmim][$MeSO_4$] and [bmim][$MeSO_4$] for the synthesis of N-acetyl-lactosamine and showed enhanced bioconversion yield as compared to that obtained in aqueous media. However, the catalytic activity of the enzyme was reduced with an increasing concentration of ILs, where the maximum amount tolerated by the enzyme was 25% (v/v) (Cantone et al., 2007).

The reactivity and stability of cellulases was investigated in eight ILs in comparison to that in aqueous medium; the results revealed that the enzyme showed improved stability and hydrolysis rate in ILs. The authors suggested that the increase in rate of hydrolysis in ILs was due to the higher viscosity of ILs and that they were not denatured at 50°C in such an environment, as compared to the aqueous medium (Bose et al., 2010).

Ionic liquid solutions are highly polar due to their ionic nature and various polarity parameters have been used to describe the solvent polarity of ILs. The polarity of ionic liquids has been reported to affect the activity of enzymes, where similar effects have been investigated for lipases. The results revealed that the catalytic activity of lipases increased with the increasing IL polarity during acylation of racemic 1-phenyl ethanol with vinyl acetate (Park and Kazlauskas, 2003). On the contrary few studies have reported no clear trend between polarity of ILs and enzyme activity (Naushad et al., 2012). The effect of alkyl chain length in the cation of ILs has also been reported to affect the enzyme activity, where it has been suggested that cations behave like surfactants in aqueous solution and an increase in the chain length can deactivate enzymes.

7.2.4 Biotransformation in supercritical fluids

7.2.4.1 Properties of supercritical fluids

The use of supercritical fluids (SCF) is a new area that is used for biotransformation, and it is expected to expand in the future (Dhake et al., 2011).

A fluid is considered supercritical when above its critical temperature and pressure, yet below the pressure to condense it into a solid. In general, supercritical fluids involve a substance that is known to be supercritical (SCF) above a singular point in a phase diagram, its critical point. Near to this critical point, minor changes in temperature or pressure may affect relatively large changes in the density or solvation ability of a supercritical fluid.

Supercritical fluid looks-like a gas with low viscosity and high diffusion characteristics and exhibit liquid density and dissolving power. Such properties are beneficial for the enhancement of mass transfer enzymatic catalyzed-reaction rates overcoming the diffusion limitations and consequently increasing the solubility of the hydrophobic substrates (Kamat et al., 1995; Mesiano et al., 1999; Yan et al., 2009; Habulin et al., 2007; Hobbs and Thomas, 2007; Jérôme and Lecomte, 2008; Dhake et al., 2011; Liu et al., 2012; van den Broek et al., 2012). In addition, the density and all related properties can be controlled with even small changes in temperature and pressure, which would result in a desired control of the reaction media (Mesiano et al., 1999; Matsuda et al., 2004; Ghanem, 2006; Hobbs and Thomas, 2007).

Because of such advantages, supercritical fluids have been used as reaction media for a wide range of biocatalytic reactions, such as esterification, transesterification (Matsuda et al., 2004; Habulin et al., 2007; Hobbs and Thomas, 2007; Shin et al., 2009; Dhake et al., 2011; van den Broek et al., 2012), polymerization (Habulin et al., 2007; Hobbs and Thomas, 2007; Jérôme and Lecomte, 2008; Kovaliov et al., 2020), acetylation, carboxylation, and transglycosylation (Matsuda et al., 2004; Habulin et al., 2007; Hobbs and Thomas, 2007; Liu et al., 2012).

In addition, the property of supercritical fluids (SCF) is used for a wide range of extractions applications, biotransformation and downstream processing steps in a single bioreactor. In addition, the relatively lower viscosity and higher diffusivity of SCFs enable enhanced mass-transfer, as compared to other reaction systems (Fontes et al. 2001). Other advantages of SCFs include, low-toxicity, lower environmental impact, easy downstream processing and recyclability (Knez et al., 2015). Carbon dioxide (CO_2) is the most commonly used supercritical fluid for biotransformation, due to its the high vapor pressure that allows complete recovery of the product that leaves no trace after the reaction, also resulting in a solvent-free reaction. The use of SCFs for biotransformation allows for an integrated production process, where the dense fluid can act as a solvent for the reaction. In general, this tunable solvation property of SCFs enable merging of biocatalytic and the downstream processing steps as a single stable process unit (Knez, 2018).

Despite the advantages associated with SCFs, the major limitation with the use of supercritical carbon dioxide (CO_2) for biocatalysis, is associated with its inability to dissolve polar compounds. In such cases, polar co-solvents, including acetone, ethanol, methanol or water may be added to the reaction system to increase its polarity and thereby solubilize the polar substrate via hydrogen bonding. These co-solvents are also known as "entrainers" and are usually added to the SCFs at moderate concentrations of < 10 mol%. Even though co-solvents may be used to increase the polarity of the reaction system, they often increase the complexity of the phase behaviour and complicate downstream processing (Knez, 2018).

7.2.4.2 Stability of enzymes

In carrying out biotransformations on a larger industrial scale, the stability of the biocatalyst and substrate selectivity, are the most important factors. In addition, for bioprocesses to be economically efficient, biocatalysts are required to exhibit enhanced stability, reusability, without losing selectivity. Although various enzymes are stable in SCFs; it is important that for each reaction system (substrate/enzyme/dense fluid) the proper biocatalysis parameters must be determined and the corresponding process conditions that affect the enzyme stability needs to be assessed. In general, the stability of enzymes is sensitive to various factors, such as pH, temperature, pressure and can often lead to their inactivation if increased or decreased beyond the optimum point. Other than this, various substrates may also block the active sites of the enzymes that results in limited catalytic activity, reduced substrate specificity or also their complete inactivation. On the contrary, substrate inhibition may be overcome by reducing the substrate concentration or by continuous addition of substrates to the reaction system; however, product inhibition is more difficult to overcome as this can be done only by continuous removal of the product from the reaction systems (Knez, 2018).

In general, conventional extraction separation can result in increased complications and hence, the use of dense gases is usually preferred, especially in case of continuous processes. It has been reported that enzyme molecules show more rigidity in dense gases, which prevents quick denaturation of the enzymes. Moreover, the use of relatively higher-pressure systems can often lead to denaturation of the biocatalyst; immobilization of enzymes has been shown to improve the enzyme stability under such extreme reaction conditions. Overall, this area is still new in terms of optimization, where it is extremely difficult to predict enzyme stability or activity in SCFs (Knez, 2018).

7.2.4.3 Effect of water content on enzyme activity

As mentioned earlier enzymes require a certain amount of water to maintain their catalytic activity in a reaction system. In fact, water content in a reaction system is one of the most important factors that influences enzyme catalytic activity. Water affects the enzyme activity by noncovalent binding to their native structure, or by disruption of hydrogen bonds, or by facilitation of reagent diffusion in a reaction system. In general, a low water content in the reaction mixture favors the optimal catalytic activity of enzymes in SCFs. Mostly, lipolytic enzymes show enhanced enzyme activity at very low water content in the reaction system, whereas a relatively higher water content may result in a decrease of enzymatic activity. This effect may result in protein structural changes under high pressure systems. Other than this, water causes hydrolysis of products and an excess of water content may lead to enzyme agglomeration, where a reduced surface area results in lower enzymatic activities. In context with this effect, hydrophobic solvents have been showed to demonstrate relatively higher enzymatic activities, where hydrophobic dense gases such as CO_2 are able to dissolve 0.3%–0.5% (w/w) of water, also depending on the temperature and pressure used (Knez, 2018). Hence, water may be added to SCFs, in order to adjust the water activity to a necessary value for optimal enzymatic activity for carrying out biocatalysis in such media. For reactions where water is a reactant, the water content must be adjusted such that it is high enough to perform a continuous reaction as well as to compensate for any losses due to solubility in SCFs. It has also been reported that use of CO_2 in reaction systems may cause acidifications in the surrounding medium, hence addition of organic or inorganic buffers can be used to control the pH value for increased enzyme stability. Salt hydrate pairs may also be used to control water activity in hydrophobic solvent, where these can have a positive effect on the initial rate of reaction and on the bioconversion rates (Knez, 2018).

7.2.4.4 Effect of pressure on enzymes

In general, the diffusivity, interphase tension, density and dielectric constant of SCFs may be varied by changing the pressure in the reaction medium. In addition, the substrate specificity and the catalytic activity of enzymes can also be modified with the change in pressure. It has been reported that most enzymes become completely deactivated in SCFs, except for lipolytic enzymes, where lipases show enhanced catalytic activity in liquid propane with an increase in pressure. This increase in enzyme activity may be due to and increased substrate solubility with an increase in pressure. However, in some cases such as ethane, fluoroform, sulfur hexafluoride or propane the pressure has a limited effect on the rate of lipase-catalyzed reactions. On the other hand, most enzymes are deactivated at high pressures and if they are active in SCFs, they exhibit enhanced stability and specificity under extreme pressures. Other than the effect on the enzyme's catalytic activity, the use of SCFs aids in an easier downstream processing, as an increase in pressure decreases the density and dielectric constant of the SCF, thereby resulting in reduced solvent power that allows an easy separation of the reaction products (Knez, 2018).

7.2.5 Solid-gas biocatalysis

Solid-gas biocatalysts employ the ability of lyophilized or immobilized enzymes to catalyze the biotransformation of gaseous substrates (Vitola et al., 2019). This type of biocatalysis offers various advantages over the conventional methods in aqueous medium. Most enzymes are more stable in their lyophilized form rather than in their dissolved state which result in enhanced thermal stability, increased yields and lower economical costs (Kulishova and Zharkov, 2017).

In addition, this type of biocatalysis avoids the use of any solvents which causes limited toxicity towards the biocatalysts as well as the formation of side chain products. Furthermore, the use of this type of reaction system also decreases the downstream processing and makes product isolation much easier, as no further extraction or purification is required (Takoua Debeche et al., 2005). The major limitations associated with biocatalysis are diffusion and mass-transfer which are completely absent in a sol/gas system, where the circulation of gases simplify the process and reduces the cost if used at an industrial scale (Kulishova and Zharkov, 2017).

The solid/gas system would allow all reaction components to establish independently the thermodynamic activity at the desired values. This would allow investigation of the effect of any additional component at a constant thermodynamic activity value (Létise et al., 2003). Lamare and Legoy (1995) showed that it would be of great importance to take into consideration the thermodynamic activity of each of the chemical components in the system (Lamare and Legoy, 1995).

Most sol/gas biocatalysis have been carried out in batch reactor systems; however, the advancement in tubular reactors with immobilized dehydrates enzymes are now most often used (Vitola et al., 2019). On the contrary a major drawback in solid/gas biocatalysis is its limited use for simple compounds with relatively lower boiling points as the highest boiling points of substrates are those of acetophenone, 1-phenylethanol, 2-octyl propionate, 3octyl propionate. Nevertheless, even with such limitations this biocatalysis approach may be used to develop ecologically cleaner industrial processes. Overall, this kind of biocatalysis opens new possibilities in biocatalysis and biotransformation, that show more control of thermodynamic parameters, including pressure, temperature, water and substrate activity (Kulishova and Zharkov, 2017).

7.3 Scale-up of enzymatic bioconversion processes

The ultimate objective in the optimization of biocatalysis of enzymes in nonconventional media is to proceed with the scale-up of the processes for their applications in industrial processes (von Weymam et al., 2008; Lv et al., 2019). In order to achieve such an objective, it is essential optimize the design of industrial processes either in batch or in continuous reactor (Pinoa et al., 2018). However, this optimization is aimed at the integration of flow reactor technology, which should lead to sustainable and higher productivity (Tamborini et al., 2017). Hence, there should be a well thought-out strategies (Osbon and Kumar, 2019) for successful processes of enzymatic transformations (Singh et al., 2019). Indeed, the process should have a synergic integration of both technical and economic advantages of enzymatic bioconversion (Nguyen and Saddler, 1991).

It is a combination of the nature of a biocatalyst and flow chemistry of the reaction media that would result in a successful industrial process. However, there are other factors that may influence the process, such as substrate/product inhibition, product recovery, mechanical mixing. The development of automation of machineries and devices for the on-line processes and product recovery are making flow-based biocatalysis much easier for the enzymatic bioconversion industrial applications.

Although the progress achieved to date in metabolic engineering and immobilization contributed greatly to the industrial applications of enzymes, there are still certain problems related to the low productivity and difficulties in scaling-up the process (Tamborini et al., 2017; Kucharska et al., 2020).

7.4 Conclusion

The enzyme-catalyzed reactions in nonconventional media, containing low water content or even free one, has resulted in an increase in the number of applications of a given enzyme in a wide range of industrial domains. For example, the expansion of catalytic activity of enzymes, that are normally classified as hydrolases, to carry out reactions for synthesis is a major benefit for such a group of enzymes. The outlined content of this chapter dealt with the definition of nonconventional- media for enzyme catalytic reactions as well the major techniques used to achieve biocatalysis, including organic solvents, solvent- free, ionic liquids, supercritical fluids, and solid-gas media. However, the concept of green chemistry is a predominant factor that should be considered in any novel approach for the development of nonconventional media. In addition, the parameters that could affect the enzyme catalytic reaction in each reported media should be considered to ensure the benefits of using these nonconventional media in bioprocesses. Progress in organic chemistry, bioengineering, immobilization, computational and modeling as well the scaling-up of the bioprocesses models to industrial levels is expected to lead to a larger and even wider range of industrial biotechnological applications of enzymes.

References

Adlercreutz, P., 1996. Modes of using enzymes in organic media. In: Koskinen, A.M.P., Klibanov, A.M. (Eds.), Enzymatic Reactions in Organic Media. Springer, New York, N.Y., pp. 9–42.

Adlercreutz, P., 2000. Biocatalysis in non-conventional media. In: Straathof, A.J.J., Adlecreutz, P. (Eds.), Applied Biocatalysis. second ed. Harwood Academic Publishers, Gordon and Breach Group, Ottawa, ON, pp. 295–316.

Adlercreutz, P., 2008. Fundamentals of biocatalysis in neat organic solvents. In: Carrea, G., Riva, S. (Eds.), Organic Synthesis with Enzymes Non-Aqueous Media. John Wiley, New York, N.Y., pp. 1–24.

Adlercreutz, P., 2013. Immobilization and application of lipases in organic media. Chem. Soc. Rev. 42, 6406–6436.

Arriagada-Strodthoff, P., Karboune, S., Neufeld, R.J., Kermasha, S., 2007. Optimization of chlorophyllase-catalyzed hydrolysis of chlorophyll in monophasic organic solvent media. Appl. Biochem. Biotechnol. 142, 263–275.

Attwood, D., Florence, A.T., 1983. Aspects of surfactant toxicity. In: Attwood, D., Florence, A.T. (Eds.), Surfactant Systems. Chapman & Hall Ltd, New York, N.Y., pp. 614–697.

Aziz, S., Dutilleul, P., Kermasha, S., 2012. Lipase-catalyzed transesterification of krill oil and 3, 4-dihydroxyphenyl acetic acid in solvent-free medium using response surface methodology. J. Mol. Catal. B: Enzym. 84, 189–197.

Bhat, S.D., Naidu, B.V.K., Shanbhag, G.V., Halligudi, S.B., Sairam, M., Aminabhavi, T.M., 2006. Mesoporous molecular sieve (MCM-41)-filled sodium alginate hybrid nanocomposite membranes for pervaporation separation of water–isopropanol mixtures. Sep. Purif. Technol. 49, 56–63.

Biasutti, M.A., Abuin, E.B., Silber, J.J., Correa, N.M., Lissi, E.A., 2008. Kinetics of reactions catalyzed by enzymes in solutions of surfactants. Adv. Colloid Interface Sci. 136, 1–24.

Bitar, M., Karboune, S., Bisakowski, B., Kermasha, S., 2004. Biocatalysis of chlorophyllase in an aqueous/miscible organic solvent medium-containing canola oil. J. Am. Oil Chem. Soc. 81, 927–932.

Bódalo, A., Bastida, J., Máximo, M.F., Montiel, M.C., Murcia, M.D., Ortega, S., 2009. Influence of the operating conditions on lipase-catalysed synthesis of ricinoleic acid estolides in solvent-free systems. Biochem. Engineer. J. 44, 214–219.

Bordusa, 2002. Proteases in organic synthesis. Chem. Rev. 102, 4817–4868.

Bose, S., Armstrong, D.W., Petrich, J.W., 2010. Enzyme-catalyzed hydrolysis of cellulose in ionic liquids: A green approach toward the production of biofuels. J. Phys. Chem. B 114, 8221–8227.

Bourg-Garros, S., Razafindramboa, N., Pavia, A.A., 1998a. Large-scale preparation of (Z)-3-hexen-1-yl acetate using *Candida antarctica*-immobilized lipase in hexane. Biotechnol. Bioengineer. 59, 495–500.

Bourg-Garros, S., Razafindramboa, N., Pavia, A.A., 1998b. Optimization of lipase-catalyzed synthesis of (z)-3-hexen-1-yl acetate by direct esterification in hexane and a solvent-free medium. Enzym. Microb. Technol. 22, 240–245.

Cantone, S., Hanefeld, U., Basso, A., 2007. Biocatalysis in non-conventional media-ionic liquids, supercritical fluids and the gas phase. Green Chem. 9, 954–971.

Capello, C., Fischer, U., Hungerbühler, K., 2007. What is a green solvent? A comprehensive framework for the environmental assessment of solvents. Green Chem. 9, 927–934.

Chaibakhsh, N., Rahman, M.B.A., Abd-Aziz, S., Basri, M., Salleh, A.B., Rahman, R., 2009. Optimized lipase-catalyzed synthesis of adipate ester in a solvent-free system. J. Industrial Microbiol. Biotechnol. 36, 1149–1155.

Chao, L., Zhou, Y.-J., Wang, N., Feng, X.-W., Li, K., Yu, X.-Q., 2010. Promiscuous protease-catalyzed aldol reactions: A facile biocatalytic protocol for carbon–carbon bond formation in aqueous media. J. Biotechnol. 150, 539–545.

Dhake, K.P., Deshmukhi, K.M., Patil, Y.P., Singhal, R.S., Bhanage, B.M., 2011. Improved activity and stability of *Rhizopus oryzae* lipase via immobilization for citronellol ester synthesis in supercritical carbon dioxide. J. Biotechnol. 156, 46–51.

Dordick, J.S., 1989. Enzymatic catalyais in monophasic organic solvents. Enzym. Microb. Technol. 11, 194–211.

Doukyu, N., Ogino, H., 2010. Organic solvent-tolerant enzymes. Biochem. Engineer. J. 48, 270–282.

Elfman-Borjesson, I., Harrod, M., 1999. Synthesis of monoglycerides by glycerolysis of rapeseed oil using immobilized lipase. J. Am. Oil Chem. Soc. 76, 701–707.

Erbeldinger, M., Halling, P.J., Ni, X., 2001. Scale-up of enzymatic solid-to-solid peptide synthesis and enzyme recovery. AIChE J. 47, 500–508.

Fernández-Fernández, M., Moldes, D., Domínguez, A., Sanromán, M.Á., Tavares, A.P.M., Rodriguez, O., Macedo, E.A., 2014. Stability and kinetic behavior of immobilized laccase from *Myceliophthora thermophila* in the presence of the ionic liquid 1-ethyl-3-methylimidazolium ethylsulfate. Biotechnol. Prog. 30, 790–796.

Fontes, N., Almeida, M.C., Barreiros, S., 2001. Biotransformations in supercritical fluids. In: Vulfson, E.N., Halling, P.J., Holland, H.L. (Eds.), Enzymes in Nonaqueous Solvents: Methods and Protocols. Humana Press, Totwa, NJ, pp. 565–573.

Gaffar, R., Kermasha, S., Bisakowski, B., 1999. Biocatalysis of immobilized chlorophyllase in a ternary micellar system. J. Biotechnol. 74, 45–55.

Gawas, S.D., Jadhav, S.V., Rathod, V.K., 2016. Solvent free lipase catalysed synthesis of ethyl laurate: Optimization and kinetic studies. Appl. Biochem. Biotechnol. 180, 1428–1445.

Ghaffari-Moghaddam, M., Eslahi, H., Aydin, Y.A., Saloglu, D., 2015. Enzymatic processes in alternative reaction media: A mini review. J. Biol. Methods 2, e25. doi: 10.14440/jbm.2015.60.

Ghanem, A., 2006. Trends in lipase-catalyzed asymmetric access to enantiomerically pure/enriched compounds. Tetrahedron 63, 1721–1754.

Gill, K.J., 2018. Encapsulation of Laccase Enzymatic Extract, from Coriolus hirsutus and its Biocatalysis in Solvent-Free Media, PhD Thesis, McGill University, Montreal, QC, Canada.

Gómez, J.L., Bastida, J., Máximo, M.F., Montiel, M.C., Murcia, M.D., Ortega, S., 2011. Solvent-free polyglycerol polyricinoleate synthesis mediated by lipase from *Rhizopus arrhizus*. Biochem. Engineer. J. 54, 111–116.
Gorman, L.A.S., Dordick, J.S., 1992. Organic solvents strip water off enzymes. Biotechnol. Bioengineer. 39, 392–397.
Goswami, D., Basu, J.K., De, S., 2012. Lipase applications in oil hydrolysis with a case study on castor oil: A review. Crit. Rev. Biotechnol. 33, 81–96.
Gupta, M.N., 1992. Enzyme function in organic solvents. Eur. J. Biochem. 203, 25–32.
Gupta, M.N., Mukherjee, J., Deepika Malhotra, D., 2013. Use of high activity enzyme preparations in neat organic solvents for organic synthesis. Universal Org. Chem., (Doi: org/10.7243/2053-7670-1-1).
Gupta, A., Khare, S.K., 2009. Enzymes from solvent-tolerant microbes: Useful biocatalysts for non-aqueous enzymology. Crit. Rev. Biotechnol. 29, 33–54.
Gupta, M.N., Roy, I., 2004. Enzymes in organic media: Forms, function and applications. FEBS J. 271, 2575–2583.
Habulin, M., Primozic, M., Knez, Z., 2007. Supercritical fluids as solvents for enzymatic reactions. Acta Chim. Slov. 54, 667–677.
Halling, P.J., 1994. Thermodynamic predictions for biocatalysis in nonconventional media: Theory, tests, and recommendations for experimental design and analysis. Enzym. Microb. Technol. 16, 178–206.
Halling, P.J., Eichhorn, U., Kuhl, P., Jakubke, H.D., 1995. Thermodynamics of solid-to-solid conversion and application to enzymic peptide synthesis. Enzym. Microb. Technol. 17, 601–606.
Hartmann, M., 2005. Ordered mesoporous materials for bioadsorption and biocatalysis. Chem. Mater. 17, 4577–4593.
Hinzmann, A., Adebar, N., Betke, T., Leppin, M., Groger, H., 2019. Biotransformations in pure organic medium: Organic solvent-labile enzymes in the batch and flow Synthesis of nitriles. Chem. Eur. 41, 6011–6916.
Hobbs, H.R., Thomas, N.R., 2007. Biocatalysis in supercritical fluids, in fluorous solvents, and under solvent-free conditions. Chem. Rev. 107, 2786–2820.
Hudson, E.P., Eppler, R.K., Clark, D.S., 2005. Biocatalysis in semi-aqueous and nearly anhydrous conditions. Curr. Opin. Biotech. 16, 637–643.
Illanes, A., Cauerhff, A., Wilson, L., Castro, G.R., 2012. Recent trends in biocatalysis engineering. Bioresour. Technol. 115, 48–57.
Illanes, A., Fernandez-Lafuente, R., Guisan, J.M., Wilson, L., 2008. Heterogeneous enzyme kinetics. In: Illanes, A. (Ed.), Enzyme Biocatalysis: Principles and Applications. Springer, New York, N.Y., pp. 155–197.
Ismail, A.A., Wafik Gouda, M., Motawi, M.M., 1970. Micellar solubilization of barbiturates. I. Solubilization of certain barbiturates in polysorbates of varying hydrocarbon chain length. J. Pharm. Sci. 50, 220–224.
Jérôme, C., Lecomte, P., 2008. Recent advances in the synthesis of aliphatic polyesters by ring-opening polymerization. Adv. Drug Deliv. Rev. 60, 1056–1076.
Jin, J.N., Lee, S.H., Lee, S.B.J., 2003. Immobilization of lipase in a mesoporous reactor based on MCM-41. J. Mol. Catal. B: Enzym. 26, 209–217.
Kamat, S.V., Beckman, E.J., Russell, A., 1995. Enzyme activity in supercritical fluids. Crit Rev. Biotechnol. 15, 41–71.
Karam, R., Karboune, S., St-Louis, R., Kermasha, S., 2009. Lipase-catalyzed acidolysis of fish liver oil with dihydroxyphenylacetic acid in organic solvent media. Process Biochem. 44, 1193–1199.
Karboune, S., Neufeld, R.J., Kermasha, S., 2005a. Immobilization and biocatalysis of chlorophyllase in selected organic solvent systems. J. Biotechnol. 120, 273–283.
Karboune, S., Safari, M., Lue, B.-M., Yeboah, F.K., Kermasha, S., 2005b. Lipase-catalyzed biosyntheis of cinnamoylated lipids in a selected organic solvent medium. J. Biotechnol. 119, 281–290.

Kaul, A., Asano, Y., 2008. Strategies for discovery and improvement of enzyme function: State of the art and opportunities. Microbial Biotechnol. 5, 18–33.
Kermasha, S., Tse, M., 2000. Biocatalysis of tyrosinase in chloroform medium, using selected phenolic substrates. J. Chem. Technol. Biotechnol. 75, 475–483.
Kermasha, S., Bao, H., Bisakowski, B., 2001a. Biocatalysis of tyrosinase in selected organic solvent media using catechin as substrate. J. Mol. Catal. B: Enzym. 11, 929–938.
Kermasha, S., Bao, H., Bisakowski, B., Yaylayan, V., 2002a. Characterization of the biocatalysis of tyrosinase in selected organic solvent media using model phenolic substrates. J. Mol. Catal. B: Enzym. 19-20, 335–345.
Kermasha, S., Dioum, N., Bisakowski, B., 2001b. Biocatalysis of lipoxygenase in selected organic solvent media. J. Mol. Catal. B: Enzym. 11, 909–919.
Kermasha, S., Dioum, N., Bisakowski, B., Vega, M., 2002b. Biocatalysis of immobilized lipoxygenase in a ternary micellar system. J. Mol. Catal. B: Enzym. 19, 305–317.
Kermasha, S., Gaffar, R., Bisakowski, B., 2000. Biocatalysis of silica gel-immobilized chlorophyllase in a ternary micellar system. Process Biochem. 35, 1103–1109.
Kermasha, S., Safari, M., Goetghebeur, M., 1995. Interesterification of butter fat by lipase from *Rhizopus niveus* in cosurfactant free microemulsion system. Appl. Biochem. Biotechnol. 53, 229–244.
Khamessan, A., Kermasha, S., 1996a. Biocatalysis of chlorophyllase from *Pheaodactylum tricornutum* in micellar ternary system containing Spans. J. Biotechnol. 45, 253–264.
Khamessan, A., Kermasha, S., 1996b. Biocatalysis of chlorophyllase in canola oil using organic solvent systems. J. Food Biochem. 20, 311–328.
Khamessan, A., Kermasha, S., Marsot, P., 1994a. Effect of polar organic solvents on the biocatalysis of chlorophyllase in a biphasic organic system. Biosci. Biotech. Biochem. 58, 1947–1952.
Khamessan, A., Kermasha, S., Marsot, P., 1995a. Biocatalysis of chlorophyllase from *Phaeodactylum tricornutum* in organic solvent media. Process Biochem. 30, 159–168.
Khamessan, A., Kermasha, S., Khalyfa, A., Marsot, P., 1993. Biocatalysis of chlorophyllase from the alga *Phaeodactylum tricornutum* in a water/miscible organic solvent system. Biotechnol. Appl. Biochem. 18, 285–298.
Khamessan, A., Kermasha, S., Marsot, P., Ismail, A., 1994b. Biocatalysis of chlorophyllase from *Phaeodactylum tricornutum* in a biphasic organic system. J. Chem. Technol. Biotechnol. 60, 73–81.
Khamessan, A., Kermasha, S., Mollimard, L., 1995b. Optimization of chlorophyllase-catalyzed hydrolytic activity in a micellar ternary system. Biotechnol. Appl. Biochem. 22, 327–343.
Khmelnitsky, Y., Levashov, A., Klyachko, N., Martinek, K., 1988. Engineering biocatalytic systems in organic media with low water content. Enzym. Microb. Technol. 10, 710–724.
Klibanov, A.M., 2001. Improving enzymes by using them in organic solvents. Nature 409, 241–246.
Klibanov, A.M., 2003. Asymmetric enzymatic oxidoreductions in organic solvents. Curr. Opin. Biotechnol. 14, 427–431.
Knez, Ž., 2018. Enzymatic reactions in subcritical and supercritical fluids. J. Supercritical Fluids 134, 133–140.
Knez, Ž., Leitgeb, M., Primožicˇ, M., 2015. Enzymatic reactions in supercritical fluids. In: Fornari, T., Stateva, R.P. (Eds.), High Pressure Fluid Technology for Green Food Processing. Springer, Berlin, Germany, pp. 185–215.
Konieczny, S., Fik, C.P., Averesch, N.L.H., Tiller, J.C., 2012. Organosoluble enzyme conjugates with poly(2-oxazoline) s via pyromellitic acid dianhydride. J. Biotechnol. 159, 195–203.
Kovaliov, M., Wright, T.A., Cheng, B., Mathers, R.T., Zhang, X., Meng, D., Szczesniak, K., Jenezyk, J., Jurga, S., Cohen-Kami, D., Page, R.C., Konekoewicz, D., Averick, S., 2020. Toward next-generation biohybrid catalyst design: Influence of degree polymerization on enzyme activity. Bioconjugate Chem. 31, 939–947.

Kragl, U., Eckstein, M., Kaftzik, N., 2002. Enzyme catalysis in ionic liquids. Curr Opin. Biotechnol. 13, 565–571.
Krishna, S.H., 2002. Developments and trends in enzyme catalysis in nonconventional media. Biotechnol. Adv. 20, 239–267.
Kucharska, K., Slupek, E., Cieslinski, H., Kaminski, M., 2020. Advantageous conditions of saccharification of lignocellulosic biomass for biofuels generation via fermentation processes. Chem. Papers 74, 1199–1209.
Kuhl, P., Elchhorn, U., Jakubke, H.D., 1995. Enzymic peptide synthesis in microaqueous, solvent-free systems. Biotechnol. Bioengineer. 45, 276–278.
Kuldamrong, W., Husson, F., Kermasha, S., 2013. Biocatalysis with hydroperoxide lyase in extracts from *Penicillium camaberti* in neat organic solvent media. Biocatal. Biotransform. 31, 94–99.
Kulishova, L.M., Zharkov, D.O., 2017. Solid/gas biocatalysis. Biochemistry (Moscow) 82, 95-85.
Kuperkar, V.V., Lade, V.G., Prakash, A., Rathod, V.K., 2014. Optimization and kinetic studies. J. Mol. Catal. B: Enzym. 99, 143–149.
Lamare, S., Legoy, M.-D., 1995. Solid/gas biocatalysis: How to fully define the system? Biotechnol. Tech. 9, 127–132.
Lamer, S., Leblanc, D., Morin, A., Kermasha, S., 1996. Biogeneration of ethyl valerate by whole cells of *Pseudomonas fragi* CRDA 037 in aqueous medium. Biotechnol. Tech. 10, 475–478.
Létise, F., Lamare, S., Legoy, M.-D., Graber, M., 2003. Solid/gas biocatalysis: An appropriate tool to study the influence of organic components on kinetics of lipase-catalyzed alcoholysis. Biochim. Biophys. Acta 1652, 27–34.
Li, Z., Yang, D., Jiang, L., Ji, J.F., Ji, H.B., Zeng, X.H., 2007. Lipase-catalyzed esterification of conjugated linoleic acid with L-carnitine in solvent-free system and acetonitrile. Bioprocess. Biosyst. Engineer. 30, 331–336.
Liu, Y., Chen, D., Xu, L., Yan, Y., 2012. Evaluation of structure and hydrolysis activity of *Candida rugosa* Lip7 in presence of sub-/super-critical CO_2. Enzym. Microb. Technol. 51, 354–358.
Llerena-Suster, C.R., José, C., Collins, S.E., Briand, L.E., 2012. Investigation of the structure and proteolytic activity of papain in aqueous miscible organic media. Process Biochem. 47, 47–56.
Lue, B.-M., Karboune, S., Yeboah, F.K., Kermasha, S., 2005. Lipase-catalyzed esterification of cinnamic acid and oleyl alcohol in organic solvent media. J. Chem. Technol. Biotechnol. 80, 462–468.
Lv, Y., Xu, S., Lyu, Y., Zhou, S., Du, G., Chen, J., Zhou, J., 2019. Engineering enzymatic cascades for the efficient biotransformation of eugenol and taxifolin to silybin and isosilybin. Green Chem. 21, 1660–1667.
Ma, H.-L., Kermasha, S., Gao, J.-M., Morales-Borges, R., Yu, X.-Z., Yi, Y., 2009. Laccase-catalyzed oxidation of phenolic compounds in organic media. J. Mol. Catal. B: Enzym. 57, 89–95.
Madamwar, D., Soni, K., 2003. Biocatalysis in non-conventional media by entrapment of the enzyme in water-restricted microenvironment. New Horizon in Biotechnology. Springer, New York N.Y., pp. 97–112.
Matsuda, T., Watanabe, K., Harada, T., Nakamura, K., 2004. Enzymatic reactions in supercritical CO_2: Carboxylation, asymmetric reduction and esterification. Catal. Today 96, 103–111.
Mb, A.R., Tajudin, S.M., Hussein, M.Z., Zalih, R.N., Rahman, R.A., 2005. Application of natural kaolin as support for the immobilization of lipase from *Candida rugosa* as biocatalsyt for effective esterification. Clay Sci. 29, 111–116.
Mesiano, A.J., Beckman, E.J., Russell, A.J., 1999. Supercritical biocatalysis. Chem. Rev. 99, 623–634.

Milasinovic, N., Knezevic-Jugovic, Z., Jakovljevic, Z., Filipovic, J., Krusic, M.K., 2012. Synthesis of n-amyl isobutyrate catalyzed by *Candida rugosa* lipase immobilized into poly(N-isopropylacrylamide-co-itaconic acid) hydrogels. Chem. Engineer. J. 181, 614–623.

Mine, Y., Fukunaga, K., Itoh, K., Yoshimoto, M., Nakao, K., Sugimura, Y., 2003. Enhanced enzyme activity and enantioselectivity of lipases in organic solvents by crown ethers and cyclodextrins. J. Biosci. Bioengineer. 95, 441–447.

Moniruzzaman, M., Nakashima, K., Kamiya, N., Goto, M., 2010. Recent advances of enzymatic reactions in ionic liquids. Biochem. Engineer. J. 48, 295–314.

Myers, D., 1988. *Surfactant Science and Technology*, in: Berg, J.C. (Ed.), VCH Publishers, New York, N.Y., pp. 153–183.

Naushad, M., ALothman, Z.A., Khan, A.B., Ali, M., 2012. Effect of ionic liquid on activity, stability, and structure of enzymes: A review. Int. J. Biol. Macromol. 51, 555–560.

Nguyen, Q.A., Saddler, J.N., 1991. An integrated model for the technical and economic evaluation of an enzymatic biomass conversion process. Bioresource Technol. 35, 275–282.

Nicolova, P., Ward, O.P., 1993. Whole cell biocatalysis in nonconventional media. J. Indust. Microbiol. 12, 76–86.

Ogino, H., Ishikawa, H., 2001. Enzymes which are stable in the presence of organic solvents. J. Biosci. Bioengineer. 91, 109–116.

Osbon, Y., Kumar, M., 2019. Biocatalysis and strategies for enzyme improvement. IntecheOpen Link, DOI: 10.5772/intechopen.85018.

Park, S., Kazlauskas, R.J., 2003. Biocatalysis in ionic liquids-advantages beyond green technology. Curr. Opin. Biotechnol. 14, 432–437.

Petersson, A.E., Gustafsson, L.M., Nordblad, M., Börjesson, P., Mattiasson, B., Adlercreutz, P., 2005. Wax esters produced by solvent-free energy-efficient enzymatic synthesis and their applicability as wood coatings. Green Chem. 7, 837–843.

Petersson, A.E.V., Adlercreutz, P., Mattiasson, B., 2007. A water activity control system for enzymatic reactions in organic media. Biotechnol. Bioengineer. 97, 235–241.

Pinoa, M.S., Rodriguez-Jasso, R.M., Micheline, M., Flores-Gallegosa, A.C., Morales-Rodriguez, Teixeir, J.A., Euiza, H.A., 2018. Bioreactor design for enzymatic hydrolysis of biomass under the biorefinery concept. Chem. Engineer. 347, 119–136.

Priyadarshini, A., Pandey, P., 2018. Enzymes in non-conventional media. In: Priyadarshini, A., Pandey, P. (Eds.), Biocatalysis and Agricultural Biotechnology: Fundamentals, Advances, and Practices for a Greener Future. first ed. CRC Press, Boca Raton, FL.

Robinson, P.K., 2015. Enzymes: Principles and biotechnological applications. Essays Biochem. 59, 1–41.

Sabally, K., Karboune, S., St-Louis, R., Kermasha, S., 2007. Lipase-catalyzed synthesis of phenolic lipids from fish liver oil and dihydrocaffeic acid. Biocatal. Biotransform. 25, 211–218.

Sabally, K., Karboune, S., Yeboah, F.K., Kermasha, S., 2005a. Lipase-catalyzed esterification of linolenyl alcohol with selected phenolic acids in organic solvent media. Appl. Biochem. Biotechnol. 127, 17–27.

Sabally, K., Karboune, S., Yeboah, F.K., Kermasha, S., 2005b. Enzymatic esterification of dihydrocaffeic acid with linoleyl alcohol in organic solvent media. Biocatal. Biotransform. 23, 37–44.

Safari, M., Kermasha, S., 1994a. Interesterification of butter fat by commercial microbial lipases in a cosurfactant free microemulsion system. J. Am. Oil Chem. Soc. 71, 969–973.

Safari, M., Kermasha, S., Lamboursain, L., Sheppard, J.D., 1994a. Interesterification of butterfat by lipase from *Rhizopus niveus* in reverse micellar systems. Biosci. Biotechnol. Biochem. 58, 1553–1557.

Safari, M., Kermasha, S., Pabai, F., Sheppard, J.D., 1994b. Interesterification of butter fat by lipase from *Mucor miehei* in microemulsion system. J. Food Lipids 1, 247–263.

Samaha, H., Kermasha, S., 1997. Biocatalysis of chlorophyllase in ternary micellar system containing Span 85 using purified and oxidized pheophytins as substrates. J. Biotechnol. 55, 181–191.

Santos, J.C., Bueno, T., Molgero da Ros, P.C., de Castro, H.F., 2007. Lipase-catalyzed synthesis of butyl esters by direct esterification in solvent-free system. J. Chem. Technol. Biotechnol. 82, 956–961.

Schimdt, A., Dordick, J.S., Hauers, B., Kiener, A., Wubbolts, M., Whiholt, B., 2001. Industrial biocatalysis today and tomorrow. Nature 409, 258–268.

Sellek, G.A., Chaudhuri, J.B., 1999. Biocatalysis in organic media using enzymes from extremophiles. Enzym. Microb. Technol. 25, 471–482.

Sheldon, R.A., 2005. Green solvents for sustainable organic synthesis: State of the art. Green Chem. 7, 267–278.

Sheldon, R.A., 2016. Biocatalysis and biomass conversion in alternative reaction media. Chem. A Eur. J. 22, 12984–12999.

Sheldon, R.A., Pereira, P.C., 2017. Biocatalysis engineering: The big picture. Chem. Soc. Rev. 46, 2678–2691.

Shin, M.H., Cheong, N.Y., Lee, J.H., Kim, K.H., 2009. Transglucosylation of caffeic acid by a recombinant sucrose phosphorylase in aqueous buffer and aqueous-supercritical CO_2 media. Food Chem. 115, 1028–1033.

Singh, N., Devi, A., Bishnol, M.B., Jaryal, R., Dahia, A., Tashyrev, O. and Hovorukha, V., 2019. Overview of the process of enzymatic transformation of biomass. Elements of Bioeconomy, Open Access, (DOI: 10.5772/intechopen.85036).

Sorour, N., Karboune, S., Saint-Louis, R., Kermasha, S., 2012a. Lipase-catalyzed synthesis of structured phenolic lipids in solvent-free system using flaxseed oil and slected phenolic acids as substrates. J. Biotechnol. 158, 128–136.

Sorour, N., Karboune, S., St-Louis, R., Kermasha, S., 2012b. Enzymatic synthesis of phenolic lipids in solvent-free medium using flaxseed oil and 3,4-dihydroxyphenyl acetic acid. Process Biochem. 47, 1813–1819.

Stamatis, H., Sereti, V., Kolisi, F.N., 2001. Enzymatic synthesis of hydrophilic and hydrophobic derivatives of natural phenolic acids in organic media. J. Mol. Catal. B: Enzym. 11, 323–328.

Stepankova, V., Bidmanova, S., Koudelakova, T., Riokpot, Z., Chaloupkova, R., Damorsky, 2013. Strategies for stabilization of enzymes in organic solvents. ACS Catal. 3, 2823–2836.

Sun, S., Song, F., Bi, Y., Yang, G., Liu, W., 2012b. Solvent-free enzymatic transesterification of ethyl ferulate and monostearin: Optimized by response surface methodology. J. Biotechnol. 164, 340–345.

Sun, J., Yu, B., Curran, P., Liu, S.Q., 2012a. Lipase-catalysed transesterification of coconut oil with fusel alcohols in a solvent-free system. Food Chem. 134, 89–94.

Takoua Debeche, T., Marmeta, C., Kiwi-Minskerb, L., 2005. Structured fiber supports for gas phase biocatalysis. Enzym. Microb. Technol. 36, 911–916.

Tamborini, L., Fernandes, P., Paradisi, F., 2017. Flow-bioreactors as complementary tools for biocatalytic process intensification. Trends Biotechnol. 36, 73–88.

Thomas, M.F., Li, L.-L., Handley-Pendleton, J.M., van der Lelie, D., Dunn, J.J., Wishart, J.F., 2011. Enzyme activity in dialkyl phosphate ionic liquids. Bioresour. Technol. 102, 11200–11203.

Tong, A.-M., Xu, J., Lub, H., Lin, W.Y., -Q, G., 2005. Construction and optimization of a monophasic organic–water system for enzymatic synthesis of *p*-nitrobenzyl-d-glucopyranosides by reverse hydrolysis. J. Mol. Catal. B: Enzym. 32, 83–88.

Tzialla, A.A., Kalogenris, E., Gournis, D., Sanakis, Y., Stamatis, H., 2009. Enhanced catalytic performance and stability of chloroperoxidase from *Caldariomyces fumago* in surfactant free ternary water–organic solvent systems. J. Mol. Catal. B: Enzym. 51, 24–35.

Ulijn, R.V., Halling, P.J., 2004. Solid-to-solid biocatalysis: Thermodynamic feasibility and energy efficiency. Green Chem. 6, 488–496.

Valivety, R.H., Halling, P.J., Macrae, A.R., 1992. Reaction rate with suspended lipase catalyst shows similar dependence on water activity in different organic solvents. Biochim. Biophys. Acta 1118, 218–222.

van den Broek, L.A.M., Boeriu, C.G., 2012. Enzymatic synthesis of oligo- and polysaccharide fatty acid esters. Carbohydr. Polym. 93, 65–72.

van Rantwijk, F., Sheldon, R.A., 2007. Biocatalysis in ionic liquids. Chem. Rev. 107, 2757–2785.

Vitola, G., Mazzei, R., Poerio, T., Barbieri, G., Fontananova, E., Buning, D., Ulbricht, M., 2019. Influence of lipase immobilization mode on ethyl acetate hydrolysis in a continuous solid-ga\z biocatalytic membrane reactor. Bioconjugate Chem. 30, 2238–2246.

von Weymam, F.N.W., Kivihariu, K.J., Jaaskelainent, S.T., Leisola, M.S.A., 2008. Scale-up of a new bacterial mannitol production process. AIChE 19, 815–821.

Vossenberg, P., Beeftink, H.H., Nuijens, T., Quaedflieg, P.J.L.M., Cohen Stewart, M.A., Trampaer, J., 2012. Performance of alcalase formulations in near dry organic media: Effect of enzyme hydration on dipeptide synthesis. J. Mol. Catal. B: Enzym. 78, 24–31.

Wang, S., Meng, X., Zhou, H., Liu, Y., Secundo, F., Liu, Y., 2016. Enzyme stability and activity in non-aqueous reaction systems: A mini review. Catalysts 6, 32, (Doi: org/10.3389/catal6020032).

Wangikar, P.P., Carmichael, D., Clark, D.S., Dordick, J.S., 1996. Active-site titration of serine proteases in organic solvents. Biotechnol. Bioengineer. 50, 329–335.

Weber, N., Mukherjee, K.D., 2004. Solvent-free lipase-catalyzed preparation of diacylglycerols. J. Agric. Food Chem. 52, 5347–5353.

Wehtje, E., Adlercreutz, P., 1997. Water activity and substrate concentration effects on lipase activity. Biotechnol. Bioengineer. 55, 709–806.

Xu, J.C., Li, W.M., Zheng, H., Lai, Y.F., Zhang, P.F., 2011. One-pot synthesis of tetrahydrochromene derivatives catalyzed by lipase. Tetrahedron 67, 9582–9587.

Yadav, G.D., Thorat, P.A., 2012. Microwave assisted lipase catalyzed synthesis of isoamyl myristate in solvent-free system. J. Mol. Catal. B: Enzym. 83, 16–22.

Yan, J.Y., Yan, Y.J., Yang, J.K., Xu, L., Liu, Y., 2009. Combined strategy for preparation of a bioimprinted *Geotrichum* sp. lipase biocatalyst effective in non-aqueous media. Process Biochem. 44, 1128–1132.

Zaharri, N.D., Othman, N., Ishak, Z.A.M., 2012. Thermal and mechanical properties of zeolite filled ethylene vinyl acetate composites. Procedia Chem. 4, 95–100.

Zaks, A., Klibanov, A.M., 1985. Enzyme catalyzed processes in organic solvents. Proc. Natl. Acad. Sci. USA 82, 3192–3196.

Zaks, A., Klibanov, A.M., 1988. Enzymatic catalysis in nonaqueous solvents. J. Biol. Chem. 263, 3194–3201.

Zhao, H., 2010. Methods for stabilizing and activating enzymes in ionic liquids-a review. J. Chem. Technol. Biotechnol. 85, 891–907.

Zoumpanion, M., Karali, M., Xenakis, A., Stamatis, H., 2006. Lipase biocatalysisin surfactant free microemulsion-like ternary systems and related organogels. J. Mol. Catal. B: Enzym. 51, 24–35.

Zoumpanion, M., Stamatis, H., Xenakis, A., 2010. Microemulsion-based organogels as matrices for lipase immobilization. Biotechnol. Adv. 28, 395–406.

CHAPTER EIGHT

Safety assessment and regulation of food enzymes

Gregory S. Ladics[a], Liuping Fan[b], Vincent J. Sewalt[c], Armin Spök[d]
[a]DuPont Nutrition & Bioscience, Wilmington, DE, United States
[b]Jiangnan University, Wuxi, China
[c]DuPont Nutrition & Bioscience, Palo Alto, CA, United States
[d]Graz University of Technology, Graz, Austria

8.1 Introduction

The industrial global enzymes market is a 3–3.5 billion-dollar industry with an annual average growth of 6%–7%. Food and feed applications account for 55%–60% of the global enzyme market with the market still growing at an estimated 6%–8% annually. The annual sale of food enzymes is approximately $440 million [Association of Manufacturers and Formulators of Enzyme Products (AMFEP), 2019, personal communication]. Globally, approximately 300 enzyme preparations are manufactured for use in the food industry with the majority produced from genetically engineered microorganisms (https://ec.europa.eu/food/safety/food_improvement_agents/enzymes/eu_list_app_en). Most enzymes are utilized in the baking and beverage (e.g., fruit juices, beer, wine), industries, as well as in starch processing and the production of dairy products.

Enzymes are proteins with highly specialized catalytic functions that occur in all living organisms. Like other proteins, enzymes are composed of amino acids; however, they differ in function in that they have the unique ability to facilitate biochemical reactions without undergoing change themselves. They are essential biological molecules that catalyze specific biochemical reactions necessary to support life. The specific reaction that each enzyme catalyzes is the basis for their classification by the International Union of Biochemistry and Molecular Biology (IUBMB system; https://www.qmul.ac.uk/sbcs/iumb/). Enzymes with a given designated activity can be from a variety of different organisms and have different amino acid sequences but a conserved active (catalytic) site sequence. These highly selective proteins play a vital role in physiological functions varying from food digestion to DNA synthesis. Enzymes are ubiquitous in fresh and processed food and are consumed by humans and animals every day in their active

Enzymes: Novel Biotechnological Approaches for the Food Industry.
http://dx.doi.org/10.1016/B978-0-12-800217-9.00008-3

state (in raw food) or as inactive proteins (in cooked or processed foods). The human body uses thousands of enzymes to perform numerous biochemical processes such as the digestion of carbohydrates, fats, and proteins. Enzymes, whether ingested or secreted endogenously, are broken down into peptides and amino acids, which are then entirely metabolized by the human body.

Since ancient times, enzymes found in nature have been used in the production of fermented foods and drinks such as cheese, vinegar, beer, and wine; leavening of bread; and meat tenderizing (e.g., juice of the papaya to soften meat). Today, nearly all commercially prepared foods contain at least one ingredient that has been made with enzymes. Some typical applications include enzyme use in the production of bakery products, sweeteners, alcoholic beverages, cheese, dairy products, egg products, soft drinks, vegetable oil, candy, and spices. Enzymes also play a significant role in nonfood applications such as laundry and dishwashing detergents, pulp and paper manufacturing, contact lens cleaner, textile processing, fuel production from renewable resources, leather tanning, de-inking of waste-paper, and yarn treatment.

Commercial enzymes are obtained from three primary sources: animal tissue, plants, and microbes. The production of enzyme preparations isolated from natural sources for use in food processing dates to the 19th century when in 1833, Anselme Payen and Jean-François Persoz, chemists at a French sugar factory, isolated "diastase," a group of enzymes that catalyzes the breakdown of starch into maltose (Armstrong, 1933). Today, "diastase" refers to any α-, β-, or γ-amylase (all of which are hydrolases) that can break down carbohydrate (Oliver, 2011). This was followed in 1874 by the Danish scientist, Christian Hansen, who extracted rennin (chymosin) from calves' stomachs for use in cheese manufacturing (Nielsen, 1994). Naturally occurring enzymes, however, are often not readily available in enough quantities for food applications or industrial use. Further, enzymes traditionally isolated from culturable microorganisms, plants, and mammalian tissues are often not well-adapted to the conditions used in modern food production methods such as high temperatures or low or high pH levels. There are also a limited number of available strains capable of growing on an industrial scale and such strains are limited in their ability to improve their expression of enzymes. Until the mid-1970s, availability of enzymes has been limited to the quantities produced in the host organism in which they were naturally derived. However, modern biotechnology techniques, often referred to as genetic engineering, developed since the mid-1970s have allowed the production of safe and efficient microbial hosts in which large quantities of commercially useful enzymes can be produced. More recently, these modern techniques have led to the development of tailored enzymes with optimized functional properties specific for their intended use. Most commercial enzymes currently available are manufactured from microorganisms that are enhanced through recombinant DNA technologies including, most recently, the use of synthetic DNA.

The first part of this review will provide background on industrial microbial enzymes and their uses, focusing on food and grain processing. The concept of safe strain lineage (SSL) as it relates to production (host) microorganisms is discussed in Section 8.2. The potential safety concerns with enzymes are then described, particularly as they relate to respiratory allergy in the occupational setting, in the third section. Safety regulations for both workers and consumers in the context of global regulatory frameworks and the agencies responsible for food enzyme safety evaluation and market authorization are discussed in the fourth part. Lastly, future challenges to industry and regulators pertaining to enzymes are summarized.

8.2 Modern biotechnology techniques and industrial enzyme production

The use of recombinant DNA technology to generate genetically engineered microorganisms (GEMs) has made it possible to manufacture novel enzymes suitable for specific food-processing conditions. Enzymes from GEMs are often identical to well-known enzymes with a history of safe use in food. The development of modern highly efficient screening techniques facilitated discovery of new enzymes from microorganisms sampled from the environment. With this approach, DNA isolated directly from the environmental samples was used to create expression libraries in *E. coli* or other suitable expression hosts (Olempska-Beer et al., 2006). When used by scientists to transfer genetic material between cells from the same or different species, microorganisms such as yeasts, bacteria and fungi with new or improved properties for industrial applications were developed. Using tools of modern biotechnology, modifications can be made more precisely compared to classical improvement techniques, with much less chance of developing unwanted secondary changes that could potentially lead to undesired effects. Given the drastically reduced cost of not only DNA sequencing but also DNA synthesis, it is now possible to design and order a synthetic gene encoding an enzyme and express it in host microorganisms that are adapted to large-scale industrial fermentation. The availability of synthetic DNA in combination with high throughput screening of candidate proteins has greatly facilitated the optimization of enzymes adapted to specific reaction conditions (e.g., pH, temperature, salt concentration) by protein engineering. With careful selection of host microorganisms, recombinant production strains can also be constructed to allow efficient production of the desired heterologous enzyme(s), remove production of unwanted enzymes such as extracellular proteases by employing the targeted knock out of host strain genes, and reduce or eliminate their potential for production of toxic secondary metabolites. By employing efficient promoters and introducing multiple copies of the enzyme-encoding gene, enzyme yield is significantly increased. This has allowed the commercialization of a wide diversity of enzymes found in nature. These include the commercialization of enzymes from unculturable microbes and microbes with enzyme yields too low to be economically viable. Of note, the chymosin first isolated by Christian Hansen in 1874 is today produced from microorganisms that contain the bovine pro-chymosin gene introduced through recombinant DNA techniques. Bovine chymosin expressed in *Escherichia coli* K-12 became the first recombinant enzyme approved for use in food by the US Food and Drug Administration (FDA) (Flamm, 1991).

Industrial production of recombinant enzymes involves several different stages: (1) establishment of a host (recipient) strain; (2) building of an expression vector; (3) transformation of the host strain; (4) identification of the top recombinant strain; and (5) characterization of the production strain. By using the tools of modern biotechnology, the enzyme industry has developed safe host organisms for producing numerous enzymes. These safe host organisms have been utilized since the early 1980s for producing different enzymes in contained manufacturing facilities. Most host strains used to develop production strains for food-processing enzymes have been derived from a relatively small number of bacterial and fungal species that include bacteria (e.g., *Bacillus subtilis*, *B. licheniformis*, *B. amyloliquefaciens*), fungi (*Trichoderma reesei*, *Aspergillus oryzae*, or *A. niger*), and yeast (*Saccharomyces cerevisiae*). These microorganisms have a long history of use as safe sources of native enzymes and an established record of proficient growth under large-scale industrial production conditions. They are also conducive to genetic modifications and recognized for their ability to secrete abundant quantities of enzymes directly into the fermentation media. The wild-type strains of several host microorganisms may produce several extracellular enzymes. These enzymes may appear in the final enzyme preparation and produce undesirable reactions in food or in the case of extracellular proteases, degrade heterologous enzymes and reduce yield. To increase enzyme yield and minimize potential for survival in the environment, protease-deficient and sporulation deficient host strains, have been constructed using modern biotechnology techniques (Olempska-Beer et al., 2006).

To produce recombinant production strains, genes encoding the selected enzyme proteins are introduced into the host strains noted above using expression vectors. An expression vector is a DNA plasmid that carries the expression cassette. Expression cassettes contain the essential genetic elements that must be present for expression of the target gene in the production microorganism. Essential components of the expression cassette include a promoter sequence, the gene encoding the desired enzyme (often with a signal peptide sequence to facilitate protein secretion), and a terminator sequence. The promoter and terminator are regulatory sequences that control the transcription of the enzyme-encoding gene in the host strain. Expression vectors may also contain DNA derived from bacterial plasmids. Typically, the most commonly used plasmids are pUB110, and the constructed plasmids pUC18, and pUC19. Plasmid pUB110 was originally isolated from *Staphylococcus aureus*. The plasmid replicates in Bacilli and has been used in the construction of expression vectors for production of enzymes in *B. subtilis* and other *Bacillus* species. (Keggins et al., 1978;). The pUC18 and pUC19 plasmids were developed for cloning in *E. coli* (Yanish-Perron et al., 1985).

The technique used to transfer vector DNA into host cells varies based on the properties of the host strain and the expression vector. For enzyme production, bacterial hosts have been transformed with vector DNA using conjugation, electroporation, DNA uptake by competent cells (known as bacterial transformation), or vector incubation with protoplasts, that is, cells from which cell walls were chemically removed. Filamentous fungi and yeasts are typically transformed by incubating DNA with protoplasts. To select transformed cells, expression vectors used in enzyme production contain a gene(s) that expresses a selectable marker. Although in the past the use of antibiotic resistance markers in microbial strains were common, these are increasingly being replaced with nutritional or metabolic markers.

8.2.1 Manufacture of microbial-derived enzymes

The manufacturing process of microbial enzymes comprises large-scale, strictly controlled (i.e., temperature, pH, nutrients and aeration) fermentation to yield high volumes of the various production strains (Fig. 8.1).

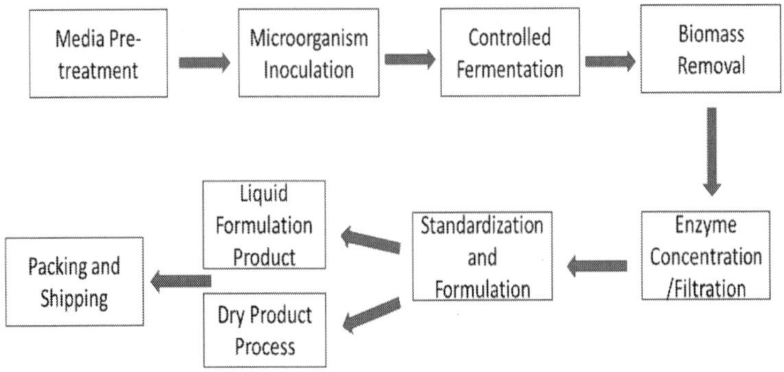

Figure 8.1 General diagram for the manufacturing process of enzymes.

Fermentation media contain nutrients and compounds that facilitate the fermentation process. Commonly used media components include dextrose, starch, corn steep liquor, ammonia, urea, soybean meal, yeast extract, and minerals such as phosphates, chlorides, or carbonates. Additional additives may include antifoaming agents and acid or alkali for pH adjustment. Enzymes either accumulate inside the cells or are secreted into the media of the fermentation tanks. In subsequent steps, cells may be lysed and the disrupted cells (or the media including the enzymes) are subjected to further refinement and purification processes using a variety of chemical, mechanical and thermal techniques (concentration, precipitation, extraction, centrifugation, filtration, and chromatography). Absence of viable production organisms is a product specification for food enzyme preparations. As the enzyme preparation does not contain the microorganism used to produce it, it is referred to as being produced *with* a microbial production organism rather than *from* the microorganism. Enzymes used in food processing and other applications are sold commercially as enzyme preparations. An enzyme preparation typically contains the enzyme of interest and additional soluble components from the fermentation (e.g., small microbial metabolites and residual amino acids and residual raw materials from the fermentation media), collectively referred to as Total Organic Solids (TOS) and several added substances such as diluents, preservatives, and stabilizers (e.g., potassium sorbate, sodium citrate, sodium sulfate, sodium chloride, sorbitol, and water). These substances are safe for use in food with little to no associated toxicity. The raw materials used in fermentation media are expected to be of appropriate purity consistent with current good manufacturing practice (cGMP) (Chaplin and Bucke, 1990; Ullmann's Encyclopedia of Industrial Chemistry, 2000). The final enzyme preparations are usually commercially marketed in granular or liquid forms. The final formulated enzyme product is assessed for compliance with specifications established for enzyme preparations in the Food Chemicals Codex (US Pharmacopeia, 2020) and JECFA (2006).

8.3 Safe strain lineage (SSL) concept

The safety of enzyme production strains continues to be a focus of attention and remains the key consideration in evaluating enzyme safety (Pariza & Johnson, 2001, 2010). The concept of developing safe strain lineages using well-characterized nonpathogenic, nontoxigenic microbial strains, particularly those with a history of safe use in food enzyme manufacture, has been adopted by the enzyme industry. As indicated above, most host strains used to develop production strains for food-processing enzymes have been derived from a relatively small number of bacterial and fungal species. Demonstrating the safety of the production strain, particularly its pathogenic and toxigenic potential, is the most critical factor in evaluating the safety of an enzyme preparation. In principle, the same safety considerations apply to enzymes derived from native and recombinant microorganisms. As described by Sewalt et al. (2016), there are two basic approaches to establish that the host strain is safe: (1) by ensuring that any newly isolated strain for enzyme production or to serve as host strain is non-pathogenic (i.e., they do not cause disease in healthy humans and animals) and non-toxigenic (i.e., "one that does not produce injurious substances at levels that are detectable or demonstrably harmful under ordinary conditions of use or exposure") or (2) by establishing a safe strain lineage for expression of desired enzyme sequences by genetic engineering techniques (Pariza and Foster 1983; Pariza and Johnson, 2001; Pariza and Cook, 2010). For human health, a literature search can be conducted to determine if the host strain has caused disease in healthy individuals. If data is lacking, then animal studies using live organisms can be conducted to demonstrate the failure to cause disease. More recently, whole genome sequence analysis has been deployed to determine if the strain is nontoxigenic by assessing the presence of toxin producing genes. A lack of toxin-encoding genes supports the conclusion of non-toxigenicity. Although neither pathogenic nor toxigenic microorganisms are intentionally used in the production of food processing enzymes, certain fungi (e.g., *A. oryzae* or *A. niger*) traditionally used as host strains for production of enzymes may produce low levels of toxic secondary metabolites, such as mycotoxins, under fermentation conditions favorable to the synthesis of these compounds (Blumenthal, 2004). Mycotoxin production by these fungi is usually minimal in single-species culture and can be further minimized or avoided by controlling the fermentation conditions and substrate selection for use in the production of enzymes. Further, through recombinant DNA methods, mycotoxin synthetic pathways can be inactivated or impaired by deletion or disruption of genes encoding enzymes involved in the mycotoxin synthesis pathway. In addition, enzyme preparations are frequently analyzed for pertinent toxins/secondary metabolites that are known to be produced by the species to which the production organism belongs or closely related species. If a new production organism is found to produce a toxin, animal studies may be conducted with relevant dosage levels to demonstrate whether the level of toxins produced are a concern relative to the exposure in the intended use. If the organism is non-toxigenic and non-pathogenic, then it is assumed that common foods or food ingredients produced from the organism, using current GMPs, are safe to consume (IFBC, 1990).

The conduct of repeated toxicological testing with enzymes derived from the same strain lineage (company specific lineages of, e.g., *T. reesei; B. licheniformis, B. subtilis, A. niger* strains) with no adverse effects allows for the establishment of a Safe Strain Lineage (SSL) (Pariza and Johnson, 2001; Pariza and Cook, 2010; Sewalt et al., 2016; Ladics and Sewalt, 2018). A SSL can be established by repeated assessment of members of the lineage using the Pariza and Johnson (2001) decision tree, which involves addressing the potential for antimicrobial activity and toxic microbial metabolites and includes a repeat dose 90-day oral toxicity study to produce a No Observed Adverse Effect Level (NOAEL) (discussed in further detail below) to determine the margin of safety (MOS) in the intended use (Fig. 8.2).

Once the SSL has been established, the decision tree allows for any strains within the lineage that are upstream from the production strains already evaluated toxicologically to be supported as safe production hosts for other enzymes, and any new enzyme preparations from the new strains in the lineage would also be considered safe, predicated on the use of safe recombinant DNA techniques and DNA sequence confirmation of the new organism. If a new branch of the lineage is developed based on random mutagenesis without conclusive whole genome sequence analysis, then it would be advised to re-establish the safety of that new branch with targeted toxin analysis (aimed at detecting toxic metabolites known to be produced by the species or genus and/or toxicology studies. The SSL concept can be applied to any new enzyme belonging to the established SSL, if that enzyme

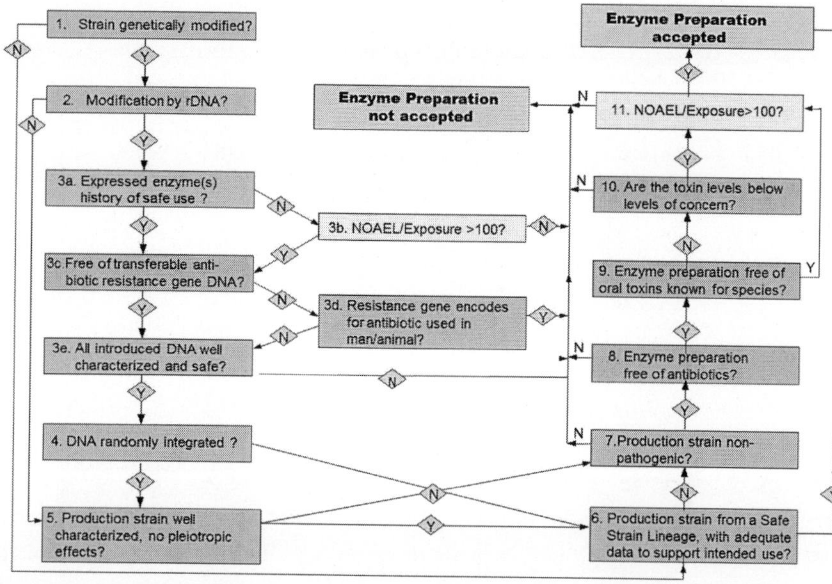

Figure 8.2 Enzyme Evaluation Decision Tree, adapted from Pariza and Johnson (2001). This decision tree serves to qualify new or existing enzymes from traditionally improved production strains or new or existing enzymes expressed in new or existing production hosts developed with recombinant DNA (modern biotechnology). The text of question 6 is updated to indicate that the available enzyme safety data for another member of the Safe Strain Lineage (SSL) needs to be examined to verify it can support exposure in the intended use. As explained in Pariza and Johnson (2001), an SSL can be established by repeated analysis using the decision tree, which generally involves oral toxicity studies. Once the SSL has been established, examination of the safe use of another food enzyme expressed with safe recombinant techniques in the same SSL can rely on the pre-existing safety data. However, the use of random mutagenesis to establish a new branch of an existing SLL requires reconfirmation of its safety, which may be accomplished by examining a whole genome sequence, by in depth analytical confirmation of absence of toxins, and preferably an oral toxicity study.

has a history of safe use, and the other pertinent questions in the Pariza and Johnson (2001) decision tree pertaining to the use of recombinant DNA techniques are satisfactorily addressed. In case a new enzyme without any history of safe use is produced by a member of an established SSL, then additional toxicology data may be needed. Therefore, the use of the SSL concept greatly decreases the need for systemic toxicity studies, resulting in reduced animal usage, while supporting any continued need in formal safety assessments to determine margins of safety (Ladics and Sewalt, 2018). The enzyme industry has introduced the concept successfully to various regulatory agencies (Sewalt et al., 2016) and in addition to its use in GRAS

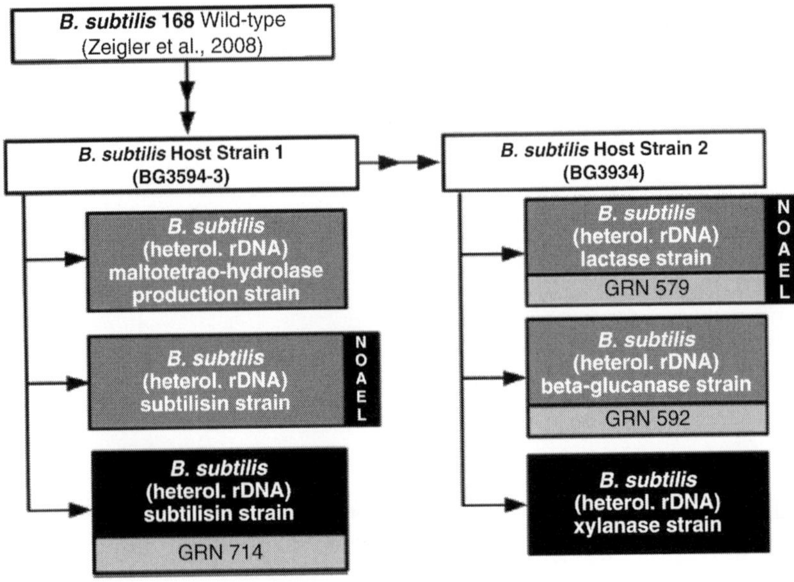

Figure 8.3 DuPont safe strain lineage for *Bacillus subtilis* (Ladics and Sewalt, 2018).

determinations, bridging data based on the SSL concept or similar considerations have been accepted, on case-by-case basis, by food authorities in Australia, Brazil, Canada, the EU and Japan.

An example of SSL is presented in Fig. 8.3, which shows three branches of a *Bacillus subtilis* lineage, each supported by at least two sets of toxicological studies. The existence of this SSL has been used to support the safety of other enzymes, such as another subtilisin enzyme produced by another member of the same SSL. The safety of this enzyme was notified to the US FDA in GRN 714 on that basis (Ladics and Sewalt, 2018).

8.4 Enzyme safety

Enzymes are proteins and proteins do not typically represent an oral hazard. Those proteins that do present a hazard are well-defined and include: plant toxins (e.g., ricin), animal toxins (e.g., prions), allergens (e.g., gluten protein, Ara H2), antinutrients (e.g., trypsin inhibitors, lectins), and pathogenic bacterial toxins (e.g., shiga toxins, botulinum) (Norton, 2001; Sandvig et al., 2010; Masilamani et al., 2012). The existence of many hazardous proteins has been documented in several publicly available databases (Zhou et al., 2007; He et al., 2008; Herzig et al., 2011; Jungo, 2012; Goodman et al., 2016). During product development, the amino acid sequence of each enzyme is screened for identity to protein toxins and allergens using in silico tools and protein databases such as AllergenOnline (http://www.allergenonline.com) and NCBI (http://www.ncbi.nlm.nih.gov) (Ladics, 2008; Ladics et al., 2011; Goodman et al., 2016; Ladics, 2018). Enzymes have also been used for decades in the treatment of both adults and children with exocrine pancreatic insufficiency. Typical enzymatic drugs (e.g., Creon® from Solvay Pharmaceuticals or Pancreas Microtabs® from Janssen/Cilaq) are a combination of the enzyme's alpha-amylase, lipase and protease, which are also used in a wide range of industrial applications. These medicinal drugs are typically administered orally at therapeutic concentrations (i.e., at concentrations where a digestive effect can be expected). Clinical trials and crossover studies verify the safe use of these compounds in patients, both in adults and children, confirming the low toxicity of the enzymes (Laake, 1980; Halm et al., 1999; Patchell et al., 2002; Konstan et al., 2004; Domínguez-Muñoz et al., 2005; Borowitz et al., 2006a,b; Keller and Layer, 2006; Saeed et al., 2007; Heubi, 2008; Konstan et al., 2008). In addition, enzymes have been widely used as digestive aids since the late 19th century. Such over-the-counter enzymes include: amylase, lipase, cellulase, invertase, papain, pepsin, bromelain, lactase, superoxide dismutase, and pancreatin (Walsh, 2005). There are few reported side effects and adverse events associated with over-the-counter digestive enzyme supplements. Those taking digestive enzymes may experience gastrointestinal upset. For the most part, those taking the amounts listed on the packages usually do not have any serious problems because over-the-counter digestive enzymes are relatively safe (Roxas, 2008; Varayil et al., 2014).

8.4.1 Food allergy potential

Proteins are an essential part of the diet (average human consumption of 70–100 g/day) (US Department of Agriculture (2008)). Since enzymes are broken down by gastric juices, digestive enzymes, and pancreatic proteolytic enzymes in the lumen of the gastrointestinal tract into simpler components such as small peptides and amino-acids, it is unlikely that exposure to enzymes commonly used in food processing would result in absorption of enzymes in toxicologically significant amounts through the gastrointestinal tract. The long-term history and well-defined presence of enzymes (microbially, plant, and animal sourced) in human food carries with them *prima facie* evidence of safety. Moreover, there are no documented cases where enzymes represent a hazard via oral exposure, and thus they are considered intrinsically safe via this route of exposure (Olempska-Beer et al., 2006;

Ladics and Sewalt, 2018). Furthermore, industrial enzymes derived from fungal, bacterial, or yeast sources used in the food industry are not known to produce food allergic reactions. Food enzymes are typically used as processing aids and are generally not present as active ingredients in consumer products. Bindslev-Jensen et al. (2006) evaluated 19 enzymes (e.g., alpha-amylases, proteases, decarboxylase, glucoamylase, lipases, glucose oxidases, pectin esterase, beta-glucanase, laccase, xylanase, and pectin lyase) used in the food industry for allergenicity. None of the 19 enzymes were found to be food allergens, and the authors concluded that ingestion of food enzymes in general is not considered to be a concern regarding food allergy. In addition, as noted above, a bioinformatic analysis comparing the amino acid sequence of the enzyme to databases of known allergens (e.g., AllergenOnline) was conducted to determine whether the enzyme is a known allergen or potentially cross-reactive to a known allergen (Ladics et al., 2011; Ladics, 2018). Taken together, these data provide further confidence that there are no food allergy concerns with the use of industrial microbial enzymes in the manufacture of foods or beverages.

8.4.2 Skin/eye irritation and respiratory allergy potential

Prolonged contact with high concentrations of proteolytic enzymes, such as proteases, has the potential for producing skin and eye irritating effects. This occurs only in cases where raw, powdered enzyme is handled and does not typically apply to encapsulated proteases intended for human ingestion. In addition, there is the well-documented potential for enzymes to produce respiratory sensitization (Type I hypersensitivity mediated by specific IgE antibodies to the enzyme) in the case of work place exposure to dusts or liquid aerosols (Pepys et al., 1969; Zachariae et al., 1981; Andersen et al., 1998; Cullinan et al., 2000; Schweigert et al., 2000; HERA 2005, 2007; Bernstein and Sarlo, 2006; Basketter et al., 2008; Basketter et al., 2012a,b). This is generally true for all enzymes regardless of their end-use. Protein induced respiratory allergy is a difficult endpoint to measure for several reasons: (1) it is a two-phase process (induction and elicitation); (2) there is no observed dose-response (difficult to identify effect limits); (3) individuals may react to different levels at induction and elicitation phases; (4) individual responses range in severity; and (5) there is currently no validated and predictive test to predict allergenicity de novo, discriminate allergenic from nonallergenic respiratory proteins, or rank-order the potency of identified protein allergens (Holsapple et al., 2006). Therefore, all enzymes are considered respiratory sensitizers and labeled accordingly. With good industrial hygiene practices, air and health monitoring, and the establishment of strict limits on air exposure to enzymes (e.g., 60 ng active enzyme protein/m^3), however, respiratory sensitization cases in the workplace can be minimized or eliminated. Outside of occupational risks, there is negligible evidence of any allergic risks to the final consumers of enzyme products. Beyond these limited cases, enzymes are intrinsically safe proteins with a low degree of toxicity and an exceptional safety record. Acute inhalation, reproduction, chronic toxicity, and carcinogenicity are not relevant for enzymes (European Commission, 2002). In agreement with the European Commission, no proteins (or enzymes) have been identified in the peer-reviewed scientific literature as reproductive, developmental, or carcinogenic compounds.

8.4.3 Dermal sensitization potential

While data support the position that industrial enzymes may be respiratory sensitizers in an occupational setting and are not likely to be food allergens, there has been some debate as to whether enzymes may produce dermal sensitization (Spök, 2006). Dermal sensitization is indicated as allergic contact dermatitis (ACD) in a clinical setting and is a cell-mediated type-IV delayed hypersensitivity. The skin sensitization potential of enzymes has been reviewed in several publications, indicating that enzymes should not be considered dermal sensitizers. (Andersen et al., 1998; HERA 2005, 2007; Basketter et al., 2008; Basketter et al., 2012a,b). While there are some reports of certain proteins inducing ACD, their ability to produce ACD appears to be quite low. A few cases of contact dermatitis have been reported in occupational settings in response to irritating enzyme preparations (*e.g.*, proteases), but this is considered a non-immunologic phenomenon (also known as irritant contact dermatitis) unrelated to ACD. Further, contact urticaria (also known as protein contact dermatitis) has been reported in occupational settings, but this is also a non-immunologic event, unrelated to ACD (Hjorth and Roed-Peterson, 1976; Smith et al., 1989; Tarvainen, 1991; Morren et al., 1993; Kanerva and Vanhanen, 1999, 2001; Basketter et al., 2008, 2015; Basketter and Raulf, 2018). The cellular mechanisms involved with ACD have been previously reviewed (Scheper and von Blomberg, 1992; Kimber, 1994, Kimber et al., 2002; Kimber and Dearman, 2002). To cause dermal sensitization, a substance must initially penetrate the *stratum corneum*, enter the epidermis and subsequently react with proteins, likely on the surface of the Langerhans cells, to form a hapten-carrier conjugate. To fulfill these requirements, a substance must be of low molecular weight (normally less than 400 Daltons) and must be capable of reacting directly or indirectly with (cell surface) protein and produce a new antigenic determinant. It is unlikely that an enzyme can be absorbed through the skin because of its high molecular weight (about 20,000–100,000 kDa) (Smith Pease et al., 2002). Thus, enzyme proteins are not able to act as classic haptens that are associated with cell-mediated delayed hypersensitivity (Divkovic et al., 2005). This is consistent with the clinical experience with enzymes used in detergent products which demonstrate an absence of dermal sensitization potential. Over a 50-year period, consumers have had skin exposure to enzymes; however, there is no evidence that this exposure has produced ACD. The lack of dermal sensitizing potential of enzymes is corroborated by evidence from controlled clinical studies together with observations attained from occupationally exposed groups (Griffith et al., 1969; Pepys et al., 1973; Zachariae, 1973; Bannan et al., 1983; White et al., 1985; Rodriguez et al., 1994; Andersen et al., 1998; Peters et al., 2001; Belsito et al., 2002; Lee et al., 2002; Cormier et al., 2004). Moreover, Basketter and Kimber (2018) indicate that evaluating enzymes in predictive animal assays previously established and validated to predict the dermal sensitization potential of low molecular weight chemicals can lead to false positive responses with enzymes. This is likely attributed to the use of adjuvants, test article occlusion, irritant dose levels, solvent based vehicles, and intradermal injections (Bergman and Broadmeadow, 1997; Festersen, 2008; Basketter et al., 2012a). Such assays include the murine local lymph node assay (LLNA) (Gerberick et al., 2005), the Buehler occluded patch test (Buehler, 1965), and the guinea pig maximization test (Cronin and Basketter, 1994). Such models were not developed nor validated to evaluate proteins (ICCVAM, 2008). Basketter and Kimber (2018) further note that such positive results with enzymes in animal models developed and validated to predict the dermal sensitization potential of low molecular weight chemicals are "without any scientific justification or predictive merit."

8.5 Industrial microbial enzymes: A comprehensive safety assessment

A comprehensive safety assessment of industrial microbial enzymes used in food applications is typically required before their introduction to the market with European and other regulators globally, requiring that toxicological evaluation is conducted for approval or registration purposes. A remark on harmonizing terminology: The EU and Japan use "risk assessment," the US and Canada "safety evaluation," and China appears to use both terms when referring to food enzyme pre-market testing by applicants. Moreover, "Safety assessment" is also used by China. Enzyme preparations are tested according to generally accepted procedures discussed in several publications and guidance documents such as the OECD (IFBC, 1990; Scientific Committee on Food (SCF), 1992; Pariza and Johnson, 2001; FDA/CFSAN, 2010; EFSA, 2016). The test material is usually the concentrated enzyme before formulation. For example, a set of toxicological studies covering in vitro genotoxicity (i.e., a test for gene-mutation in bacteria and a test for chromosomal aberrations in human peripheral blood lymphocytes) and systemic toxicity (i.e., 90-day oral studies in rats) is required for inclusion of food enzymes in the Community positive list (European Commission, 2008; EFSA, 2009a,b). Similarly, these Toxicology studies are also required for registration purposes in other countries such as China and Brazil. Additionally, a bioinformatic analysis comparing the amino acid sequence of the enzyme to databases of known toxins and proteins (http://www.uniprot.org; http://www.uniprot.org/program/Toxins) is conducted to determine whether the enzyme is a protein toxin. Consequently, enzyme producers have subjected their products to extensive safety and toxicity testing that has been reviewed by global regulatory bodies as well as JECFA. The safety evaluation of food processing enzymes from recombinant microorganisms have been extensively discussed in the literature (IFBC, 1990; Jonas et al., 1996; Pariza and Johnson, 2001, Olempska-Beer et al., 2006; Sewalt, 2017; Ladics and Sewalt, 2018) and in guidance documents issued by regulatory authorities. According to Pariza and Johnson (2001), a 100-fold Safety Margin compared to the No Observable Adverse Effect Level (NOAEL) established in a 90-day oral toxicity study provides an acceptable level of safety for food use.

Because of various global regulatory requirements, the enzyme industry has conducted numerous genotoxicity and 90-day oral toxicity studies in rodents for registration purposes. Several hundred mutagenicity studies on bacterial (i.e., Ames studies) and mammalian cells (e.g., chromosomal aberration studies) using a variety of enzymes (e.g., amylase, glucanase, lipase, protease, xylanase, cellulase, catalase, glucoamylase, glucose isomerase, lactase, phytase, and pullulanase) derived from multiple species (e.g., *Trichoderma reesei; Bacillus licheniformis, Bacillus subtilis*) have been performed for product approvals and registrations (Pariza and Johnson, 2001; Pariza and Cook, 2010; Ladics and Sewalt, 2018; https://ec.europa.eu/food/sites/food/files/safety/docs/fs_food-improvement-agents_enzymes-applications.pdf). Several enzymes were also evaluated for genotoxicity in the in vivo Mammalian

Erythrocyte Micronucleus test. No positive findings were observed in any of the genotoxicity studies. Further, >225 90-day oral gavage or feeding studies with enzymes were performed and submitted to the European Food Safety Authority (EFSA) with no reported adverse findings, including on the bone marrow (https://ec.europa.eu/food/sites/food/files/safety/docs/fs_food-improvement-agents_enzymes-applications.pdf). The lack of adverse effects observed on the bone marrow in 90-day oral gavage or feeding studies is not unexpected. Our digestive systems have evolved to convert proteins, including enzymes, are digested into amino acids and peptides for incorporation into our bodies. As a result, parent enzymes would not be transported to the bone marrow following oral exposure. Repeat dose oral studies are typically conducted as the enzyme expressed as Total Organic Solids (TOS). TOS includes the enzyme itself, as well as other organic material that originated from the production organism and enzyme processing. TOS is derived experimentally by TOS (%) = 100 − % Ash − % Water − % diluents. The safe level derived from toxicological studies (i.e., the NOAEL) and the exposure in the intended use are expressed in mg TOS/kg body weight/day. These studies also confirm that any remaining microbial metabolites and fermentation materials, to the extent present in enzyme products as TOS (Pariza and Foster, 1983), also lack toxicity. This is an important observation, as microbial enzymes used in food processing are normally sold as enzyme preparations that not only contain a specific enzyme(s) but also other metabolites of the production strain. Additionally, this reflects that the safety of the enzyme production organism is the primary consideration in evaluating the safety of an enzyme preparation (Pariza and Johnson, 2001; Pariza and Cook, 2010).

Taken together, the weight-of-evidence of the above toxicological data, along with the numerous regulatory approvals for food enzymes around the world, demonstrate that enzymes are not mutagenic or clastogenic, nor toxic by repeat oral administration. Therefore, the continued routine practice of performing repeat dose oral toxicity and genotoxicity studies on enzyme preparations is questionable and establishing general health-based guidance values (e.g., ADI values) for enzymes may be considered. The derivation of health-based guidance values involves the identification of appropriate reference points using modeling analysis such as the Benchmark Dose and the establishment of uncertainty factors to be used in human risk assessment. An important criterion for our assertion that general health-based guidance values for common food enzymes be established is the need to select and use suitable non-toxigenic microbial production strains. Importantly, if the enzyme protein is well characterized, the production organism has been established as a safe strain, and the introduced DNA and rDNA techniques used to modify the organism are well established and appropriate for food use, it is highly unlikely that the enzyme preparation will pose any safety risks.

8.5.1 Enzyme exposure assessment

Exposure to enzyme products is minimal as recommended use levels are low, generally less than 0.2% (wt/wt) or 2000 parts per million. Enzymes are often either not carried over to food as consumed or are inactivated during cooking or baking. Therefore, any amounts of enzyme carried over into the final food would be at *de minimis* levels. Humans consume approximately 1.8 kg of food/day. If it is assumed that 10% of all food ingested/day contains 0.2% of a microbial derived enzyme product (worst-case), then the amount of microbial derived enzyme consumed would be 360 mg/day or 5 mg/kg bw/day. The average human consumption of protein in the diet is 70–100 g/day. Thus, of the 100 g/day of protein ingested, the 360 mg of enzyme would only represent 0.36%. For comparison, in 90-day oral gavage studies in rodents aimed at determining the maximum safe level, the NOAEL is typically ≥1000 mg/kg bw/day (i.e, the limit dose) of enzyme product (Ladics and Sewalt, 2018). Further, enzymes have an outstanding track record in the United States Generally Recognized as Safe (GRAS) Notification process, with 98% of enzyme GRAS notices receiving a "No Questions" response letter from the US FDA (Sewalt et al., 2016) and in the Joint FAO/WHO Expert Committee on Food Additive evaluation process, with the more than seventy-five completed food use enzyme evaluations all resulting in an acceptable daily intake (ADI) "not specified" (http://apps.who.int/food-additives-contaminants-jecfadatabase/search.aspx). JECFA defines ADI "not specified" as used to refer to "a food substance of very low toxicity which, on the basis of available data (chemical, biochemical, toxicological, and other) and the total dietary intake of the substance arising from its use at the levels necessary to achieve the desired effect and from its acceptable background in food, doesnot, in the opinion of the (JECFA), represent a hazard to health (Sewalt et al., 2016). Consequently, such enzymes can be used *quantum satis*.

A food process-based methodology is used to estimate dietary exposure to food enzymes. Exposure assessment of food enzymes relies on two inputs: (1) the enzyme uses levels in foods typically expressed in TOS and (2) the consumption data of the corresponding foods. There are several different methods for calculating human exposure to food enzymes from worst-case estimations to refined methods designed to evaluate actual exposure. The selection of the method typically depends on several factors, including the purpose of the assessment and the availability of food consumption data. Historically, due to the absence of appropriate food consumption data, the Budget method, originally developed for setting maximum use levels for food additives, was proposed as a screening tool in the assessment of food enzymes (Hansen, 1979; Douglas et al., 1997; EFSA, 2011b; Silano 2015). For the Budget method, the levels of consumption of foods and beverages considered are the maximum physiological levels of consumption. For non-milk beverages, the daily consumption of 0.1 L/kg bw is used and corresponds to 6 L/day for a 60 kg adult. For solid foods, levels of 100 kcal/kg and 50 kcal/kg are applied for children and adults, respectively. Using 60 kg as a default body weight for adults, the maximum lifetime energy intake of 50 kcal/kg bw/day for solid foods corresponds to 25 g/kg bw/day. In the absence of analytical or residue data, the concentration of enzyme in foods and beverages is the maximum application rate of the enzyme product. The proportions of solid foods and beverages that may contain the enzyme are set arbitrarily. A default that is often used for European assessments is 12.5% for solid foods and 25% for non-milk beverages. If the enzyme is used for a wide range of solid foods, then the default may be set at 25%. Therefore, the Budget method overestimates the theoretical maximum daily intake (TMDI) level of consumption of enzymes from foods and beverages.

The TMDI may also be based on national food consumption survey data provided by the US Department of Agriculture, 2008. The USDA provides United States specific consumption data for each subcategory, such as dairy products, juices, red meats, poultry, eggs, etc. (https://www.ers.usda.gov/data-products/food-availability-per-capita-data-system/). The data serve as proxies for actual consumption at the national level. The food availability data include estimates for over 200 commodities, including individual fruits, vegetables, grains, added sugars and sweeteners, dairy products, nuts, meat, poultry, and seafood. These consumption data, together with enzyme use levels in various foods, are used to calculate human exposure levels to enzymes. Similarly, EFSA has recently been developing a Comprehensive European Food Consumption Database (Comprehensive Database) which will be the source of consumption data in the EU going forward (https://www.efsa.europa.eu/en/food-consumption/comprehensive-database) (EFSA, 2011a). The database contains detailed data for various EU countries. The database plays a key role in the evaluation of the risks related to possible hazards in food in the EU and allows estimates of consumers' exposure to such hazards. EFSA's Food Enzyme Intake Model (FEIM) is a tool for estimating chronic dietary exposure to food enzymes used in various food processes (https://www.efsa.europa.eu/en/applications/foodingredients/tools). It has been developed on the summary statistics of food consumption data collected from Member States (stored in the Comprehensive Database). Each release uses the most recent consumption data from the Comprehensive Database. In applications submitted according to Regulation (EC) No 1332/2008, applicants provide recommended use levels of food enzymes and a description of intended conditions of use. The enzyme use levels are expressed at the raw material level, and the intended uses are described by giving examples of commonly consumed foods and/or food processes. FEIM comprises process-specific calculators, which allow estimation of dietary exposure to food enzymes used in individual food manufacturing processes. Exposure results are reported at mean and high level (e.g., 95th or 99th percentiles) for different population groups (e.g., infants, toddlers, adults, etc.) in different countries (European Commission, 1998; WHO, 2009). To date, EFSA has completed the FEIM-baking and FEIM-brewing tools for public use.

8.5.2 Enzyme margin of safety

Once the potential human exposure to the food enzyme has been estimated based on TOS, a safety margin can be calculated by using the dose level based on TOS at the NOAEL, typically obtained from a 90-day oral gavage or feeding study, divided by the estimated human exposure. There should be no adverse effect at a dose that is 100 times the estimated mean human exposure. This is established on the 100-fold safety factor applied to food ingredients by US FDA and other published methodology (US-FDA, 1997; 21 CFR 170.22, 2015). Recently, however, EFSA has applied a 200-fold safety factor to food ingredients, such as enzymes.

8.6 Regulation of food enzymes in the European Union

8.6.1 Harmonized statutory legislation

In the EU, the food enzyme legislation is its final phase of its transition from different national legislations of Member States to a fully harmonized system—this section, therefore, focuses on the EU regulation only. Information regarding national requirements still in force for the next few years may be found in Spök (2006) and websites by the national food safety authorities in Denmark and France. In addition, national requirements in Britain may evolve in connection with Britain's departure from the EU (aka Brexit).

Placing on the market of food enzyme in the EU requires approval according to Regulation 1332/2008 (European Parliament and the Council, 2008a), last updated by Regulation 1056/2012 (European Commission, 2012b) and Regulation 1331/2008 establishing a common authorization procedure for food enzymes, food additives, and food flavorings (European Parliament and the Council, 2008b). In terms of approval requirements, this legislation does not distinguish between food enzymes used as additives and those used as processing aids. In terms of labeling only enzymes that are functional in the final food product need to be specified on the food label (European Commission, 2011). Specific guidance to differentiate food enzymes used as processing aids and as functional ingredients is provided by a European Commission Guidance Document (European Commission, 2014). Excluded from the scope of harmonized legislation are enzymes intended for direct human consumption, for example, for nutritional or digestive purposes, and enzymes used in the production of food additives or processing aids. Microbial cultures traditionally used in the production of food (e.g., cheese, wine), which may incidentally produce enzymes but are not specifically used to produce them, are not considered food enzymes.

Applicants must conduct a risk assessment and compile the safety relevant information and information on the technological need into a dossier. For certain types of enzymes made with non-genetically engineered microorganisms, it is possible for applicants to establish joint dossiers (European Commission, 2012a); however, EFSA has recently clarified its requirement for strain-specific enzyme safety data even for joint dossiers. The technical and procedural requirements for the risk assessment and the application process are specified in the Implementing Regulation 234/2011 (European Commission, 2011) most recently amended by Commission Implementing Regulation 562/2012 (European Commission, 2012a). Procedural requirements are further detailed in a European Commission Guidance document (European Commission, 2018). Criteria for health safety evaluation and what is expected to be included in the dossiers are further detailed in an EFSA Guidance Document (EFSA, 2009a,b) and in an Explanatory Note (EFSA, 2014). An overview of the safety information required is provided in Table 8.1.

Microorganisms to which QPS status (Qualified Presumption of Safety; list provided in Ricci et al., 2017) has been assigned do not need to undergo

Table 8.1 Overview of safety relevant information recommended by EFSA to be included in EU food enzyme dossiers.

Topic	Endpoint	Type of information/ method if specified
Enzyme identification, source, manufacturing, application		
	Identification	• Enzyme names, internationally accepted specification, principal and side activities, chemical composition, stability
	Source and Manufacturing	• Microorganisms: taxonomic identity on strain level according to OECD taxonomic guidance • GMM: type of genetic modification, genetic stability; absence of production strain and DNA in the food enzyme; specific guidance provided in EFSA (2011) • Monitoring of production strain • History of use and possible QPS status • Non-QPS strains: pathogenicity, toxigenicity; bacteria: absence of antimicrobial resistance genes via e.g., whole-genome sequencing (if positive absence of DNA and viable production strain in the food enzyme) • Manufacturing process
	Application	• Reaction and fate in food • Technical need • Conditions of use
	Existing evaluations/ authorizations	
Toxicity assessment		
	Genotoxicity in vitro (2 of 3 studies required)	• Bacterial reverse mutation assay (OECD 471) • Mammalian micronucleus test (OECD 487) • Mammalian chromosomal aberration test (OECD 473)
	Genotoxicity in vivo (a Additional studies if 1 or 2 in vitro tests are positive)	• Transgenic rodent cell gene mutation assay (OCED 488) • Mammalian erythrocyte micronucleus test (OECD 474) • Comet assay (OECD 489)
	Systemic toxicity	• Repeated dose 90-day oral toxicity study in rodents (OECD TG 408)
Possible waivers	If derived from edible parts of non-GMO plants and animals not posing health problems, QPS microorganisms.Validity of studies of same or closely related microorganisms decided case-by-case.	

(*Continued*)

Table 8.1 Overview of safety relevant information recommended by EFSA to be included in EU food enzyme dossiers. (*Cont.*)

Topic	Endpoint	Type of information/ method if specified
Allergenicity assessment		
	Allergenicity of source organism	
	Homology with known allergens	
	In case of concerns	Specific guidance for a step-wise approach provided in EFSA (2006, 2011) and Naegeli et al. (2017) including, e.g., in-vitro digestibility tests and serum-screening
Exposure assessment		
	Exposure of consumers	Food Enzyme Intake model (FEIM): novel exposure assessment tool developed by EFSA, based on the EFSA Comprehensive European Food Consumption Database, food enzyme use and concentration, and the fate of the enzyme in the production process – see (EFSA, 2018; Silano et al., 2016)

The focus of this overview is on food enzymes derived from micro-organisms; QPS: qualified presumption of safety. Source: EFSA, 2019; 2014; 2011b; EFSA, 2009a,b, Naegeli et al., 2017, Silano et al., 2016; Tlustos and Liu, 2019).

a full risk assessment with strain specific toxicology data; however, qualification of individual strains under QPS status for the microbial species requires verification that the strain does not produce toxins known to be produced by certain members of the species and the strain does not possess acquired resistance to antimicrobials of importance. For joint dossiers, the European Association of Manufacturers and Formulators of Enzyme Products (AMFEP) has established a framework (AMFEP, 2014). Unfortunately, EFSA recently issued guidance requiring strain-specific safety data for each strain included in a joint dossier, which may pose difficulty to some manufacturers who do not have such strain-specific data.

Each dossier is evaluated by the Panel on Food Contact Materials, Enzymes and Processing Aids (CEP) of the European Food Safety Authority (EFSA), which is composed of scientists across the EU with expertise in chemical risk assessment, toxicology, epidemiology, chemistry, exposure assessment, food technology and microbiology (https://www.efsa.europa.

eu/en/panels/cep). The members of the CEP, the minutes of their review meetings, and the final opinions are publicly available at the EFSA website (EFSA, no publication year indicated). After evaluation the dossier is transferred to the EU Working Party Governmental Experts on Food Enzymes alongside with the EFSA opinion.

The Working Party – composed of Member State representatives - decides on the approval by voting. In their decisions, they consider the opinion of EFSA, the technological need, and if there is a potential to mislead consumers. Hence, the evaluation of the safety dossiers is conducted by EFSA as part of its risk assessment responsibility, while the final decision on approval resides with the EU Commission as part of its risk management mandate.

Until the deadline for the first wave of applications in March of 2015, the European Commission received 304 food enzyme applications (European Commission, 2016) essentially mirroring the number of enzymes on the EU market at that point in time. The vast majority of these enzymes are produced with microorganisms (ca. 93%), with over halve of those produced with five microbial species (*Aspergillus niger, A. oryzae, Trichoderma reesei, Bacillus subtilis, B. licheniformis)* (Roncancio, 2019). Sixteen joint dossiers were submitted.

At the time of the writing of this chapter (November 2019), some 50 evaluations were completed by EFSA. Once the evaluations of all applications are completed, the food enzyme list will be published by the European Commission in a public Union list of approved enzymes only then will the EU harmonized legislation be fully in force. Until this date, most enzymes – those used as processing aids—are still regulated by national legislation of EU Member States. France and Denmark are the only Member States with national pre-market approval systems for food enzymes. Some Member States have established rules regulating the use of food enzymes in specific food production processes (e.g., dairy and bread). These vertical legislations differ from country to country as far as the number and type of permitted food enzymes in various applications is concerned. Two food enzymes used as food additives, lysozyme and invertase, were previously approved by the EU harmonized legislation on food additives.

8.6.2 Challenges

One challenge is the work load involved in reviewing over 300 dossiers. Another is the fact that requirements seem to be shifting even while EFSA is evaluating the initial wave of dossiers. This includes requirements for exposure assessment (shifting from the Budget method to consumption database models, several of which still need to be developed almost 5 years after the initial submission deadline for existing enzyme preparations). Based on experiences gained in assessing submitted dossiers, the CEP noted that the method for exposure assessment originally designed to assess food additives, was not universally appropriate for food enzymes. The alternative methodology developed by the CEP uses actual food consumption data for the assessment of dietary exposure to food enzymes. To put this model into practice, the following information needs to be compiled for each food enzyme application: enzyme concentrations applied, the fate of the enzyme in processing food, and consumption data from national dietary food consumption databases. Completing the generic data requirements for the Food Enzyme Intake Model (FEIM) is anticipated for 2020 (Silano et al., 2016; Tlustos and Liu, 2019). Further changing requirements include (1) acceptable margin of exposure (from 100–200, and most recently even 300), (2) recombinant DNA detection threshold (from company-specific to a firm LOD of 10 ng/g), and (3) the recent requirement of strain-specific safety data even for joint dossiers. This is likely to affect the time schedule for fully getting the EU harmonized system to work. Industry initially expected completion of food enzyme dossier evaluations and publication not later than 2021. EFSA's estimate seems to be more cautious (not until 2023) provided that average evaluation time is 9–10 months and only ~50 dossiers have been completed by the time of the writing (2019) (Roncancio, 2019).

However, even this less optimistic estimate does not account for the 300 new food enzyme applications expected to enter the market between 2015 and 2023 (both new enzyme preparations and extension of use for existing enzyme preparations. These enzymes can legally be marketed under national legislation until the establishment of an European Community (EC) positive list (estimated to be 2023) and, without some transitional measure, would need to be withdrawn from the market until positively reviewed by EFSA and approved by the EC. Drawing on the experience from the review of the first tranche of dossiers, the time-period for reviewing another 300 dossiers might extend to many more years. Industry sees, therefore, an urgent need for transitional measures in order to prevent business disruption (AMFEP, 2019).

As many enzymes are products of modern biotechnology, including recombinant DNA technology, their GMO status in the EU is briefly discussed here as well. Food and feed enzyme products produced by genetically engineered microorganisms (aka GMMs in the EU) are outside the scope of the Genetically Modified Food and Feed Regulations in the EU and do not trigger GMO labeling of food if (1) they are used as processing aids or (2) they are produced by fermentation in accordance with the requirements in the contained use directive (90/219/EEC as amended by 98/81/EC) and additional discussion reflected in COM report 2006-626 on the implementation of Regulation 1829-2003. As the enzyme preparation does not contain the GMM used to produce it, it is referred to as being produced *with* a GMM rather than *from* a GMM, and these terms are used to differentiate between enzyme preparations that are not subject to GMO authorization and labeling requirements (all food-grade enzyme preparations) and those that do not qualify for such exemption. This requirement is independent from the EFSA dossier requirement of demonstrating absence of recombinant DNA in enzyme preparations undergoing review for placement on the Union list of approved enzymes expected by 2023.

8.7 Regulation of food enzymes in the United States

US market access for food enzymes, like any other food ingredient, is based on the legal definition of "food additive" and its exemption, "GRAS", which stands for Generally Recognized As Safe. The legal definitions of "food additive" and "GRAS" have been in place since 1958, when the US Congress enacted the Food Additives Amendment to the Food, Drug & Cosmetics Act (FD&CA). The 1958 amendment required that the US Food & Drug Administration (FDA) establish a regulation prescribing the conditions under which a food additive may be safely used for use in human or animal food. Amongst these is the requirement for formal food additive approval by FDA upon their review of a Food Additive Petition (FAP) filed by industry. When reviewing an FAP, FDA also takes any other available information into account and, in effect, FDA formally undertakes the safety evaluation that forms the basis for a food additive approval, including setting product specifications and limitations of use, as needed to support safety. From an enforcement perspective, FDA has full authority to take enforcement actions to stop distribution of food containing any unapproved food additive, as such food is considered adulterated under the FD&CA.

Section 201(s) of the FD&CA defines a "food additive" as "any substance the intended use of which results or may reasonably be expected to result, directly or indirectly, in its becoming a component or otherwise affecting the characteristics of any food. .. if such substance is not Generally recognized, among experts qualified by scientific training and experience to evaluate its safety, As having been adequately shown through scientific procedures (or, in the case of a substance used in food prior to January 1, 1958, through either scientific procedures or experience based on common use in food) to be Safe under the conditions of its intended use ..."

As applicable to food enzymes, an enzyme that is demonstrated to be GRAS under the conditions of its intended use is not a "food additive" and therefore it is not subject to mandatory premarket review by FDA's Center for Food Safety and Applied Nutrition (CFSAN). However, when the GRAS criteria are not met, the use of the enzyme in food (i.e., if the enzyme is expected to become a component of food or to affect the characteristics of the food) is subject to premarket approval by CFSAN as mandated by the FD&CA. As food enzymes have an established safety record, GRAS status is the main exemption for food enzymes to the requirement of obtaining formal food additive approval.

Contrary to several other jurisdictions (such as Canada, the EU today, and countries that follow Codex) yet like Australia/New Zealand, processing aid status of a food enzyme is of no concern in determining whether food additive approval is required, unless the use of the enzyme in food processing does not result in the enzyme protein becoming part of the food at all. For example, given extensive purification of distilled spirits, an enzyme used in potable alcohol manufacture is not considered a food additive in the US, and therefore strictly speaking would also not need to be GRAS for that specific use (potable alcohol), provided that data are available to show absence in the final food. However, the enzyme may still need GRAS status, as in potable alcohol manufacture the inactive enzyme protein ends up in the distiller grains co-product, an animal feed ingredient that is under jurisdiction of FDA's Center for Veterinary Medicine.

FDA established a GRAS regulation in part 170 of Chapter 21 of the Code of Federal Regulations for human food. The GRAS regulation defined eligibility criteria for GRAS, including for the use of scientific procedures and experience based on common use. The criteria were updated a few times (e.g., to include that applicable information and experience from both inside and outside the US contributes to GRAS), most recently in 2016 (81 FR 54960) as part of formalizing the provisional GRAS Notification rule in a formal "GRAS final rule."

The criteria for eligibility for classification as GRAS as established in 21 CFR 170.30(a) through (c) for a substance intended for use in human food are outlined here:

1. General recognition of safety may be based only on the views of experts qualified by scientific training and experience to evaluate the safety of substances directly or indirectly added to food. The basis of such views may be either (1) scientific procedures or (2) in the case of a substance used in food prior to January 1, 1958, through experience based on common use in food. General recognition of safety requires common knowledge throughout the scientific community knowledgeable about the safety of substances directly or indirectly added to food that there is reasonable certainty that the substance is not harmful under the conditions of its intended use (see § 170.3(i)).
2. General recognition of safety based upon scientific procedures shall require the same quantity and quality of scientific evidence as is required to obtain approval of a food additive. General recognition of safety through scientific procedures shall be based upon the application of generally available and accepted scientific data, information, or methods, which ordinarily are published, as well as the application of scientific principles, and may be corroborated by the application of unpublished scientific data, information, or methods.

A "GRAS Panel" may be convened for evaluating whether the available scientific data, information, and methods establish that a substance is safe under the conditions of its intended use in food (81 FR 54960 at 54964). A GRAS panel is not required to demonstrate that a substance is GRAS under the conditions of its intended use but is one mechanism that the regulated industry has used to demonstrate that the safety of a substance under the conditions of its intended use is generally recognized by qualified experts (81 FR 54960 at 54975). In fact, the enzyme industry has utilized a GRAS panel for its food enzyme GRAS determinations only in the minority of GRAS determinations notified to FDA/CFSAN (Hanlon et al., 2017) as it deployed various other very effective ways endorsed by the FDA to support the public recognition element, including publication in peer-reviewed journals of the data, information, or methods used to establish the safety of food enzymes and their production organisms (Sewalt et al., 2017). Experts on microbial food enzyme safety have published several papers on the topic of safety evaluation of enzymes starting with Pariza & Foster (1983) outlining an initial decision tree based on the premise that the safety of the microbial source is of utmost important to the safety of any enzyme, followed by a more elaborate decision tree for enzymes from recombinant DNA derived production organisms Pariza & Johnson, 2001), and a review of the safety of common production organisms (Olempska-Beer et al., 2006). The

Pariza decision tree has stood the test of time since its publication: it was more recently extended to animal food (Pariza & Cook, 2010) and industry has since long incorporated the decision tree, including the concept of Safe Strain Lineage, in its standard practices for enzyme evaluation to meet the GRAS criteria (Sewalt et al., 2016). In addition, the history of safe use in food has been well documented for many common food enzymes (most recently reviewed by Sewalt et al., 2016), as has the safety of commonly used production organisms such as *A. niger, A. oryzae, T. reesei, B. subtilis and B. licheniformis* (e.g., Olempska-Beer et al., 2006), inlcuding the safety of company-specific lineages of *T. reesei* (van Dijck et al., 2003), *B. subtilis* (Ladics and Sewalt, 2018), and *B. licheniformis* (Sewalt et al., 2018). These two key aspects of the Pariza decision tree (history of safe use of the expressed enzyme and the use of production organisms from a safe strain lineage) support the general recognition of safety for food enzymes when assessed in the overall publicly accepted safety evaluation methodology outlined in the Pariza and Johnson (2001) decision tree.

The GRAS final rule formalized a voluntary notification procedure initially established in 1997 under a proposed rule (62 FR 18938), under which any person may notify FDA of a conclusion that a substance is GRAS under the conditions of its intended use. FDA/CFSAN filed its first GRAS notice in 1998 and has filed 873 GRAS notices as of October 29, 2019, of which 109 (12.5%) were for food enzymes. Indeed, enzymes are the largest and arguably the most successful individual category of food substances that have been GRAS notified to FDA (Hanlon et al., 2017), with success rate in the GRN process (as measured by receiving a No-Questions Letter from FDA) of approximately 95% versus 78% for all other food substances.

Establishing GRAS status as the legal basis for marketing an enzyme used in food is not a formal approval, as it does not require formal review by an authority, but rather it requires demonstration of general recognition. The responsibility of demonstrating general recognition of safety is with the developer and/or importer of the substance, and a so-called independent GRAS conclusion, if reached based on considerations according to the guidelines, is sufficient. FDA has published data guidelines for food enzymes (https://www.fda.gov/regulatory-information/search-fda-guidance-documents/guidance-industry-recommendations-submission-chemical-and-technological-data-food-additive-petitions), and it is of note that these recommendations apply to both GRAS Notices and Food Additive Petitions. The difference between a GRAS Notice and a Food Additive Petition and practical implications for food enzymes are outlined in Table 8.2 below.

Table 8.2 Difference between FAP and GRN for food enzyme preparations.

Item	Food Additive Petition (FAP)	GRAS Notice (GRN)
Frequency of use for enzymes	Rare	Common
Review time	Multiple years (2–7)	6 ± 2 months
Primary responsibility for safety	Food and Drug Administration	Developer or importer selling the enzyme preparation
Size of dossier/ level of detail of information	Full dossier, with all supporting studies	Summary of the data and information to support safety, leveraging various general publications on safety of production organism, safety of common food enzymes, and safety evaluation methodology
Transparency/ Data confidentiality	Evaluation not published; often contains confidential business information (CBI)	Pivotal data, information or methods needs to be in public domain, corroborative data may be confidential – rarely needed for human food enzymes
Mandatory/voluntary filing	Mandatory unless GRAS	Voluntary
Use restrictions/ manufacturing conditions	As defined/limited by FDA	Defined by submitter based on intended use.
Toxicology data	New data usually required (bridging data acceptable by exeption)	New toxicology may not be required, predicated on availability of substitute data applicable from use of Safe Strain Lineage

As discussed above, GRAS status, not processing aid status impacts the US regulatory approval requirements for enzyme preparations present in food. Processing aid status of the enzyme used to manufacture the food does exempt US food manufacturers from required inclusion of the enzyme on the finished food label, per 21 CFR 101.100 (a)(3). This exemption is common to all jurisdictions discussed in this paper, although the definition of "processing aid" differs from one jurisdiction to another. A comparison of processing aid definitions will follow later in this paper. If an enzyme is included in the ingredient statement on the label of the finished food (such as cheese), it often is stated as "enzyme" or "enzymes" although it may also list a more specific enzyme name based on activity.

Processing aid status also impacts the emerging US requirement for Bioengineering (BE) Disclosure on consumer food packaging, the US version of GMO labeling, which is managed by the US Department of Agriculture (USDA). The BE rule was finalized in December 2018 (83 FR 65814) with implementation for most of industry beginning January 2020 (small manufacturers: January 2021), and mandatory enforcement from January 1, 2022 onward. Processing aids and other incidental additives are out of scope of the definition of Bioengineered substances. Detailed implications of the BE rule for food enzymes remain to be clarified in follow-up guidance from the USDA. However, among the inputs to produce microbial enzymes, the production host and fermentation ingredients can be considered incidental additives in the final enzyme preparation, unlike the formulation ingredients. Therefore, for consumer food and dietary supplement products in which the enzyme is an active ingredient, BE disclosure will likely need to be considered only for enzyme formulation ingredients as triggered by the BE status of the enzyme formulation ingredients, and only if such formulations ingredients have a function in the consumer product. More information is available from the USDA website (https://www.ams.usda.gov/rules-regulations/be).

8.8 Regulation of food enzymes in Canada

Health Canada regulates food enzymes as food additives that need to undergo formal review before approval, unless they are a processing aid. To aid in decision making, Health Canada issued a *Policy for Differentiating Food Additives and Processing Aids* in 2014.

The regulatory definition of "food additive" is the primary basis for differentiating food additives and processing aids and is defined in section B.01.001 of the Food and Drug Regulations as "any substance the use of which results, or may reasonably be expected to result, in it or its by-products becoming a part of or affecting the characteristics of a food", but does not include certain nutrients, agricultural chemicals, spices and flavorings, food packaging materials, and animal drugs for food production animals.

Although there is no mandatory premarket review of food processing aids, Health Canada's Food Directorate has for many years conducted premarket assessments of food processing aids, upon voluntary request by industry, to determine the acceptability of proposed uses from a food safety perspective. One can obtain a Letter of No Objection from Health Canada in response to a submission based on A Guide for Preparing Food Processing Aid Submissions. Health Canada will generally consider that a significant removal step in/after the food manufacture process is indicative of not becoming a significant part of a food, while more strictly adhering to the criterion of "not affecting the characteristics of a final food" to issue an opinion that an enzyme is used as a processing aid.

Examples of enzyme processing aid uses in Health Canada's Policy include the use of an enzyme in the production of a food ingredient derived from starch, or to assist with extracting juice from fruit, or to facilitate the mashing process in manufacturing a distilled beverage. By and large, one would expect that most enzymes used in commodity processing of grain (and refined derivatives like starch, sugars, oil, protein), fruit (to extract juice), plant and animal protein sources (to isolate protein), and in brewing would be considered processing aids, unless directly affecting the characteristics of a food. Conversely, most enzymes used in baking and dairy applications would be food additives under the Health Canada policy (as those processes often don't include an enzyme removal step and the characteristics of the resulting food may be impacted), unless used in the production of an ingredient, such as production of galacto-oligosaccharide from lactose, which is used itself as an ingredient in food. To our knowledge, seven distinct enzyme uses have been reviewed and determined to be processing aid uses by Health Canada, but this list is unlikely an exhaustive list of processing aid uses. Enzymes regarded to be processing aids by Health Canada still need to be assessed by the manufacturer or seller to meet suitable food-grade specifications, and not to lead to toxicological, nutritional, microbiological, or other safety concerns. This assessment of safety for their intended use in food processing is not unlike a due diligence determination in the EU. In practice, a self-GRAS determination that satisfies all requirements in the US will also satisfy this requirement in Canada and pre-FIAP EU.

To gain formal marketing authorization for a new food enzyme used as a food additive, one files a comprehensive safety dossier with Health Canada following the Guide for the Preparation of Submissions on Food Additives, followed by Health Canada's review, which takes 12–18 months. Within 6 months after completion of the safety review, the enzyme is added to the Permitted List of Food Enzymes (https://www.canada.ca/en/health-canada/services/food-nutrition/food-safety/food-additives/lists-permitted/5-enzymes.html), and "incorporated by reference" in the marketing authorization for food additives that may be used as food enzymes. This Marketing

Authorization procedure for enzymes. The Authorization procedure for enzymes was implemented in 2012 with the intent to replace Table V of Division 16 of the Food and Drug Regulations, and greatly streamlined the regulatory approval process. In case of new uses for an enzyme already on the list, the "incorporation by reference" procedure takes 2–3 months.

Health Canada created a transition guide to provide stakeholders with further information on the Lists of Permitted Food Additives (such as the List of Permitted Food Enzymes) as well as guidance on how to interpret and use these lists. Enzymes are included in the list by activity and source (e.g., microbial production organism). The microbial source is not strain-specific, with non-genetically engineered microbes listed at the species level and genetically engineered microbes with indication of the plasmid used for transformation. Discussion is ongoing between industry and Health Canada regarding the relevance of the plasmid reference in the list, the removal of which would simplify the listing.

8.9 Regulation of food enzymes in Australia/New Zealand

The Food Safety Authority of Australia and New Zealand regulates enzymes as processing aids as covered in Standard 1.3.3, clauses 15–17. Enzymes use as processing aids are not exempt from approval requirements, as Standard 1.3.3 prohibits the use of processing aids in food unless there is a specific permission within this Standard via listing in the tables behind clauses 15–17. Processing aids are defined in this Standard in clause 1 as:

1. the substance is used in the processing of raw materials, foods, or ingredients, to fulfil a technological purpose relating to treatment or processing, but does not perform a technological function in the final food; and
2. the substance is used in the manufacture of a food at the lowest level necessary to achieve a function in the processing of that food, irrespective of any maximum permitted level specified.

Enzymes may be used as processing aids in food manufacturing only if the enzyme is derived from the corresponding source or sources specified in the Tables pertaining to clauses 15–17. To get a new enzyme listed in the Table, a comprehensive submission is needed to allow for a safety evaluation by FSANZ.

For enzymes derived from a microbial source, a listed microbial source organism may contain additional copies of genes from the same organism; that is, FSANZ does not differentiate between nongenetically engineered production organisms and self-cloned ones, as the latter do not contain novel DNA or novel protein. However, FSANZ considers protein-engineered enzymes to be novel and thus not identical to the native sequence, and it has started to list protein engineered enzymes with a specific designation as "variant enzyme."

The stance taken by FSANZ on protein-engineered enzymes impacts food labeling requirements: enzymes from self-cloned organisms used in food as processing aids do not need to be included on the food label as they are not considered novel; whereas protein-engineered enzymes produced in the same original host, even with the smallest possible change of 1 amino acid, do trigger the resulting food processed with the subject enzyme to bear a GMO label in most jurisdictions in Australia.

8.10 Codex/JECFA

Many countries in emerging markets rely on Codex guidelines, JECFA specifications, and/or JECFA evaluation for their management of food enzymes. According to Codex Alimentarius, processing aids do not require formal evaluation or approval. In the 44th Session of the Codex Committee on Food Additives (CCFA) held in Hangzhou, China in March 2012, CCFA agreed to establish an electronic Working Group, led by New Zealand and co-chaired by China, to develop criteria for the entry of substances into a database for processing aids (http://www.fao.org/tempref/codex/Meetings/CCFA/ccfa45/fa45_18e.pdf). An international processing aid (IPA) database has been established and is currently managed by China. (https://ipa.cfsa.net.cn/login?task=index_pro). The list is established based on the principles contained in Codex Committee on Food Additives (CCFA) "Guidelines on Substances used as Processing Aids (CAC/GL 75-2010)" (https://tinyurl.com/CAC-GL-75-2010). The Codex processing aid definition is quite broad and like the EU definition:

"Any substance or material, not including apparatus or utensils, and not consumed as a food ingredient by itself, intentionally used in the processing of raw materials, foods or its ingredients, to fulfil a certain technological purpose during treatment or processing and which may result in the non-intentional but unavoidable presence of residues or derivatives in the final product."

In addition to the definition itself, Codex Guideline (CAC/GL 75-2010) outlines several requirements regarding its function, manufacture/use, and safety/quality:

- *Function*: needs to have (a) technological function(s) in processing of raw materials, foods, or ingredient, whereas any residues should not perform a technological function
- *Manufacture*: shall be prepared and used under the conditions of GMP (lowest level to accomplish its function, residues/derivatives to be reduced to the extent reasonably achievable, and should not pose a health risk) and general principles of Food Hygiene (CAC/RCP 1-1969)
- *Safety*: to be demonstrated by the supplier or user, including an assessment of any unintended or unavoidable residues resulting from its use under GMP
- *Quality*: should be food-grade, conforming to applicable specifications recommended by Codex (e.g., JECFA, 2006) or responsible (inter)national bodies (e.g., FCC, 2004) and in compliance with microbiological criteria established in CAC/GL 21-1997.

Although processing aids are not required to undergo formal evaluation or approval in most countries that follow Codex, the IPA list is not considered a positive list but rather an informational tool. It is still possible to submit an enzyme used as processing aid in food for evaluation by the Joint Evaluation Committee for Food Additives (JECFA). The JECFA process includes the generation of a tox monograph and risk assessment for the food enzyme in the application, and a final JECFA specification. To date, all enzyme evaluations by JECFA have concluded that it is not necessary to establish an upper limit of exposure in the form of an Acceptable Daily Intake (ADI). The lack of need for an ADI is testimony to the safety and low risk of common food enzymes produced by common production organism under GMP against JECFA (2006) specifications.

The lack of risk associated with food enzymes has reduced JECFA's overall priority of enzymes versus food additives in the JECFA evaluation process, and a completed enzyme evaluation will not lead to an entry in the Codex General Standard for Food Additives (GSFA). This has led to some confusion around the concept of individual countries "following Codex guidelines." One the one hand, most countries follow the Codex guidelines for processing aids, including exemption of formal approval predicated upon manufacturing, use, safety and quality considerations such as GMPs and adherence to the JECFA (2006) specifications for food enzymes; on the other hand, a few countries do require national approval of enzymes used as processing aids, whereas others like Thailand have expected or required JECFA approval before allowing entry into their regulatory process.

In addition, several countries have started to explore inclusion of GMO considerations in their oversight of food enzymes, including Brazil, China, India, Indonesia, and Thailand. This has resulted in several information exchange events between industry, academic experts, and regulating agencies regarding the use and oversight of biotechnology in strain construction resulting in genetically engineered microorganism (GEMs) for the manufacture of enzymes and other food ingredients (Hanlon and Sewalt, 2020).

8.11 Japan

The Japanese food enzyme preparations are managed by the Ministry of Health, Labor and Welfare (MHLW), and they are regulated as food additives. Food enzyme preparations must be approved before they can be produced and sold, and there are nearly 70 enzyme preparations authorized for use in food. The Japanese government attaches great importance to the quality and safety of food enzyme preparations and has given great attention to detail in its oversight of the production, operation, and use of food enzyme preparations (Xiao-dan, 2018; Luo, 2014).

The agencies in charge of overall food safety in Japan mainly include the Food Safety Committee in charge of food risk assessment and the Ministries of Health, Labour, and Welfare (MHLW) and Agriculture, Forestry and Fisheries (MAFF) in charge of food risk management (Liu, 2010).

In June 2003, Japan promulgated the basic law on food safety. Based on the new concept of food safety supervision, risk assessment, and risk management were clearly separated. The newly established food safety committee has become Japan's professional and independent food risk assessment agency (not unlike EFSA in the EU). Food safety management in Japan mainly depends on the supervision conducted by the local administrative departments in the metropolitans, prefectures, and counties, as well as local agricultural bureaus of MAFF.

8.11.1 Food safety committee

The food safety committee (FSC) has seven non-governmental members as its top decision-making body, who are appointed by the Prime Minister with the approval of the National People's Congress for a term of 3 years. In total, 200 non-governmental special committee members participate on assessment panels, with a 3-year term of office. Three separate assessment panels exist, for chemical risk assessment (including enzymes), bioassessment (including microorganisms), and food risk assessment (including genetically modified food, fertilizers, and novel food). MAFF and MHLW undertake food risk management, risk communication and promotion of food safety research at the farm stage and industrial food processing, respectively.

8.11.2 Japanese food safety regulations

In Japan, there are three main laws related to food safety and standards, the basic law on food safety, the food hygiene law, and the Japanese agricultural and forestry standards (JAS) law. The basic law on food safety sets out the principles for establishing a food safety system and the role of the food safety committee, an assessment body for food risks. According to the food hygiene law, the Ministry of Health, Labour and Welfare of Japan shall act as the food risk management agency to ensure food safety and hygiene. The purpose of Japanese agricultural and forestry standard law is to formulate and popularize Japanese agricultural and forestry standards, improve product quality, and seek to rationalize production, use, or consumption. According to the law, the Ministry of Agriculture and Forestry carries out risk management of agricultural products and food, and they are mainly responsible for food quality labeling and genetically modified food (Luo, 2014).

8.11.2.1 Japanese food enzyme preparation regulations

In Japan, there are three legal documents concerning the regulations on the administration of food enzyme preparations: food hygiene law, implementation of regulations on food hygiene law, and 101 provisional measures on the revision of food hygiene law and nutrition promotion law. The Ministry of Health and Welfare of Japan is responsible for the implementation of the food hygiene law and the standardized management of food enzyme preparations according to the law. The law applies to both the Japanese domestic products and imported products, which stipulate in article 6 that "only allow sales, production, import, processing, use, or display, for the purpose of sales gained based on the drug and the opinion of the food hygiene council does not constitute a hazard to human health designated food enzymes (not including the natural seasonings and usually for diet also can be used for preparation of enzyme preparation) or food containing such enzymes."

The regulations on the implementation of the food hygiene law list the specified food enzyme preparations and make detailed provisions on the labeling requirements, approval process, import declaration, and product inspection of food enzyme preparations. Although the inspection results of the official laboratories of the exporting countries approved by Japan are regarded as the same as those issued by the Japanese quarantine station, the inspection of the composition specifications of imported food enzyme preparations must be conducted according to the inspection methods specified in the food hygiene law. As far as approved composition, nearly 70 food enzymes are allowed in Japan, based on a required formal safety assessment of food enzymes and. hygienic limits of food enzyme preparations a limit standard of the amount of enzyme preparations (contained in the final food preparation). In addition, Japan has clear requirements on the labeling of food enzyme preparations, such as the basic requirements of labels, exemption from labeling requirements, food enzyme preparations that should be labeled, and method of labeling. Detailed information related to food enzyme preparation should be provided when the product is imported to speed up the customs clearance process of exported food in Japan. Food enzymes are subject to food additive public regulations, also known as the Japanese food additive standards based on the technical regulations of Japanese food hygiene law. A variety of testing methods are specified, and quality standards are set for about 400 food additives. The first edition of the Japanese food additive public agreement was issued in 1960, and since then eight revisions were issued, with the latest revised (ninth) edition of the Japanese food additive public regulations issued in 2017.

8.11.3 General approval procedure of enzyme preparations in Japan

Applicants who apply for the designation of a new food enzyme preparation or for the modification of an approved food additive should apply to the MHLW, submitting the required information. The application is reviewed by the secretariat office, and MHLW will initiate procedures to seek the opinions of the council on drug affairs and food health if the minister considers it necessary to hear the opinions of the council on the application. The council may, when necessary, request further information from the applicant. If, after discussion, the council considers it reasonable to designate and amend the application, it will submit a confirmation report to MHLW, who will decide in accordance with procedures, including amending the enforcement provisions of the food hygiene law. Although the standard review cycle (from when MHLW receives an application to when they decide to revise a regulation) is 1 year, this does not account for situations when incomplete dossiers need to be updated and resubmitted. In fact, the process can take several years especially when the council requests additional studies to support the safety of the new food additive. It has been a recent observation that approval times for common food enzymes from common production organisms has decreased substantially, especially in case of enzymes or production organisms that are like those previously reviewed.

The required information to be submitted in a food additive dossier is similar to what is needed for GRAS status or approval in the EU with a few differences, and include the source of the enzyme, detailed information of the development process, biochemical characteristics, quality specifications and standards, and, effectiveness data, stability data, and a description of effects of food enzyme preparations on the major components of the final food. Safety data required include animal toxicity studies under GLP conditions, daily intake information based on use rate and food intake information from the National Nutrition Examination Survey and other relevant surveys. When evaluating the safety of food enzyme preparations, the daily intake of food additives should be compared with the allowable daily intake obtained through toxicity studies. If food additives are consumed in conjunction with other similar food additives, their safety in combination should also be considered and evaluated. In case the applicant considers it necessary to establish a limit standard for the use of food enzyme preparations, the applicant shall put forward suggestions. If the applicant does not consider it necessary to restrict the use of food enzyme preparations, enough evidence should be provided.

Table 8.3 Below contains links to Japanese food safety related websites. The enzyme preparations approved in Japan can be found under the first link.

8.12 China

In China, enzyme preparations used in food industry must meet the relevant national quality standards in production and use and are managed as food additives. The current health standards for the use of food additives (GB2760-2014) stipulate the requirements for the use of food enzyme preparations in food. Commercial enzyme products allow the addition of ingredients that are easy to store and use. As enzyme preparations are managed as food additives, the administrative provisions on food additives in the food safety law, its supporting laws, regulations, and rules are applicable to food enzyme preparations (Wang, 2014).

The competent departments involved in enzyme preparation mainly include the State Administration for Market Regulation, the National Health Commission, and the Ministry of Agriculture and Rural Affairs. The production and circulation of food enzymes used in the food industry

Table 8.3 Japan food additives related websites.

Number	URL	Site name
1	http://www.ffcr.or.jp/zaidan/FFCRHOME.nsf/pages/e-foodadditives	Japan Food Additive Use Standard and Food Additive Code
2	http://www.mhlw.go.jp/topics/yunyu/ihan/index.html	Japan's Ministry of Health, Labour and Welfare announced the detention of imported food
3	http://www.fsc.go.jp/english.index.html	Food Safety Commission of Japan
4	http://www.mhlw.go.jp/english	Japan's Ministry of Health, Labour and Welfare
5	http://www.maff.go.jp/	Japan's Ministry of Agriculture, Forestry, and Fisheries
6	http://www.caa.go.jp/en/	Consumer Bureau of Japan
7	http://www.fferor.jp/zaidan/ffcrhome.nsf/pages/e-foodadditives	Japan Food Chemistry Research Foundation
8	http://www.jetro.gojp/en/reports/regulations	Japan Foreign Trade Organization
9	http://www.jsaorjp/defaultenglish.asp	Japan Standards Association

are mainly regulated by the State Administration for Market Regulation which has been merged with the State Administration for Industry and Commerce, General Administration of Quality Supervision, Inspection and Quarantine, and the food and drug administration. The National Health Commission is responsible for the approval and safety evaluation of new enzyme preparations, acceptance of the new enzyme preparations in accordance with the procedures, organization of experts on the safety and technological necessity of enzyme preparations, and other materials for technical review. These are done according to the results of the technical review to decide on approval. As for enzyme preparations produced by genetically engineered microorganisms (also known as GMMs and referred to as "transgenic" in China), according to the current relevant regulations, the safety of the production strain should be evaluated first, and then the safety and technological necessity of enzyme preparations should be evaluated by the National Health Commission (Table 8.4).

China's enzyme preparation industry has developed rapidly in recent years. According to the latest statistics of China biofermentation industry association in 2019, the production value of domestic enzyme preparation

Table 8.4 Relevant food additive laws and regulations in China.

Category	Department	Laws and regulations
Comprehensive law	NPC (National People's Congress)	Food safety law of the People's Republic of China
Listing of new varieties	National Health and Family Planning Commission Ministry of Agriculture and Rural Affairs	Measures for the administration of new varieties of food additives. Provisions on the declaration and acceptance of new varieties of food additives.
Production link	State Administration for Market Regulation National Health and Family Planning Commission	Provisions on the supervision and administration of the production of food additives. Health standards for food additive manufacturers Detailed rules for implementing the replacement (issuance) of production license for food additive products Food safety toxicology assessment procedures General guidelines for food additive labeling
Sector of using	State administration for market regulation National health and family planning commission	Measures for the administration of food production licenses. Measures for the hygiene control of food additives Food recall management measures

in 2018 was 3.3 billion Yuan, about 500 million US dollars, accounting for about 10% of the global market. The production volume and the output value increased by 4.3% and 10%, respectively (Zhang, 2019). However, there are only about 50 kinds of enzyme preparations approved in China, which is far behind those in Europe and the United States. China is not far behind the foreign countries in terms of biotechnology and human resources. Therefore, using these advantages to develop new enzyme preparations and applications, as well as giving new technologies and products more opportunities for development is important for the future development of the biological industry (Zhang, 2019). However, the development of regulations has lagged the pace of technological progress, limiting the application of new enzyme varieties in industry.

The main law under which the safety of food additives is managed today is the 2010 Food Safety Law of the People's Republic of China, as amended in 2018, along with the Measures for the Administration of New Food Additives issued in 2010. With the implementation of these laws and regulations, the management of food enzyme preparations is increasingly standardized, and the objective and scientific understanding of food enzyme preparations by the society and the public is gradually enhanced, which ensures stable and normal development of the production and management of food enzyme preparations.

In the current "food safety law," there are nine provisions directly on food additives. Chapter iii, "hygiene of food additives," clearly points out that the production, marketing, and use of food additives must comply with the hygiene standards for the use of food additives and hygiene management measures. Food additives that do not conform to the hygiene standards and hygiene control measures shall not be marketed or used. Article 44 and article 45 also stipulate corresponding administrative penalty provisions for acts violating the requirements of food additive production, marketing, use, and labeling in the food hygiene law (Li and Wang, 2004).

Provisions on food enzyme preparations in the food safety law of the People's Republic of China:

The food safety law of the People's Republic of China has 104 articles; Article 37, 38, 39, 40, 50, and 52 are directly related to the administration of food additives (The food safety law of the People's Republic of China was last amended in 2018). These articles deal with the approval requirements for the production of food with new food raw materials or for the production of a new variety of food additive or food-related product, including requirements for risk assessment, a production licensing system, and downstream food manufacturing requirements (Table 8.5).

8.12.1 Provisions on the declaration and acceptance of new varieties of food additives

In China, new enzyme preparations are subject to the approval and safety evaluation of new food additives. Approval process to produce new varieties of enzyme preparation unit or the use of the unit in accordance with the above application materials is handled by the National Health Committee, in accordance with the procedures for acceptance. The committee organizes experts on enzyme preparation technology to evaluate safety and technology, and based on the result of the technical review, it decides whether to approve or not the new enzyme preparations. As for enzyme preparations produced by transgenic microorganisms, according to the current relevant regulations, the Ministry of Agriculture and Rural

Table 8.5 Development of regulation of enzyme preparation.

Time	Department	Administrative regulations
1965	The State Council	Trial regulations on food hygiene administration
1973	Ministry of Health	Measures for the hygiene control of food additives
1979	The State Council	Regulations of the People's Republic of China on food hygiene
1982	NPC (National People's Congress)	Food hygiene law of the People's Republic of China (trial)
1995	NPC	Food hygiene law of the People's Republic of China
2002	Ministry of Health	Health standards for food additive manufacturers
2009	NPC	Food safety law of the People's Republic of China
2010	Ministry of Health	Measures for the administration of new varieties of food additives
2018	NPC	Food safety law of the People's Republic of China (amended in 2018)

Affairs (MARA) first evaluates the safety of transgenic strains, and then the National Health Commission evaluates the safety and technological necessity of enzyme preparations. Data for review by the NHC mainly includes basic information of the enzyme preparation (including name, classification, physical and chemical characteristics, etc.), the source of the enzyme preparation and process control data, the use of the enzyme preparation, information on enzymes safety evaluation (including the source of enzymes and safety evaluation data of enzyme preparation product), enzyme preparation process (function and principle in food production and processing, the effect with and without food additives, data, etc.), quality specification of enzyme preparation, three product inspection report, enzymes approved for use as reference in other countries, sexual information, etc. (Zhang, 2019).

China's food enzyme preparation standards mainly include the following:

In December 2014, the National Health and Family Planning Commission promulgated the standards for the use of food additives (GB 2760-2014). This standard defines the principles for the use of food additives, the types of food additives allowed to be used, the range of use, and the maximum amount of use or residue (Table 8.6).

Table 8.6 Chinese standards for food enzyme preparations.

Sequence number	Standard number	Standard name
1	GB 2760-2014	Standards for the use of food additives
2	GB1886.174—2016	Enzyme preparation for food industry
3	GB15193-2014	Food safety toxicology evaluation procedures and methods
4	GB7718-2019	General guidelines for labeling prepackaged foods

Food Safety National Standard "enzyme preparation for food industry" GB1886.174 —2016

This standard replaces GB 25594-2010 on January 1, 2017. Compared with GB25594-2010, the main changes of this standard are as follows: more terms and definitions of enzyme activity are provided; increased product classification, as well as physical and chemical requirements; the methods for the determination of enzyme activity of some enzyme preparations are given in the appendix. Limits for metal pollutants and microbial indexes of enzyme preparations are shown in Tables 8.7 and 8.8. This standard is applicable to enzyme preparation for food industry permitted as stipulated by GB 2760.

In addition, no viable production organism should be detected in enzyme preparation produced by microbes of gene recombination technology.

Table 8.7 Metal limits.

Item	Index	Test method
Pb (mg/kg) ≤	5	GB5009.12
As (mg/kg) ≤	3	GB/5009.12

Table 8.8 Microbiological limits.

Item	Index	Test method
CFU/g (mL) ≤	50,000	GB 4789.2
MPN/g (mL) ≤	30	GB 4789.3
E. coli /25g (mL)	not detectable	GB/T 4789.38
Salmonella/25g (mL)	not detectable	GB 4789.4

Food safety toxicology evaluation procedures and methods (GB15193-2014) consists of 26 standard food toxicology system in China, including one food safety toxicology evaluation procedures, one item of food toxicology laboratory practice, one item of health guidance value of formulation, one item of content processing method, one of the "three" material processing method, one item of pathology examination technical requirements, and 20 specific food toxicology test methods, such as acute oral toxicity test, genetic toxicity test, and teratogenic test, reproductive toxicity test, dynamic test, chronic toxicity experiment, etc. The evaluation covers food and its raw materials, food additives, new food raw materials, irradiated food, food-related products, and food contaminants.

In the evaluation, information and regulations from international organizations may be used to determine the tests to be carried out. For other food additives, toxicological test items should be specified in detail according to different situations. In China, it usually takes about a year to review and approve new food additives and half a year to expand the scope of use. However, this has not been the case for enzymes produced with modern biotechnology (i.e., recombinant micro-organism), for which an approval process was not in place for approximately 9 years, between 2010 and 2018. Review of food enzymes produced with modern biotechnology resumed in 2018 and, as of the writing of this paper, are now awaiting final approval publication.

8.12.2 General rules for prepackaged food labeling (GB7718-2019)

The standard requires that the labeling of food enzyme preparations in food should use the product name or type name specified in GB 2760. If the food additive contained in compound ingredients is less than 25% of the total amount of food, in accordance with the principle of introduction stipulated by GB2760 and does not play a technological role in the final product, it is not required to be labeled. In addition, Chinese food enzyme preparation related websites are shown in Table 8.9.

8.13 Trends and challenges pertaining to industrial microbial enzymes

Based on the overview presented earlier, several challenges regarding both the further development/harmonization of food enzyme regulations and safety evaluation can be recognized. As demonstrated in the previous sections of this chapter, there are differences in the regulatory processes to

Table 8.9 Chinese food enzyme preparation related websites.

Number	URL	Site name
1	http://www.samr.gov.cn/	China General Administration for Market Regulation
2	http://www.sac.gov.cn/	Website of Standardization Administration of China
3	http://www.tbtsps.cn/pages/tradez/IndexTrade.action	China National Standards and Regulations Research Center
4	http://jckspj.customs.gov.cn/	China National Import and Export Food Safety Administration website
5	http://www.cfaa.cn	China Food Additives and Ingredients Association website
6	http://www.nhc.gov.en/sps	Food Safety Standards, Monitoring, and Assessment Bureau, China Health Commission
7	http://www.cfsa.net.cn	China National Center for Food Safety Risk Assessment

which food enzymes are subjected in various jurisdictions, and some of these will continue in the future given that each jurisdiction will continue to have its own laws and regulations. For example, some countries exempt food enzymes from formal approval requirement if they are used as processing aids, whereas others, including Australia, China, Japan, and the United States, regulate food enzymes as food additives regardless of how they are used (note that in the United States, all substances present in food are considered Food Additives, unless they are GRAS). Now that the EU is implementing the FIAP regulation for all enzymes, even enzymes used as processing aids need to undergo formal approval. In contrast, the Canada approval process continues to differentiate between enzymes used as food additives and enzymes used as processing aids, although it also requires industry to conduct a safety assessment for enzymes used as processing aids, not unlike an independent GRAS conclusion in the United States. Processing aid status, however, consistently determines in all reviewed countries whether a food enzyme needs to be included in the ingredient statement of the consumer food label. It would be inconsistent with virtually other jurisdictions globally if this were to change in the EU based on the implementation of the FIAP regulation.

In addition, differences exist in the requirements for the safety evaluation of food enzymes among different countries, some of which relate to differing viewpoints regarding the safety of different techniques to improve the microbial production organisms. For example, the US FDA has always taken the approach to focus on the safety of the resulting enzyme product, whereas the emphasis in several other jurisdictions has historically been more focused on concerns around the use of in vitro gene technology in developing enzyme products. Interestingly, as experience in dossier review increases over time along with the mounting weight-of-evidence on enzyme safety (Ladics and Sewalt, 2018) and safe practices of modern biotechnology as applied to enzyme development and enzyme evaluation have become standard (Sewalt et al., 2016), most agencies now realize that modern biotechnology is more precise than traditional ways of genetic modification. This realization combined with the use of well-documented safe microbial hosts to produce enzymes (whether via QPS status or the SSL concept) is changing the paradigm for safety evaluation globally. For example, EFSA recently clarified that new toxicology data may not be required if relevant bridging data is available for a similar enzyme produced in a different QPS strain, but not if random mutagenesis (rather than recombinant DNA technology) was used to improved that new strain. This is exactly how industry is also managing study selection according to the SSL concept, with a new branch developed via mutagenesis from an existing SSL needing reconfirmation of safety (Ladics and Sewalt, 2018).

Although food enzymes have become more generally recognized as safe to consumers ingesting the food produced with these enzymes, the rate at which this recognition occurs differs between regions of the world. For example, the early requirement for genotoxicity studies was suggested according to precautionary principles, considering insufficient general evidence of safety, and well before the advent of modern biotechnology enabled the development of standardized safe production hosts. Various agencies differ in the rate at which they allow knowledge building to influence their adherence to strict requirements such as required genotoxicity studies, with these studies not being required to achieve GRAS status, but still required to satisfy approval requirements in most other countries. Similarly, different jurisdictions may differ in their viewpoint on accepting the weight-of evidence on protein engineering as a safe means to improve functionality of food enzymes, although such views may be influenced more by public perception rather than actual safety evidence. Hence, due to the differing rates of acceptance of the weight-of-evidence of safety data, differences in the required toxicology endpoints have emerged. An inter-agency review and discussion of the current weight-of-the-evidence of safety data pertaining to food enzymes and enzymes in general, and the adoption of a standard evaluation methodology not unlike the Pariza and Johnson (2001) decision tree designed for both naturally selected, traditionally modified (by mutagenesis) and recombinant production stains, would be instrumental in clarifying the relevance of specific endpoints and test methods required in regulatory guidance documents for product approvals. The current decision tree emphasizes safe transformation methods and characterization of the inserted DNA, while allowing for concepts such as history of safe use (of the enzyme) and safe strain lineage to dictate the need for new toxicology data to set a NOAEL. Most enzymes are produced by microorganisms that are regularly used for the manufacture of enzymes and that pertain to safe strain lineages. Repeated toxicological evaluation of enzymes from the same production organism pertaining to the same lineage has resulted in the establishment of multiple SSLs. Data further indicate that there are no concerns for oral toxicity of enzymes in general nor genotoxicity (Ladics and Sewalt, 2018). Therefore, the continued routine practice of performing subchronic oral and genotoxicity studies on enzyme preparations is questionable and establishing general health-based guidance values for common enzymes produced by common production organisms may be considered. Uniformity in safety evaluations would be improved and unnecessary animal testing reduced. Reduction in animal testing is an important animal welfare target and can be accomplished without compromising food safety for common food enzymes produced in QPS micro-organisms and non-QPS microorganism belonging to a well-documented SSL. More uniform safety evaluation procedures across jurisdictions would also facilitate mutual recognition of approval status (such as is currently the case between Denmark and France) and even unilateral adoption of food enzymes formally assessed as safe and acceptable elsewhere. Of note are early indications that Turkey and Russia will adopt enzymes reviewed by EFSA, once these are placed on the EU positive list, as suitable for use in their jurisdiction as well. To facilitate regulators in emerging markets to make informed decisions regarding acceptability of food enzymes in their country, the enzyme industry has collated a global list of food enzymes, their GRAS status in the US (per 21 CFR or GRAS-notified to FDA), and their approval status in key countries including Australia/New Zealand, Canada, Denmark, France, and Japan. This list is available upon request from the Enzyme Technical Association (ETA, https://www.enzymetechnicalassociation.org/) or AMFEP (https://amfep.org/).

Notwithstanding the lack of food allergenicity in relation to microbial enzymes, further research and data into the etiology of how enzymes act as respiratory allergens in the work-place are needed. Currently, all enzymes (regardless of their source or the use of modern biotechnology) are considered respiratory sensitizers and labeled as such on their respective material safety data sheets. The current state-of-the-science does not differentiate between nonallergenic/weakly allergenic enzymes and those moderate/strongly allergenic enzymes via the respiratory route of exposure nor allow for a rank-order of potency. Research involving in silico, in vitro, and/or in vivo methods is required to establish predictive and validated assays to evaluate the respiratory sensitization potential of enzymes in general. The ETA and AMFEP provide information on the safe handling of enzymes in various industrial applications, including food processing.

Finally, as the use of enzymes potentially expands into new areas of commercialization, such as cosmetics and personal care products, there is a risk of exposure to the broader, general population. With such new applications of enzymes, it is important to investigate how consumers may be exposed, the level of exposure, and the subsequent risk for sensitization. Such evaluations may include clinical allergy studies where real-world exposures are simulated and subjects evaluated for IgE specific antibody to the enzyme(s) in question. Such studies would be important given previous clinical findings of IgE sensitization to personal care products containing protease enzyme (Kelling et al., 1998; Sarlo et al., 2004).

Acknowledgment

The authors thank Kees Broekhuizen and Fred Wondergem for valuable comments on sections of the manuscript.

Relevant websites

https://ec.europa.eu/food/safety/food_improvement_agents/enzymes_en
http://www.efsa.europa.eu/en/topics/topic/food-enzymes
http://www.cfaa.cn.
http://www.jetro.go.jp/en/reports/regulations/.
http://www.ffcr.or.jp/zaidan/FFCRHOME.nsf/pages/e-foodadditives.
http://down.foodmate.net.
http://www.samr.gov.cn.
http://www.chinanutri.cn.
http://www.moa.gov.cn.
http://www.caa.go.jp/en/pdf/110817_orginization.pdf.

http://www.fscgo.jp/english/aboutus/organization of the food-safety commission_el.html. http://www.maff.go.jp/e/pdf/09ep9.pdf. http://www.mhlw.go.jp/english/org/detail/index.html. http://www.jetro.go.jp/en/reports/regulations/pdf/foodext2010e.pdf.

References

AMFEP, 2014. Terms of reference for joint food enzyme dossiers: Amfep/14/33-21 FEB 2014. Available from: https://amfep.org/_library/_files/amfep-14-33-joint-food-enzymes-dossiers-tors.pdf.

AMFEP, 2019a. Regulatory aspects on food enzymes. Available from: https://amfep.org/about-enzymes/regulatory/food-enzymes-regulations/.

Andersen, P.H., Bindslev-Jensen, C., Zacharia, H., Masbech, H., 1998. Skin symptoms in patients with atopic dermatitis using enzyme- containing detergents: A placebo-controlled study. Acta Derm. Venereol. 78, 60–62.

Armstrong, E.F., 1933. Enzymes: A discovery and its consequences. Nature 131, 535–537.

Bannan, E.A., Griffith, J.F., Nusair, T.L., Sauers, L.J., 1983. Skin testing of laundered fabrics in the dermal safety assessment of enzyme containing detergents. J. Toxicol. Cutaneous Ocular Toxicol. 11, 327–339.

Basketter, D.A., Kimber, I., 2018. Are skin sensitization test methods relevant for proteins? Regul. Toxicol. Pharmacol. 99, 244–248.

Basketter, D.A., Raulf, M., 2018. Industrial enzymes, in: Kanerva's Occupational Dermatology, third ed. Springer, New York, pp. 1–5.

Basketter, D.A., English, J.S., Wakelin, S., White, I.R., 2008. Enzymes, detergents and skin: Facts and fantasies. Br. J. Dermatol 158, 1177–1181.

Basketter, D.A., Brockhuizen, C., Firldeend, M., Kirkwood, S., Kluin, C., Mathieu, S., Rodriguez, C., 2012a. Enzymes in cleaning products: An overview of toxicological properties and risk assessment/management. Regul. Toxicol. Pharmacol. 64, 117–123.

Basketter, D.A., Berg, N., Kruszewski, F.H., Sarlo, K., Concoby, B., 2012b. The toxicology and immunology of detergent enzymes. J. Immunotox. 9, 320–326.

Basketter, D.A., Kruszewski, F.H., Mathieu, S., Kirchner, D.B., Panepinto, A., Fieldsend, M., Siegert, V., Bames, F., Bookstaff, R., Simonsen, M., concoby, B., 2015. Managing the risk of occupation allergy in the enzyme detergent industry. J. Occup. Environ. Hyg. 12, 431–437.

Belsito, D.V., Fransway, A.F., Fowler, Jr., J.F., Rietsohel, R.L., Storrs, F.J., Netheroott, J.R., 2002. Allergic contact dermatitis to detergents: A multicenter study to assess prevalence. J. Amer. Acad. Dermatol. 46, 200–206.

Bergman, A., Broadmeadow, A., 1997. An overview of the safety evaluation of the *Thermomyces lanuginosis* xylanase enzyme (SP 628) and the *Aspergillus aculeatus* xylanase enzyme (SP 578). Food Addit. Contam. 14, 389–398.

Bernstein, J., Sarlo, K., 2006. Enzymes, in: Bernstein, D.I., Chan-Yeung, M., Malo, J.L. (Eds.) Asthma in the Workplace,, Taylor and Francis, London, U.K., pp. 377–392.

Bindslev-Jensen, C., Skov, P.S., Roggen, E.L., Hvass, P., Brinch, D.S., 2006. Investigation on possible allergenicity of 19 different commercial enzymes used in the food industry. Food Chem. Toxicol. 44, 1909–1915.

Blumenthal, C.Z., 2004. Production of toxic metabolites in *Aspergillus niger, Aspergillus oryzae*, and *Trichoderma reesei*: Justification of mycotoxin testing in food grade enzyme preparations derived from the three fungi. Regul. Toxicol. Pharmacol. 39, 214–228.

Borowitz, D., Goss, C.H., Stevens, C., Hayes, D., Newman, L., O'Rourke, A., Konstan, M.W., Wagener, J., Moss, R., Hendeles, L., Orenstein, D., Ahrens, R., Oermann, C.M., Aitken, M.L., Mahl, T.C., Young, K.R., Dunitz, J., Murray, F.T., 2006a. Safety and preliminary clinical activity of a novel pancreatic enzyme preparation in pancreatic insufficient cystic fibrosis patients. Pancreas. 32, 258–263.

Borowitz, D., Goss, C., Stevens, C., Hayes, D., Newman, L., O'Rourke, A., Konstan, M., Wagener, J., Moss, R., Hendeles, L., Orenstein, D., Ahrens, R., Oermann, C., Aitken, M., Mahl, T.C., Young, K.R., Dunitz, J., Frederick Murray, F.T., 2006b. Study of a novel pancreatic enzyme replacement therapy in pancreatic insufficient subjects with cystic fibrosis. J. Pediatr. 149, 658–662.

Buehler, E.V., 1965. Delayed contact hypersensitivity in the guinea pig. Arch. Dermatol. 91, 171–177.

Chaplin, M., Bucke, C., 1990. Enzyme Technology, Cambridge University Press. Available from: http://www.sbu.ac.uk/biology/enztech/.

Cormier, E.M., Sarlo, K., Scott, L.A., MacKenzie, D.P., Payne, N.S., Carr, G.J., Smith, L.A., Gua-Lim, F., Bunag, F.G., Vasunia, K., 2004. Lack of type 1 sensitization to laundry detergent enzymes among consumers in the Philippines: Results of a 2-year study in atopic subjects. Annal Allergy Asthma Immunol. 92, 549–557.

Cronin, M.T., Basketter, D.A., 1994. Multivariate QSAR analysis of a skin sensitization database. SAR QSAR Environ. Res. 2, 159–179.

Cullinan, P., Harris, J.M., Newman Taylor, A.J., Hole, A.M., Jones, M., Bernes, F., Jolliffe, G., 2000. An outbreak of asthma in a modern detergent factory. Lancet 356, 1899–1900.

Divkovic, M., Divkovic, M., Pease, C.A., Gerberick, G.F., Basketter, D.A., 2005. Hapten-protein binding: from theory to practical application in the in vitro prediction of skin sensitization. Contact Dermat. 53, 189–195.

Domínguez-Muñoz, J.E., Iglesias-Garcia, J., Iglesias-Rey, M., Figueiras, A., Vilarino-Insua, M., 2005. Effect of the administration schedule on the therapeutic efficacy of oral pancreatic enzyme supplements in patients with exocrine pancreatic insufficiency: a randomized, three- way crossover study. Aliment. Pharmacol. Therap. 21, 993–1000.

Douglas, J.S., Barraj, I.M., Tennant, D.R., Long, W.R., Chaisson, C.F., 1997. Evaluation of the budget method for screening food additive intakes. Food Addit. Contam. 14, 791–802.

EFSA, 2006. Guidance document for the risk assessment of geneticallymodified plants and derived food and feed by the Scientific Panel on Genetically Modified Organisms (GMO). EFSA Journal 4(4), 99–105. https:// doi.org/10.2903/j.efsa.2006.99.

EFSA, 2009a. Guidance of the Scientific Panel of food contact material, enzymes, flavourings and processing aids (CEF) on the Submission of a Dossier on food enzymes for safety evaluation. EFSA J 7 (8), 1305. doi: 10.2903/j.efsa.2009.1305.

EFSA, 2009b. European Food Safety Authority- Guidance of EFSA prepared by the scientific panel of food contact material, enzymes, flavourings and processing aids on the submission of a dossier on food enzymes. EFSA J. 1305, 1–26.

EFSA, 2011a. Use of the EFSA comprehensive European food Consumption Database in exposure assessment. EFSA J. 9(3), 2097.

EFSA (2013). Investigation of the state of the art on identification of appropriate reference points for the derivation of health-based guidance values (ADI, AOEL and AAOEL) for pesticides and on the derivation of uncertainty factors to be used in human risk assessment. Question number: EFSA-Q-2011-00322. EFSA Supporting Publications 10(4), 1–169. doi: 10.2903/sp.efsa.2013.EN-413.

EFSA 2014. Explanatory Note for the Guidance of the Scientific Panel of Food Contact Materials, Enzymes, Flavourings and Processing Aids (CEF) on the Submission of a Dossier on Food Enzymes: Technical Report of EFSA. EFSA supporting publications EN-689. Available from: https://efsa.onlinelibrary.wiley.com/doi/epdf/10.2903/sp.efsa.2014.EN-689.

EFSA, 2016. Guidance of the Scientific Panel of Food Contact Material, Enzymes, Flavourings and Processing Aids (CEF) on the Submission of a Dossier on Food Enzymes for Safety Evaluation. Available from: www.efsa.euopa.eu/en/efsajournal/pub/1305.

EFSA, 2018. Annex B-Process-specific technical data used in exposure assessment of food enzymes: Annex Adopted: 05 June 2018. Available from: https://efsa.onlinelibrary.wiley.

com/action/downloadSupplement?doi=10.2903%2Fj.efsa.2016.4581&file=efs24581-sup-0001-Annex_B.pdf.

EFSA, 2019. Technical meeting with stakeholders on applications for food enzymes: Brussels, Belgium, 19 June 2019 to 20 June 2019. Available from: https://www.efsa.europa.eu/en/events/event/190619-0.

European Commission, 1998. Development of methodologies for the monitoring of food additive intake across the European Union. Development of methodologies for the monitoring of food additive intake across the European Union, Task 4.2-Final Report submitted by the Task Coordinator 16 January 1998. Volume European Commission, Brussels, Belgium, pp. 1–45 Annex A-J.

European Commission, 2002. Collection of Information on Enzymes. ISBN 92-894-4218-2. Luxembourg: Office for Official Publications of the European Communities.

European Commission 2008. Food enzymes. Regulation (EC) No 1332/2008 of the European Parliament and of the council of 16 December establishing a common authorization procedure for food additives, food enzymes and food flavourings. Official J L354/1, 1. 31/12/2008. Available from: http://eur-lex.europa.eu/legal content/EN/TXT/PDF/?uri=CELEX:32008R1332&from=EN.

European Parliament and the Council, 2008b. Regulation (EC) No 1331/2008 of the European Parliament and of the Council of 16 December 2008 establishing a common authorisation procedure for food additives, food enzymes and food flavourings.

European Commission 2011. Commission Regulation (EU) No 234/2011 of 10 March 2011 implementing Regulation (EC) No 1331/2008 of the European Parliament and of the Council establishing a common authorisation procedure for food additives, food enzymes and food flavourings Text with EEA relevance.

European Commission 2012a. Commission Implementing Regulation (EU) No 562/2012 of 27 June 2012 amending Commission Regulation (EU) No 234/2011 with regard to specific data required for risk assessment of food enzymes Text with EEA relevance.

European Commission 2012b. Commission Regulation (EU) No 1056/2012 of 12 November 2012 amending Regulation (EC) No 1332/2008 of the European Parliament and of the Council on food enzymes with regard to transitional measures.

European Commission 2014. Guidance Document on Criteria for Categorisation of Food Enzymes: Version 1, 24 February 2014. Available from: https://ec.europa.eu/food/sites/food/files/safety/docs/fs_food-improvement-agents_enzymes-guidance-categorisation.pdf.

European Commission 2016. Food enzyme applications submitted to the Commission within the legal deadline (from 11 September 2011 to 11 March 2015): Version 4. Updated on 25 July 2016. Available from: https://ec.europa.eu/food/sites/food/files/safety/docs/fs_food-improvement-agents_enzymes-applications.pdf.

European Commission 2018. Practical guidance for applicants on the submission of applications on food additives, food enzymes and food flavourings: Version 12. Updated on 7 March 2018. Available from: https://ec.europa.eu/food/sites/food/files/safety/docs/fs_food-improvement-agents_guidance-auth-proc.doc.

European Parliament and Council 2008a. Regulation (EC) No 1332/2008 of the European Parliament and of the Council of 16 December 2008 on food enzymes and amending Council Directive 83/417/EEC, Council Regulation (EC) No 1493/1999, Directive 2000/13/EC, Council Directive 2001/112/EC and Regulation (EC) No 258/97.

EFSA, 2011b. Guidance on the risk assessment of genetically modified microorganisms and their products intended for food and feed use. EFSA J. 9 (6), 2193, 10.2903/j.efsa.2011.2193.

Festersen, U., 2008. Alternative application route in the LLNA provides crucial environmental enrichment and broadens the usability of vehicles. Altern. Animal Exp. (ALTEX) 14, 433–436.

Flamm, E.L., 1991. How FDA approved chymosin: A case history. Biotechnology 9, 349–351.

Food Chemicals Codex (FCC), 2004. Institute of Medicine of the National Academies, fifth ed. The National Academies Press, Washington, DC.

Gerberick, G.F., Ryan, C.A., Kern, P.S., Schlatter, H., Dearman, R.J., Kimber, I., Patlewicz, G.Y., Basketter, D.A., 2005. Compilation of historical local lymph node data for evaluation of skin sensitization alternative methods. Dermat. 16, 157–202.

Goodman, R.E., Ebisawa, M., Ferreira, F., Sampson, H.A., van Ree, R., Vieths, S., Baumer, J.L., Bohl, B., Lalithambika, S., Wise, J., Taylor, S.L., 2016. AllergenOnline: A peer-reviewed, curated allergen database to assess novel food proteins for potential cross-reactivity. Mol. Nutr. Food Res. 60, 1183–1198.

Griffith, J.F., Weaver, J.E., Whitehouse, H.S., Poole, R.L., Newmann, E.A., Nixon, G.A., 1969. Safety evaluation of enzyme detergents, oral and cutaneous toxicity, irritancy and skin sensitization studies. Food Cosmetics Toxicol. 7, 581–593.

Halm, U., Loser, C., Lohr, M., Katschinski, M., Mossner, J., 1999. A double-blind, randomized, multicentre, crossover study to prove equivalence of pancreatin minimicrospheres versus microspheres in exocrine pancreatic insufficiency. Aliment. Pharmacol. Therap. 13, 951–957.

Hanlon, P.R., Sewalt, V.J., 2020. Genetically Engineered Microorganisms: Uses in modern food production and their regulatory oversight. Crit Rev Food Sci Nutr (published online). doi: 10.1080/10408398.2020.1749026.

Hanlon, P.R., Frestedt, J., Magurany, K., 2017. GRAS from the ground up: Review of the Interim Pilot Program for GRAS notification. Food Chem Toxicol. 105, 140–150.

Hansen, S.C., 1979. Conditions for use of food additives based on a budget method for an acceptable daily intake. J. Food Protect. 42, 429–434.

He, Q.Y., He, Q.Z., Deng, X.C., Yao, L., Meng, E., Liu, Z.H., Liang, S.P., 2008. ATDB: A unidatabase platform for animal toxins. Nucleic Acids Res. 36, D293–D297.

HERA, 2005. Human and Environmental Risk Assessment on Ingredients of Household Cleaning Products -amylases, Cellulases and Lipases. Available from: www.heraproject.com.

HERA, 2007. Human and Environmental Risk Assessment on Ingredients of Household Cleaning Products - Subtilisins (Proteases), Edition 2.0. Available from: www.heraproject.com.

Herzig, V., Wood, D.L.A., Newell, F., Chaumeil, P.A., Kass, Q., Binford, G.J., Nicholson, G.M., Gorse, D., King, G.F., 2011. ArachnoServer 2.0, an updated online resource for spider toxin sequences and structures. Nucleic Acids Res. 39, D653–D657.

Heubi, J.E., 2008. Zentase, a novel pancreatic enzyme product (Pep), is effective in mild, moderate, and severe exocrine pancreatic insufficiency (Epi). Gastroenterol. 134, A583–A584.

Hjorth, N., Roed-Peterson, J., 1976. Occupational protein contact dermatitis in food handlers. Contact Dermatitis. 2, 28–42.

Holsapple, M.P., Jones, D., Kawabala, T.T., Kimber, I., Sarlo, K., Selgradee, M.J.K., Shah, J., Woolhiser, M.R., 2006. Assessing the potential to induce respiratory hypersensitivity. Toxicological Sci. 91, 4–13.

ICCVAM (Interagency Coordinating Committee for the Validation of Alternative Methods) 2008. Validation status of new version and application of the murine local lymph node assay: A test method for assessing the allergic contact dermatitis potential of chemicals and products, May, 2008, Available from: http://iccvam.niehs.nih.gov/docs/immunotox_docs/LLANPRPRept2008.pdf.

International Food Biotechnology Council (IFBC), 1990. Safety evaluation of foods and food ingredients derived from microorganisms in Biotechnologies and Food: Assuring the safety of foods produced by Genetic modification. Regul. Toxicol. Pharmacol. 12, S114–S128.

JECFA (Join FAO/WHO Expert Committee on Food Additives). 2006. General Specifications and Considerations for Enzyme Preparations Used in Food Processing. Compendium of Food Additive Specifications, volume 1-3. FAO JECFA Monographs. AVailable from: http://www.fao.org/ag/agn/jecfa-additives/docs/enymes_en.htm.

Jonas, D.A., Anbtignac, E., Antoine, J.-M., Classen, H.G., Hugggett, A., Knudsen, I., Mahler, J., Ockhuizen, T., Smith, M., Teuber, M., Walker, R., De Vogel, P., 1996. The safety assessment of novel foods. Guidelines prepared by ILSI Europe Novel Food Task Force. Food Chem. Toxicol. 34, 931–940.

Jungo, F., 2012. The UniProtKB/Swiss-Prot Tox-Prot program: a central hub of integrated venom protein data. Official J. Int. Soc. Toxicol. (TOXICON) 60, 551–557.

Kanerva, L., Vanhanen, M., 1999. Occupational protein contact dermatitis from glucoamylase. Contact Dermatitis. 41, 171–173.

Kanerva, L., Vanhanen, M., 2001. Occupational allergic contact urticaria and rhino-conjunctivitis from a detergent protease. Contact Dermatitis. 45, 49–51.

Keggins, K.M., Lovett, P.S., Duvall, E.J., 1978. Molecular cloning of genetically active fragments of *Bacillus* DNA in *Bacillus subtilis* and properties of the vector plasmid pUB110. Proc. Natl. Acad. Sci. USA. 75, 1423–1427.

Keller, J., Layer, P., 2006. Are monolithic enteric-coated enzyme preparations effective in pancreatic exocrine insufficiency? A multicentre, double blind, placebo controlled cross-over trial. Gastroenterol. 130, A517–A521.

Kelling, C.K., Bartolo, R.G., Ertel, K.D., Smith, L.A., Watson, D.D., Sarlo, K., 1998. Safety assessment of enzyme-containing personal cleansing products: Exposure characterization and development of IgE antibody to enzymes after a 6-month use test. J. Allergy Clin. Immunol. 101, 179–187.

Kimber, I., 1994. Cytokines and regulation of allergic sensitization to chemicals. Toxicol. 93, 1–11.

Kimber, I., Dearman, R.J., 2002. Allergic contact dermatitis: The cellular effectors. Contact Dermatitis. 46, 1–5.

Kimber, I., Basketter, D.A., Gerberick, G.F., Dearman, R.J., 2002. Allergic contact dermatitis. Int. Immunopharm. 2, 201–211.

Konstan, M.W., Stern, R.C., Trout, J.R., Sherman, J.M., Eigen, H., Wagener, J.S., Duggan, C., Wohl, M.E.B., Colin, P., 2004. Ultrase MT12 and Ultrase MT20 in the treatment of exocrine pancreatic insufficiency in cystic fibrosis: safety and efficacy. Aliment. Pharmacol. Ther. 20, 1365–1371.

Konstan, M.W., Liou, T.G., Strausbaugh, S.D., Ahrens, R., Kanga, J.F., Graff, G.R., Moffettt, K., Millard, S.L., Nasr, S.Z., Simeon, E., Spenard, J., Grondin, J., 2008. Efficacy and safety of Ultrase (R) MT20 in treating pancreatic insufficiency in cystic fibrosis. Gastroenterol. Res. Pract. 134, A228–A229.

Laake, K., 1980. Enzyme drugs. Side Eff. Drugs Annu. 5, 222–225.

Ladics, G.S., 2008. Current Codex guidelines for assessment of potential protein allergenicity. Food Chem. Toxicol. 46, S20–S23.

Ladics, G.S. (2018). Protein allergy and GMOs. In Comprehensive Toxicology, Vol. 11, 3rd edition, C.A. McQueen (Ed.), Oxford: Elsevier Ltd., Oxford, U.K., pp. 638-667.

Ladics, G.S., Sewalt, V., 2018. Industrial microbial enzyme safety: what does the weight-of-evidence indicate? Regul. Toxicol. Pharmacol. 98, 151–154, https://doi.org/10.1016/j.yrtph.2018.07.016.

Ladics, G.S., Cressman, R.F., Herouet-Guicheney, C., Herman, R.A., Privalle, L., Song, P., Ward, J.M., McClain, S., 2011. Bioinformatics and the allergy assessment of agricultural biotechnology products: Industry practices and recommendations. Regul. Toxicol. Pharmacol. 60, 46–53.

Lee, M.Y., Park, K.S., Hayashi, C., Lim, H.H., Lee, K.H., Kwak, I., Laurie, R.D., 2002. Effects of repeated short-term skin contact with proteolytic enzymes. Contact Dermatitis. 46, 75–80.

Li, X., Wang, M., 2004. The review of regulations and standards on food additive worldwide. Chinese J. Food hygiene. 4, 308–312.
Liu, C. 2010. Research on Regulation of Food Safety in Japan. Jilin University.
Luo, P., 2014. Review on administration, regulations and standards of food additives in Japan and South Korea. Chinese Food Additives. 7, 88–93.
Masilamani, M., Cummins, S., Shreffler, W., 2012. Determinants of food allergy. J. Allergy Clin. Immunol. 32, 11–33.
McKenzie, T., Hoshino, T., Tanaka, T., Sueoka, N., 1986. The nucleotide sequence of pUB110: Some salient features in relation to replication and its regulation. Plasmid. 15, 93–103.
McKenzie, T., Hoshino, T., Tanaka, T., Sueoka, N., 1987. Correction. A revision of the nucleotide sequence and functional map of pUB110. Plasmid. 17, 83–85.
Morren, M.A., Janssens, V., Doom-Gossens, A., van Hoeyveld, E., Cornelis, A., De Wolf-Peeters, C., Heremans, A., 1993. Alpha-amylase, a flour additive: An important cause of protein contact dermatitis in bakers. J. Amer. Acad. Dermatol. 29, 723–728.
Naegeli, H., Birch, A.N., Casacuberta, J., Schrijver, A., de Gralak, M.A., Guerche, P., Jones, H., Manachini, B., Messéan, A., Nielsen, E.E., Nogué, F., Robaglia, C., Rostoks, N., Sweet, J., Tebbe, C., Visioli, F., Wal, J.-M., Eigenmann, P., Epstein, M., Hoffmann-Sommergruber, K., Koning, F., Lovik, M., Mills, C., Moreno, F.J., van Loveren, H., Selb, R., Fernandez Dumont, A., 2017. Guidance on allergenicity assessment of genetically modified plants. EFS2 15, 493, (10.2903/j.efsa.2017.4862c).
Nielsen, T.K., 1994. Enzyme applications (industrial), in: Kirk-Othmer (Ed.), *Encyclopedia of Chemical Technology*, vol. 9, fourth ed., John-Wiley & Sons, Inc., New York, N.Y., pp. 567–620.
Norton, S., 2001. Toxic effects of plants. In: Klaassen, C.D. (Ed.), Casarett and Doull's Toxicology: The Basic Science of Poisons. sixth ed. McGraw-Hill, New York, N.Y., pp. 96–976.
Olempska-Beer, Z.S., Merker, R.I., Ditto, M.D., DiNovi, M.J., 2006. Food-processing enzymes from recombinant microorganisms-a review. Regul. Toxicol. Pharmacol. 45, 144–158.
Oliver, G., 2011. *The Oxford Companion to Beer*. Oxford: Oxford University Press, Oxford, U.K., p. 49.
Pariza, M.W., Cook, M., 2010. Determining the safety of enzymes used in animal feed. Regul. Toxicol. Pharmacol. 56, 332–342.
Pariza, M.W., Foster, E.M., 1983. Determining the safety of enzymes used in food processing. J. Food Protect. 46, 453–468.
Pariza, M.W., Johnson, E.A., 2001. Evaluating the safety of microbial enzyme preparations used in food processing: Update for a new century. Regul. Toxicol. Pharmacol. 33, 173–186.
Patchell, C.J., Desai, M., Weller, P.H., Macdonald, A., Smyth, R.L., Bush, A., Gilbody, J.S., Duff, S.A., 2002. Creon 10,000 Minimicrospheres vs. Creon 8,000 microspheres: An open randomised crossover preference study. J. Cyst. Fibros. 1, 287–291.
Pepys, J., Longbottom, J.L., Hargreave, F.E., Faux, J., 1969. Allergic reactions of the lungs to enzymes of *Bacillus subtilis*. Lancet 1, 1181–1184.
Pepys, J., Wells, I.D., D'Souza, M.F., Greenberg, M., 1973. Clinical and immunological responses to enzymes of *Bacillus subtilis* in factory workers and consumers. Clin. Allergy 3, 143–160.
Peters, G., Johnson, G.Q., Golembiewski, A., 2001. Safe use of detergent enzymes in the workplace. Appl. Occup. Environ. Hyg. 16, 389–395.
Ricci, A., Allende, A., Bolton, D., Chemaly, M., Davies, R., Girones, R., Herman, L., Koutsoumanis, K., Lindqvist, R., Nørrung, B., Robertson, L., Ru, G., Sanaa, M., Simmons, M., Skandamis, P., Snary, E., Speybroeck, N., Ter Kuile, B., Threlfall, J., Wahlström, H., Cocconcelli, P.S., Klein, G., Prieto Maradona, M., Querol, A., Peixe, L., Suarez, J.E., Sundh, I., Vlak, J.M., Aguilera-Gómez, M., Barizzone, F., Brozzi, R., Correia, S., Heng, L., Istace, F., Lythgo, C., Fernández Escámez, P.S., 2017. Scientific Opinion on the update of the list of QPS-recommended biological agents intentionally added to food or feed as notified to EFSA. EFSA J. 15, 222. doi: 10.2903/j.efsa.2017.4664.

Rodriguez, C., Calvin, G., Lally, C., LaChapelle, J.M., 1994. Skin effects associated with wearing fabrics washed with commercial laundry detergents. J. Toxicol. Cutaneous Ocular Toxicol. 13, 39–45.
Roncancio, C., 2019. Withdrawal and updating dossiers. EFSA workplan for the existing dossiers. Technical meeting with stakeholders on applications for food enzymes 19-20 June 2019, Brussels, Belgium.
Roxas, M., 2008. The role of enzyme supplementation in digestive disorders. Alter. Med. Rev. 13, 307–313.
Saeed, Z., Wojewodka, G., Marion, D., Guilbault, C., Rodzioch, D., 2007. Novel pharmaceutical approaches for treating patients with cystic fibrosis. Curr. Pharmaceut. Des. 13, 3252–3263.
Sandvig, K., Torgersen, M.L., Engedal, N., Skotland, T., Iversen, T.G., 2010. Protein toxins from plants and bacteria: probes for intracellular transport and tools in medicine. FEBS Lett. 584, 2626–2634.
Sarlo, K., Adamson, G.M., Hollis, V.L., Innis, J.D., Babcock, L.s., Kirchner, D.B., 2004. Development of allergic antibody to an enzyme in a body lotation: Results of an 18-month clinical study. J. Immunotoxicol. 1, 71–77.
SCF. 1992. Scientific Committee on Food (SCF), Guidelines for the Presentation of Data on Food Enzymes (April 11, 1991); Report of the Scientific Committee on Food: Twenty-seventh series, Catalogue No. EUR 14181, 1992, pp. 13–22.
Scheper, R.J., von Blomberg, B.M.E., 1992. Cellular mechanisms in allergic contact dermatitis. In: Rycroft, R.J.G., Menne, T., Frosch, P.J., Benezra, C. (Eds.), Textbook of Contact Dermatitis. Springer-Verlag, Berlin, Germany, pp. 11–27.
Schweigert, M.K., Mackenzie, D.P., sarlo, K., 2000. Occupational asthma and allergy associated with the use of enzymes in the detergent industry: A review of the epidemiology, toxicology, and methods of prevention. Clin. Exp. Allergy. 30, 1511–1518.
Sewalt, V.J., 2017. Performance enzymes for food ingredients at the BIO World Congress—enabling biotechnologies and supporting capabilities join in a model for successful commercialization of food ingredients. Ind. Biotechnol. 13, 219–220.
Sewalt, V., Shanahan, D., Gregg, L., La Marta, J., Carillo, R., 2016. The generally recognized as safe (GRAS) process for industrial microbial enzymes. Ind. Biotechnol. 12, 295–302.
Sewalt, V., La Marta, J., Shanahan, D., Gregg, L., Carillo, R., 2017. Letter to the Editor regarding "GRAS from the ground up: Review of the Interim Pilot Program for GRAS notification" by Hanlon et al. 2017. Food Chem. Toxicol. 107 (Pt A), 520–521.
Sewalt, V.J., Reyes, T.F., Bui, Q., 2018. Safety evaluation of two α–amylase enzymes from *Bacillus licheniformis* expressing an α–amylase gene from Cytophaga species. Regul. Toxicol, Pharmacol. 98, 140–150.
Silano, V., 2015. Exposure assessment of food enzymes. EFSA J. 14, 4581–4590.
Silano, V., Bolognesi, C., Castle, L., Cravedi, J.-P., Fowler, P., Franz, R., Grob, K., Gürtler, R., Husøy, T., Kärenlampi, S., Mennes, W., Milana, M.R., Penninks, A., Smith, A., Tavares Poças, M.d.F., Tlustos, C., Wölfle, D., Zorn, H., Zugravu, C.-A., Arcella, D., Liu, Y., Engel, K.-H., 2016. Exposure assessment of food enzymes: Statement adopted 14 September 2016. EFSA J. 14, 249, (Doi: 10.2903/j.efsa.2016.4581).
Smith, D.J., Toby Mathias, C.G., Greenwald, D.J., 1989. Contact dermatitis from *B. subtilis*-derived protease enzymes. Contact Dermatitis. 20, 58–59.
Smith Pease, C.K., White, I.R., Basketter, D.A., 2002. Skin as a route of exposure to protein allergens. Clin. Exp. Dermatol. 27, 296–300.
Spök, A., 2006. Safety regulations of food enzymes. Food Technol. Biotechnol. 44, 197–209.
Tarvainen, K., 1991. Occupational allergies due to cellulase, xylanase, and α-amylase. Z. Hautkr. 66, 964–967.
Tlustos, C., Liu, Y., 2019. Dietary Exposure Assessment of Food Enzymes. Technical meeting with stakeholders on applications for food enzymes 19-20 June, Brussels, Belgium.
Ullmann's Encyclopedia of Industrial Chemistry, 2000. Electronic Version, sixth ed. Wiley-VCH, Weinheim.

US Department of Agriculture, 2008. Department of Agriculture, Agricultural Research Service, Nutrient Intakes from Food: Mean Amounts and Percentages of Calories from Protein, Carbohydrate, Fat, and Alcohol, One Day, 2005-2006. Available from: www.ars.usda.gov/ba/bhnrc/fsrg.

US-FDA., 1997. Food and Drug Administration. Substances Generally Recognized as Safe Proposed Rule. Fed Reg 1997;62(74):18938-964. Available from: https://federalregister.gov/a/97-9706.

US-FDA. 2010. Center for Food Safety and Applied Nutrition. Guidance for Industry: Enzyme Preparations: Recommendations for Submission of Chemical and Technological Data for Food Additive Petitions and GRAS Notices. Available from: http://www.fda.gov/FoodGuidances.

US Pharmacopeia, 2020. Monograph: enzyme preparations. In: Food Chemicals Codex, 12th Edition. The United States Pharmacopeial Convention, Rockville, Maryland.

van Dijck, P., Selten, G., Hempenius, R., 2003. On the safety of a new generation of DSM Aspergillus niger enzyme production strains. Regulatory Toxicology and Pharmacology 38, 27–35.

Varayil, J.E., Bauer, B.A., Hurt, R.T., 2014. Over-the-counter enzyme supplements: What a clinician needs to know. Mayo Clinic Proc. 89, 1307–1312.

Walsh, G., 2005. Directory of Therapeutic Enzymes, in: McGrath, B.M., Walsh, G. (Ed.), CRC Press, New York, N.Y. pp. 278–279.

Wang, H., 2014. The situation and problems of domestic and international management of food additives. Chinese Food Additive 1, 68–72.

White, I.R., Lewis, J., El Alami, A., 1985. Possible adverse reactions to an enzyme-containing washing powder. Contact Dermatitis. 13, 175–179.

WHO, 2009. (World Health Organization): Dietary exposure assessment of chemicals in food. Principles and methods for the risk assessment of chemicals in food. Environmental Health Criteria 240. FAO/WHO. International Programme on Chemical Safety (IPCS). Geneva: WHO. Available from: http://www.who.int/ipcs/food/principles/en/index.html.

Xiao-dan, L., 2018. Food additives safety regulations in US, EU, Japan and suggestions to China. World Agric. 4, 62–67.

Yanish-Perron, C., Vieira, J., Messing, J., 1985. Improved M13 phage cloning vectors and host strains: Nucleotide sequences of the M13mp18 and pUC19 vectors. Gene. 33, 103–119.

Zachariae, H., 1973. Occupational enzyme dermatitis. Results of patch testing with alcalase. Act. Dermato-Venereologica. 53, 145–148.

Zachariae, H., Hoegh-Thomsen, J., Witmeur, O., Wide, L., 1981. Detergent enzymes and occupational safety. Observations on sensitization during Esperase production. Allergy 36, 513–516.

Zeigler, D.R., Pragai, Z., Rodriguez, S., Chevreux, B., Muffler, A., Albert, T., Bai, R., Wyss, M., Perkins, J.B., 2008. The origins of 168, W23, and other *Bacillus subtilis* legacy strains. J. Bacteriol. 190, 6983–6995.

Zhang, J., 2019. Regulatory management of enzyme preparation for food industry. Biotechnol. Industry Technol. 3, 83–90.

Zhou, C.E., Smith, J., Lam, M., Zemla, A., Dyer, M.D., Slezak, T., 2007. MvirDB: A microbial database of protein toxins, virulence factors and antibiotic resistance genes for biodefense applications. Nucleic Acids Res. 35, D391–D394.

21 CFR 170.22, 2015. Safety factors to be considered. In: Code of Federal Regulations Title 21, Subchapter B --Food for Human Consumption, Part 170 --Food Additives, Subpart B --Food Additive Safety, Section 22. Office of the Federal Register National Archives and Records Administration (NARA)/US Government Publishing Office (GPO), 2015, Washington, DC, pp. 13 -15. Available from: www.gpo.gov/fdsys/pkg/CFR-2011-title21-vol3/pdf/CFR-2011-title21-vol3-sec170-22.pdf.

CHAPTER NINE

Selected industrial enzymes

Selim Kermasha[a], Michael N.A. Eskin[b]
[a]McGill University, Montreal, Quebec, Canada
[b]University of Manitoba, Winnipeg, Manitoba, Canada

9.1 Introduction

The food and beverage industry have become dependent on the use of enzymes as they perform their tasks efficiently within minutes or seconds. Without them, the myriad of biochemical reactions catalyzed would require hundreds of years to complete (Dalby, 2003; Otten and Quax, 2005). To meet the increasing demands by food industries, animal feed producers, cosmetic and pharmaceutical industries, as well as detergent, textile, leather, and paper processors, a continuous and dependable supply of enzymes is crucial (Mohammadi Panah et al., 2015). This is provided by industrial enzymes which represent a global market estimated between \$5000 million and \$5500 million in 2016, where the food and feed applications account for 55%–60% of the total enzymes market. It is important to note that the latter is growing at an estimated 6%–8% annually (Guerrand, 2017).

The milder reaction conditions under which enzymes operate, their specificity, as well as lower environmental and physiological toxicity makes them very acceptable adducts (Bommarhus and Paye, 2013; Madhaven et al., 2017; Roy and Prasad, 2018). The development of immobilized enzymes, described in Chapter 6 is an effective way of retaining their catalytic activity as well as making them reusable (DiCosimo et al., 2013; Bezerra et al., 2016; Cao et al., 2016).

The first successful commercial use of enzymes in food processing was in the cheese-making process, since then enzymes have been used extensively in brewing, meat tenderization, baking, and protein hydrolysis. In addition to their use as aids in food processing, enzymes also play an essential role in the production of natural ingredients, additives, and nutraceuticals (Singh and Kumar, 2019). Some enzymes used in the food industry, however, do not have the same activity when they are exposed to extreme conditions of high temperatures, pH, and low-water activity. This led to the development of enzymes with greater resistance to extreme conditions enabling them to operate more effectively and economically (Flores-Gallegos et al., 2019).

Enzymes: Novel Biotechnological Approaches for the Food Industry.
http://dx.doi.org/10.1016/B978-0-12-800217-9.00009-5

New developments in enzyme engineering, together with more acceptance by consumers of recombinant DNA technology, will go a long away to ensuring cleaner and greener processing of foods (Singh and Kumar, 2019).

The use of chemicals around the globe has increased tremendously, with deterimental affects to the health of populations Consequently, there is an urgent need for the world to replace these noxious chemicals with environmentally friendly products for the betterment of life on this planet (Razzaq et al., 2019). As a result, green processes using enzymes are gaining prominence as key factors for maintaining the quality and safety of the environment. Enzymes, green catalysts, play a prominent role in the food industry because they are essential tools in many of the processing stages in the production of food products. Enzymes require less energy and exhibit high specificity, and the development of novel enzymes is consistent with the green trend in the food industry (Nair and Jayachandran, 2019).

This chapter will highlight a select number of enzymes currently used by the food industry to ensure the optimal production of nutritious and safe food products for the consumer. While the majority of the enzymes used in the food industry are hydrolases, there are also an increasing number of oxidoreductases, isomerases and lyases.

9.2 Hydrolases and their uses in selected application

This class of enzymes, as defined by the International Union of Biochemistry and Molecular Biology, are biological catalysts that catalyze the hydrolytic cleavage of such bonds as C-O, C-N, and C-C as well as phosphoric anhydride bonds. As mentioned in Chapter 2, the common name for some of these enzymes reflects the name of their substrate with the added suffix ase. Hydrolases are classified according to the major substrates hydrolysed, and are grouped into carbohydrases, proteinases, and lipases.

9.2.1 Carbohydrases

Carbohydrases are a diverse group of enzymes involved in both the hydrolysis and synthesis of carbohydrates. They include a great variety of enzymes as discussed below. The majority of these enzymes are obtained from microorganisms with a preference for extracellular enzyme-producing microorganisms, which not only simplifies their processing but also lowers their costs (Tufvesson et al., 2010; Simpson et al., 2012). Because of their wide application, carbohydrases are also immobilized for use by the food industry using simple, efficient, and inexpensive techniques (Contesini et al., 2013).

A more detailed discussion of enzyme immobilization can be found in Chapter 6.

9.2.1.1 Amylases

Amylases are a group of enzymes that catalyze the hydrolysis of starch, the major storage carbohydrate in plants (Tester et al., 2004). Storage starch is produced in the amyloplasts as discrete granules with a distinct morphology in different plants (Bertoft, 2017). These are composed primarily of two major polysaccharides, amylose and amylopectin which account for 20% and 80% of the starch granule, respectively. Both components consist of linear chains of α-(1,4)-linked D-glucose residues interconnected through α-(1,6)-glucosidic linkages that form branches in these polymers (Fig. 9.1). While amylose is primarily considered a linear polymer, it may contain a few such branches (Hizikuri et al., 1981). In sharp contrast, however, amylopectin is extensively branched with the α-(1,6) glucosidic linkages that account for around 5% of the total links (Esmaeili and Noorolahi, 2017).

These are an important group of starch degrading enzymes widely found in both plants and microorganisms (Moller and Svensson, 2016). For example, amylases from plants and microorganisms have been used for many years by both the brewing and baking industries (Saini et al., 2017; Park et al., 2018). One of the most widely studied enzymes in this group is α-amylase (EC 2.2.1.1,), an extracellular enzyme that degrades

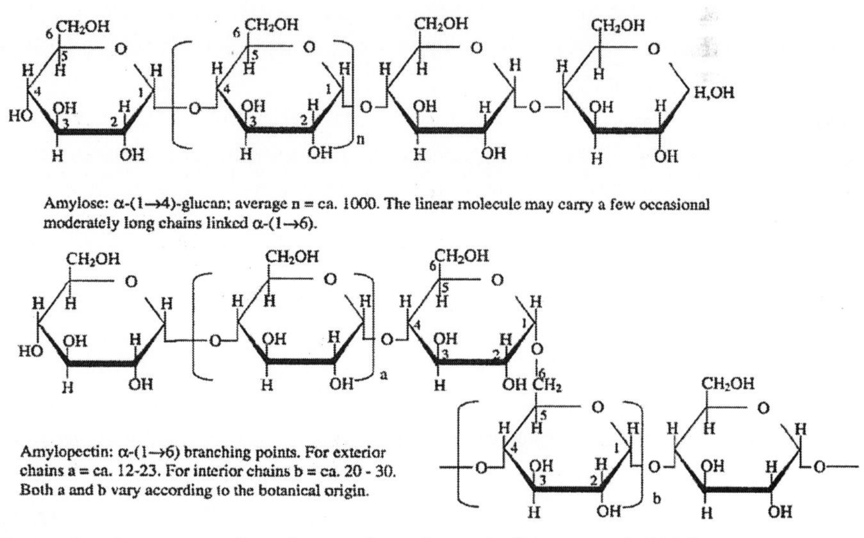

Figure 9.1 Structures of amylose and amylopectin (Tester et al., 2004).

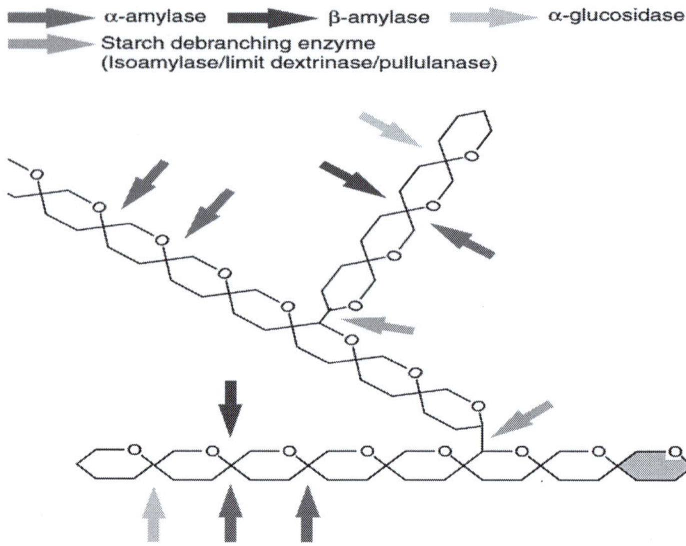

Figure 9.2 Starch degrading enzymes including α-amylase, β-amylase, and pullanase (Moller and Svesson, 2016).

α-1,4-glucosidic linkages in an endo-manner releasing oligosaccharides (Zubeyde et al., 2008). The main commercial source of α-amylase is fungal (*Aspergillus oryzae*). The low level or absence of α-amylase in wheat flour from mechanically harvested grain necessitates supplementation with this enzyme in the baking process. During the fermentation period α-amylase produces dextrins from the damaged starch granules. These dextrins are then catalyzed by β-amylase (EC 3.2.1.2), an exo-hydrolase enzyme, that works from the nonreducing end of the polysaccharide releasing successive maltose units (Fig. 9.2). These maltose units provide the fermentable sugars for yeast in the production of bread.

9.2.1.2 Amyloglucosidase

Amyloglucosidase (AMG EC 3.2.1.3) or glucoamylase, together with α-amylase are the two main classes of amylases identified among microorganisms (John, 2017), This is an exo-enzyme that hydrolyzes the α-1,4-glycosidic bonds from the non-reducing end of the linear amylose, yielding β-D-glucose (Fig. 9.2). Amyloglucosidase is used for the production of ethanol syrup and is also used to enhance the barley mash used in beer production (Aiyer, 2005; Zambare, 2010). Its economic importance is related to its effectiveness to produce β-D-glucose from starch. Because of the high demand for this enzyme, glucoamylase is produced industrially from the filamentous fungi, *Aspergillus* spp. and *Rhizopuss* sp. via submerged fermentation from a production medium composed of 20% corn and 2.5% corn steep liquor at 30–35°C (Nigam and Singh, 1995). The commercial demands for this enzyme have been solved using a combination of chemical mutagenesis and genetic engineering (Kumar and Satyanatayana, 2009).

9.2.1.3 Pullulanase

Pullulanase (EC. 3.2.1.41), an extracellular carbohydrase, was discovered by Bender and Wallenfels (1961) to be produced by the mesophilic organism, *Klebsiella pneumonia.* It has since become an indispensable enzyme in starch processing. Pullulanase is also known as the debranching enzyme as it hydrolyses α-1.6-glycosidic branch chain linkages (Fig. 9.2). In addition, this enzyme attacks α-1,4-glycosidic linkages of polysaccharides making it an invaluable agent for studying the structures of oligo and polysaccharides (Ling et al., 2009). There are two types of pullulanases, type-1 and type-2, with the former only hydrolysing α-1, 6 glycosidic linkages while the latter hydrolyses both α-1,4 and α-1,6 glycosidic linkages (Krigshauser and Liebl, 2000). The high temperature stability of pullalanases from *Klebsiella pneumonia* and *Bacillus* species make them ideally suited for the industrial processing of starch (Malakar et al., 2010).

Pullulanase is primarily used for the production of high glucose (30%–50% glucose; 30%–40% maltose) or high maltose (39%–50% maltose; 6%–10% glucose) corn syrups (Van Der Maarel et al., 2002; Ben Messaoud et al., 2002). Together with glucoamylase and β-amylase, pullulanase plays an important role in the saccharification of starch.

9.2.1.4 β-Glucanase

(1,3) β-D-Glucanases (Fig. 9.3) effectively depolymerise (1→3)(1→6)-β-D-glucans producing degraded products with biological and functional properties suitable as pharmaceuticals or as prebiotics in foods and beverages (Pang et al., 2005; Kadi and Crouzet, 2008; Barbosa et al., 2010). The enzyme is present in two forms, exo-β-D-glucanase (EC. 3.2.1.58) and endo-β-glucanase (EC 3.2.1.39). One bacterial source of these enzymes is *Trichoderma* species, which was reported by Giese et al. (2011a,b) to be produced in *Trichoderma harzainum* RifaiPAMB-86. Together with xylanase, β-glucanase improved nutrient utilization of poultry fed cereal base diets (Zarghi, 2018).

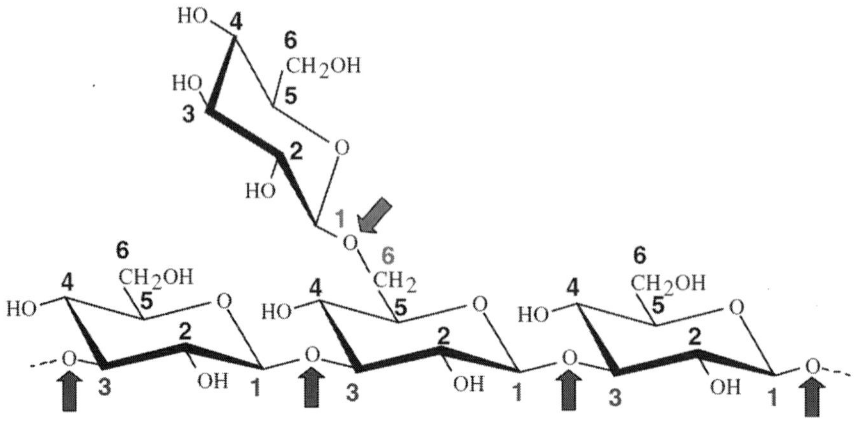

Figure 9.3 Proposed target sites for attack by β(1,3)-glucanases and β(1,6)-glucanases against a β-(1,3)(1,6)-branched glucan (Martin et al., 2007).

9.2.1.5 Xylanase

Xylanases are microbial enzymes that break down the hemicellulose, xylan. This complex polysaccharide has a main backbone chain composed of β-xylopyranose residues joined by β-1,4-glycosidic bonds (Fig. 9.4). Depending on the particular plant it has different sugar moieties attached including arabinose and glucuronic acid (Beg et al., 2001). It is the second

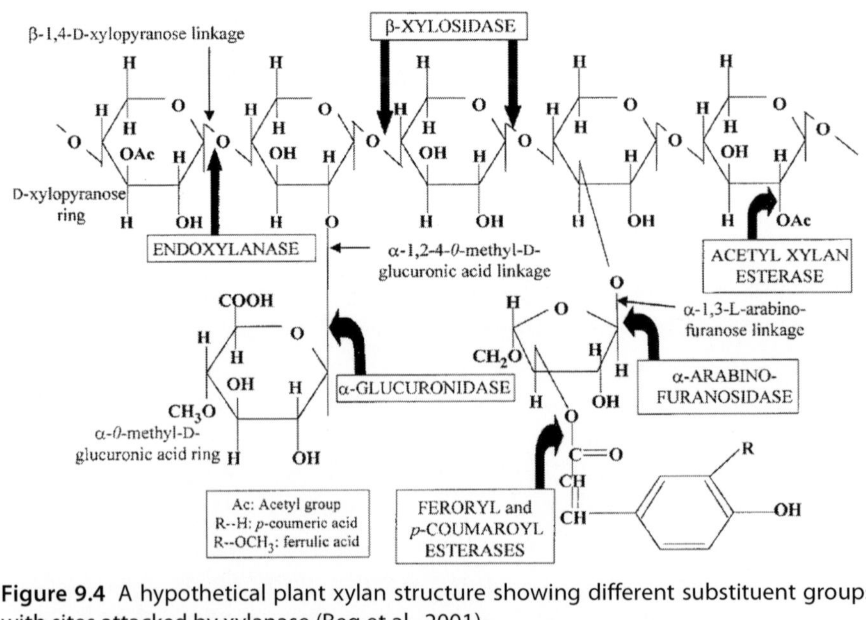

Figure 9.4 A hypothetical plant xylan structure showing different substituent groups with sites attacked by xylanase (Beg et al., 2001).

most abundant polysaccharide in plant cell walls after cellulose. The bioconversion of xylan, an important source of lignocelluloses material, has applications in both the food and paper making industries (Harris and Ramalingam, 2019). Xylanase (endo-1,3-β-D-xylanohydrolase, EC. 3.2.1.8) has also been reported in bacteria, fungi, actinomycetes, and yeast. While it has an important role to play in the pulp and paper industry it has also been shown to improve the digestibility of animal feed stocks, modify cereal-based material, convert lignocelluloses and agro-wastes to ferementable products as well as play a role in the clarification of fruits juices and degumming of plant fibers (Kapoor et al., 2001; Kuhad and Singh, 1993; Virupakshi et al., 2005).

9.2.1.6 Invertase

Invertase, an enzyme from yeast, has been used for almost a century to produce cast cream centres in the confectionery industry. It is one of the simplest carbohydrases in nature (Kotwal and Shankar, 2009). It cleaves the disaccharide sucrose, α-D-glucopyanosyl-β-D-fructofuranoside, into its two constituent monosaccharides, glucose and fructose (Fig. 9.5). The name is derived from the inversion of the plane of polarized light for sucrose, which is slightly dextrorotatory, to strongly levorotatory due to the release of fructose. There are two classes of invertase, each cleaving sucrose at a different point in the molecule (Myrback, 1960; Fukiu et al., 1974). One is α-D-glucoside glucohydrolase (EC 3.2.1.20), which attacks the sucrose molecule from the glucose side of the oxygen glycosidic ring while the other attacks from the fructose end and is β-D-fructofuranoside fructohydrolase (E.C. 3.2.1.26).

Commercially, invertase is obtained from several sources although the chief source is *Saccharomyces cerevisiae* commonly called baker's yeast. The enzyme exhibits high activity over a wide pH range of 3.5–4.5 with an optimum at around 4.5. Invertase, a glycoprotein, is stable at 50°C with a maximum activity at 55°C (Kulshrestha et al., 2013). It is used to produced cream

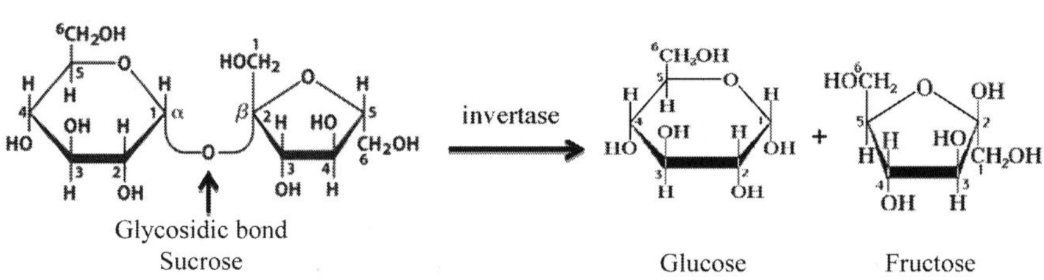

Figure 9.5 Hydrolysis of sucrose by invertase (Nadeem et al., 2015).

and liquid centers by liquefying creams or fondant-containing materials. Creams are biphasic systems of a syrup (liquid) fraction and a sugar (crystalline) fraction. The fluidity of the cream is determined by the concentration of the syrup fraction. Invertase hydrolyses the sucrose crystalline fraction to a solid content of 82% at which point enzyme activity is terminated. Fortunately, this sugar level will not support microbial fermentation (Eskin, 1990).

9.2.1.7 Lactase

Lactase or β-D-galactosidase (EC 3.2.1.23) catalyzes the hydrolysis of lactose (Fig. 9.6) to glucose and galactose (Richmond et al., 1981). It is used widely to remove lactose from milk products for those suffering from lactose intolerance. The removal of lactose is beneficial as it improves the sweetness, solubility, flavor, and digestibility of dairy products (Grosova et al., 2007). The enzyme is widely distributed in microorganisms, plants and animal organs but the major industrial sources are fungal sources *Aspergillus* sp. and *Kluyveromyces* sp. (Hussain, 2010).

Lactose intolerance is a pathophysiological condition resulting from an insufficiency of the enzyme "lactase" in the jejunum (Suri et al., 2019). Accumulation of lactose containing products can cause intestinal digestion and colonic fermentation resulting in abdominal pain and distension, borborygmi, flatulence, and diarrhea (Deng et al., 2015). Since animal milk is high in calcium, eliminating milk from the diet of children and adults can lead to a deficiency in calcium which could result in less dense and more fragile bones (Hodges et al., 2019). Early work led to the development of a powdered lactase that can be added to milk prior to consumption (McCormick, 1976). This was followed by a more stable and convenient liquid form of lactase (Dahlqvist et al., 1977; Scrimshaw and Murray, 1988). This caused a change in the viscosity and whiteness of the lactose-free milk compared to regular milk. In addition, the release of D-galactose and D-glucose increases sweetness while heating can produce undesirable flavors via the Maillard reaction (Onwulata et al., 1989: Harju et al., 2012). To overcome these problems, Zhang and Zhong (2018) used solid/oil/water (S/O/W) emulsions

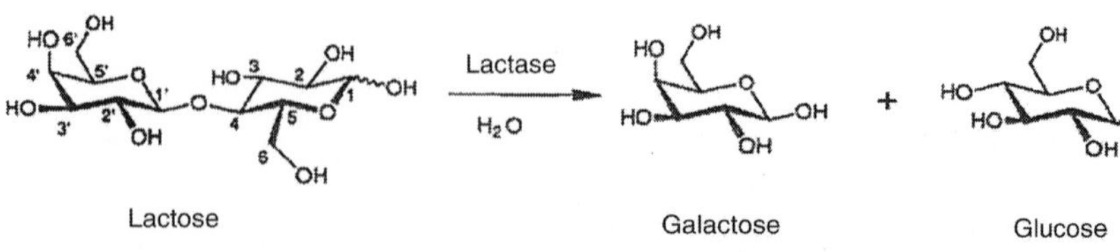

Figure 9.6 Hydrolysis of lactose by lactase (Riviera-Sagredo et al., 1992).

to encapsulate spray-dried lactase powder to control lactose hydrolysis in milk, The results looked very promising but further research was needed to improve the loading of lactase in the capsule as well as determine its effect on the sensory properties of the milk.

9.2.1.8 Pectinases

Pectinases or pectin esterases are a group of three distinct types of enzymes pectin methyl esterases, polygalacturonases and pectin lyases, each differing in their mode of action. These enzymes essentially break down complex polysaccharides in plant tissues into simpler molecules such as galacturonic acid (Tapre and Jain, 2014). The primary substrates for these enzymes are collectively referred to as pectic substances and contain the basic polymer α-1,4-D-polygalacturonic acid. Pectin, a colloid, is itself composed of a polygalaturonic acid chain in which about two thirds of the carboxyl groups are esterified with methanol (Kertesz, 1951). Pectic enzymes either depolymerize or de-esterify pectic substances.

9.2.1.8.1 Pectin methylesterase (PME)

Pectin methyl esterase (PME) (EC 3.1.1.11) removes the methoxyl groups from pectin by de-esterification to form pectic acid (Fig. 9.7B). Pectin methylesterase can be classified into Type I and Type II based on their protein

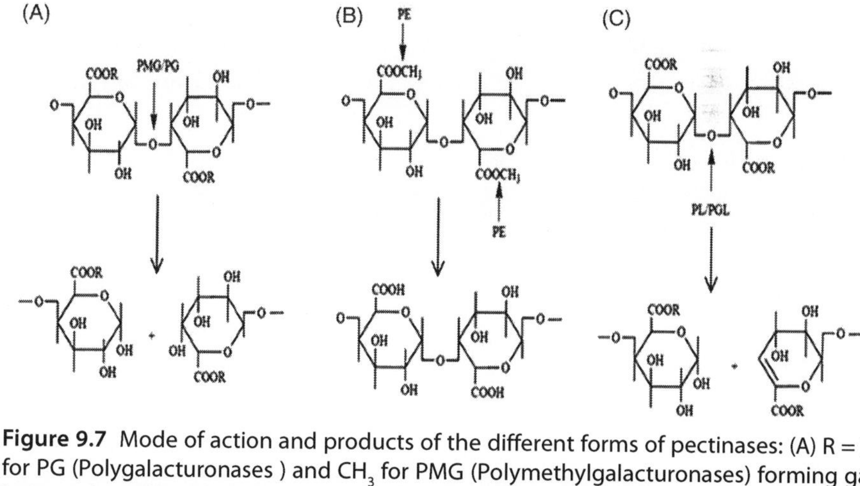

Figure 9.7 Mode of action and products of the different forms of pectinases: (A) R = H for PG (Polygalacturonases) and CH_3 for PMG (Polymethylgalacturonases) forming galacturonic acid by PG and saturated methoxylated galacturonide by PMG; (B) PE (Pectin esterase) forming pectic acid; (C) R = H for PGL (Polygalacturonate lyase) and CH_3 for PL (Pectin Lyase) forming unsaturated galacturonic acid by PG and unsaturated methoxylated galacturonide by PL (Garg et al., 2016).

(PRO) domain (Kohli et al., 2015). Type I PMEs have a 1-3 PRO domains (500–900 amino acids, 52–105 kDa) with 2 or 3 introns while Type II PMEs (250–400 amino acids, 27–45 kDa) contained no PRO domain but had five or six introns (Micheli, 2001; Tian et al., 2006). PMEs associated with bacteria and fungi have structures close to the Type II sequence.

9.2.1.8.2 Polygalacturonases (PG) and polymethylgalcturonases (PMG)

Polygalacturonases basically hydrolyse the α-1,4-glycosidic linkages in poylgalacturonic acid releasing D-galacturonic acid units. Polymethylgalacturonases, on the hand, primarily hydrolyses the α-1,4-glycosidic linkages of highly esterified pectin yielding 6-methyl-D-galacturonate (Jayani et al., 2005). As shown in Fig. 9.7A, both types can act as endo- or exo-enzymes. The former, endo PG (EC.3.2.1.15) and endo-PMG randomly catalyze the hydrolysis of the substrate, while exo PG (EC 3.2.1.67) and exo-PMG act from the nonreducing end of the substrate cleaving mono-galacturonate or digalacturonate units (Rombouts and Pilnik, 1980; Eskin, 1990; Kashyap et al., 2001).

Pectinases are used extensively in fruit and vegetable processing particularly for removing the pulp (Kashyap et al., 2001; Chaudhri and Suneetha, 2012; Khan et al., 2013). Consumers prefer low viscosity and clear juices, which requires the removal of the peels by liquefaction (Kareen and Adebowale, 2007; Makky and Yusoff, 2015). These enzymes have also been used in the clarification of wine, grape juice, and cider (Kibra, 1982; Revilla and Ganzalez-san Jose, 2003; Jayani et al., 2005). Removal of pectins in the cell walls of tea leaves by pectinases accelerates fermentation and is important for the development of color and aroma (Praveen and Suneetha, 2014). Degradation of mucilage surrounding the coffee bean by pectinolyic microorganisms is important in coffee fermentation (Murthy and Nardo, 2011). For the production fruit juice concentrates and tomato puree or ketchup, however, cloud stability is very important. To maintain cloud stability requires the complete inactivation of pectinase activity (Eskin, 1990).

A particular industrial environmental benefit of pectinase is that, together with xylanase, it reduces the amount of chlorine needed for biobleaching by the paper and pulp industries (Bajpai, 1999; Dhimann et al., 2009).

9.2.2 Proteases

9.2.2.1 Generality

Proteases make up the largest single family of enzymes and accounts for 2% of the human genome (Schilling and Overall, 2008; Marnett and

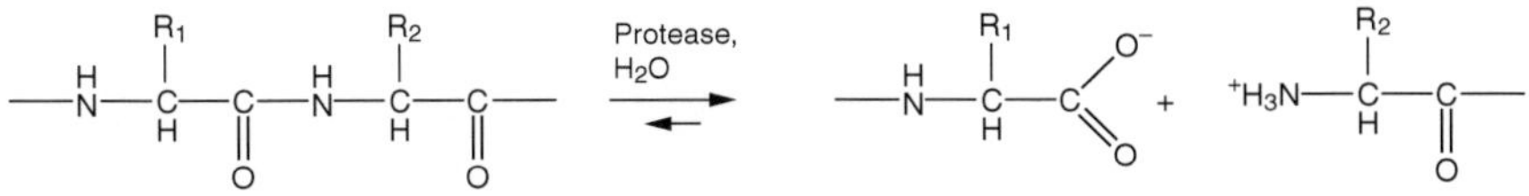

Figure 9.8 Protease enzymatic reaction (Wikipedia).

Craik, 2005). These enzymes, are also named peptidases or proteinases, catalyze the breakdown of protein macromolecules into smaller polypeptides or single amino acids. Their action cleaves the peptide bonds between amino acid residues of the protein chains (Fig. 9.8).

Although the hydrolysis of a peptide bond is thermodynamically favored, it is still a slow reaction without the presence of a protease enzyme. This is due to the fact that the peptide bonds are very stable because their resonance structure forms a partial double bond. In addition, the specificity of the peptide bond hydrolyzed by these enzymes is high.

Proteases were first grouped into 84 families in 1993 according to their evolutionary relationship. They are currently classified according their catalytic mechanisms into six types, aspartate, cysteine, glutamate, metallo, serine, and threonine (Lopez-Otin and Bond, 2008). While these enzymes all hydrolyze peptide bonds, they have different action mechanisms and participate in a vast array of metabolic reactions (Stryer et al., 2019).

9.2.2.2 Types of proteases

Proteases exhibit a broad pH optimum range and have wide commercial applications. They can be classified (Shen and Chou, 2009) into six major sub-classes, serine, threonine, cysteine, aspartate, metallo- and glutamate proteases. Below, are the summarized outlines of the mechanisms of the most important groups of proteases.

9.2.2.2.1 Serine proteases

Serine hydrolases are one of the largest and important known enzyme classes that include approximately 200 enzymes or around 1% of the genes in the human proteome. These enzymes are characterized by the presence of a nucleophilic serine in their active site, which is implicated in the hydrolysis of their substrates. Serine proteases represent over one third of all known proteolytic enzymes in which the serine residue creates a nucleophilic amino acid that cleaves the peptide bond (Rubio et al., 2003). Serine proteases can be found in both eukaryotes and prokaryotes and can be divided into major groups, based on their structure, chymotrypsin-like or trypsin-like and subtilisin-like activities.

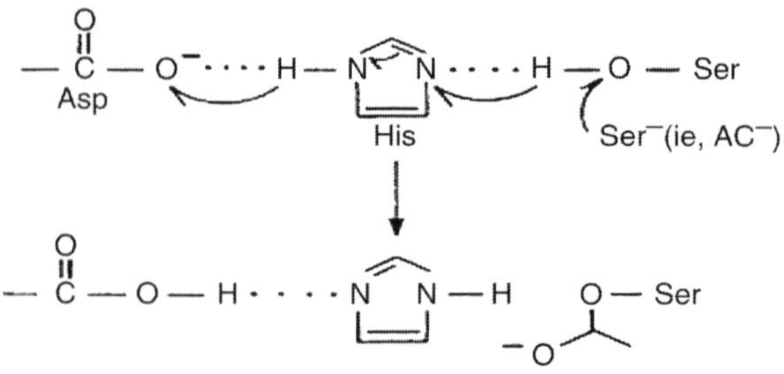

Figure 9.9 Serine protease mechanistic action (Wikipedia).

Serine proteases are alkaline enzymes (Gupta et al., 2002, Beg and Gupta, 2003) that function as endo- or exopeptidases by cleaving peptide bonds in the inner region of the peptide chains or from the terminal end. They exhibit a wide range of optimum pH values ranging from 7.0 to 11.0 (Hedstrom, 2002). In addition, they can also be classified as amino- or carboxy- peptidases depending on the specific site of catalytic action. This group of enzymes are characterized by having a serine residue at the active site (Rubio et al., 2003). For example, chymotrypsin, cleaves peptide bonds selectively on the carboxyl terminal side of the large hydrophobic amino acids such as tryptophan, tyrosine, phenylalanine, and methionine (Fig. 9.9).

Other proteases use different mechanisms in their mode of peptide bond hydrolysis, although the objective in each case is to generate a nucleophile that attacks the peptide carbonyl group.

9.2.2.2.2 Cysteine proteases

Cysteine proteases, also known as thiol proteases, are enzymes that degrade proteins (Verma et al., 2016). These enzymes share a common catalytic mechanism that involves a nucleophilic cysteine thiol in a catalytic triad or dyad set of organization (Fig. 9.10).

Cysteine proteases are commonly found in fruits, such as papaya, pineapple, fig, and kiwifruit. A good example of cysteine protease is papain,

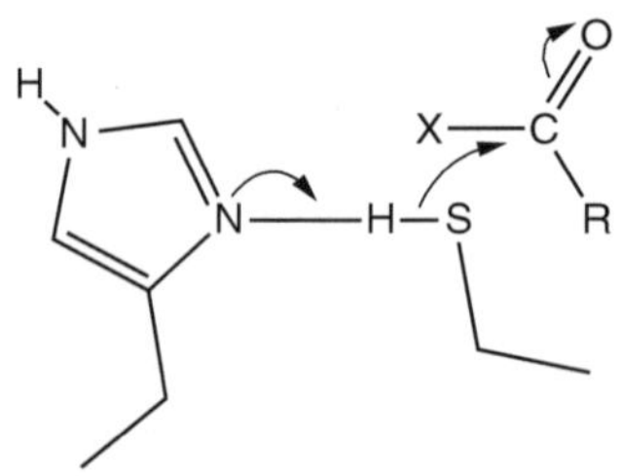

Figure 9.10 Cysteine protease (Wikipedia).

which is found in the papaya fruit, The catalytic mechanism used by this enzyme for the hydrolysis of the peptide bond involves the activation of a cysteine residue by a histidine residue, both are present in the active site of the enzyme. The outcome of the synergistic presence of both amino acids, cysteine and histidine, at the active site of the enzyme is the formation of a powerful nucleophile that can attack the carbon present in the carbonyl group next to the peptide bond.

9.2.2.2.3 Aspartate proteases

Aspartyl proteases are a type of proteolytic enzymes classified as endonucleases (Mamo and Assefa, 2018). They are characterized by having a catalytic triad of three coordinated amino acids Asp- Thr- Gly at the active site (Fig. 9.11). Most of aspartate proteases are found as monomeric enzymes consisting of two domains. Aspartyl proteases are found in vertebrates, plants, plant viruses as well as in retroviruses. They are important in healthy human biological systems for regulating blood pressure and digestion. They are also used as enzymes in food systems (Shahenvaz et al., 2019).

9.2.2.2.4 Metalloproteases

Metalloproteases are a group of enzymes that breakdown proteins normally found in the extracellular matrix proteins or in other words in the spaces between cells in tissues. They are responsible for the degradation of most extracellular matrix proteins, such as collagen, during organogenesis, growth, and normal tissue turnover (Bogyo, 2006). These enzymes need zinc or calcium atoms to proceed with their catalytic activity, and they are hence referred to as metalloproteases (Waschkowitz et al., 2009). Indeed, metalloproteases contain an active site that has a bound metal ion, which is almost always zinc (Fig. 9.12). The mechanism of catalysis of metalloproteases involves activation of a water molecule by the metal ion to carry out a nucleophilic attack on a carbonyl peptide bond. A base is present to help deprotonate the metal-bound water (Tallant et al., 2010; Cerda-Costa and Gomis-Ruth, 2014).

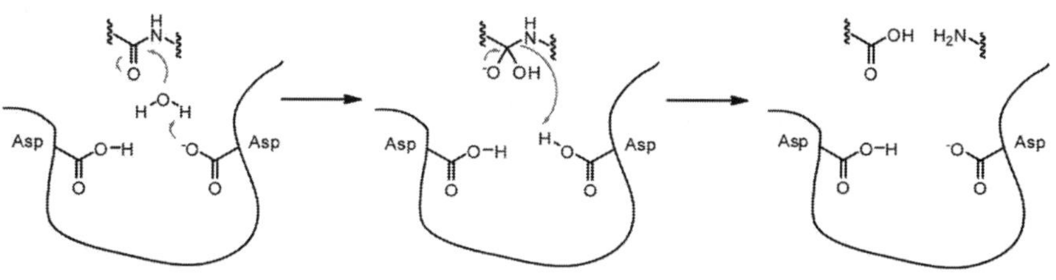

Figure 9.11 Aspartyl protease.

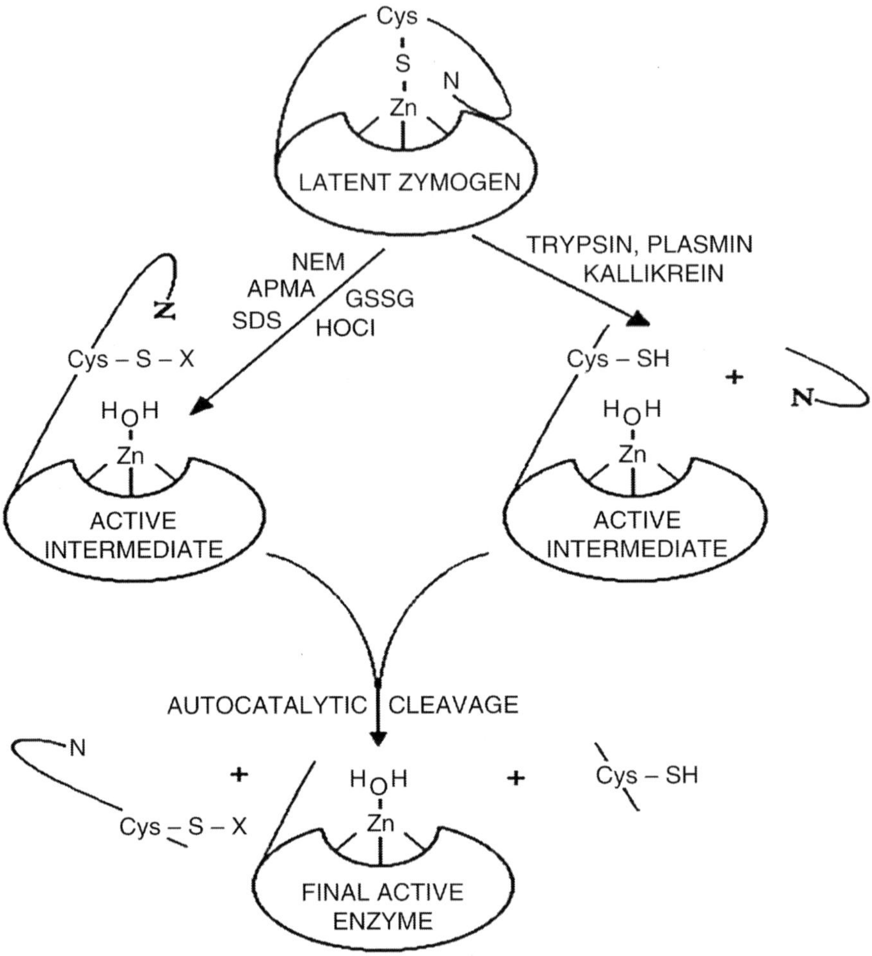

Figure 9.12 Cysteine switch mechanism for activation of metalloproteinases (Chakraborti et al., 2003).

There are two major types of metalloproteases, metalloendopeptidases, and metalloexopeptidases, of which the most common investigated ones are carboxypeptidases A and B, as well as thermolysin (Jablonska-Typuc et al., 2016).

The most common zinc metalloproteases, include the digestive enzymes carboxypeptidases and various matrix metalloproteases (MMPs) secreted by cells as well as one lysosomal protease. Metalloproteases have a role in degrading extracellular matrix during tissue remodeling, cell signaling the release of cytokines or growth factors through cleavage of proteins (Alaiyad and Conant, 2019).

9.2.2.3 Applications of proteases in food industry

Proteases (Flores-Gallegos et al., 2019) are multifunctional hydrolases involved in human metabolism in both healthy and diseased individuals (Carlos López-Otín and Bond, 2008). They are used extensively by the food industries (Mohammadi Panah et al., 2015; Tavano et al., 2018). The use of certain proteases in traditional food processing goes back centuries. For example, chymosin in the rennet extract was recovered from the stomach of calves and used for the production of cheese. Similarly, papain from the leaves and unripe fruit of papaya has been used for the tenderization of meat (Bekhit et al., 2014). These traditional food processing methods have led to the development of various food applications for a wide range of proteases from many sources.

Microbial proteases represent one of the three largest groups of industrial enzymes and account for about 60% of the total global sale of enzymes (Sousa et al., 2016). They are preferred because they exhibit most of the characteristics required for their biotechnological applications (Mamo and Assefa, 2018).

Microbial enzymes are generally preferred to plant sources because of their desirable properties and suitability for industrial use (Gupta et al., 2002). The diversity of microbial sources, with various pH values and enzyme specificities, provide a wide range of industrial applications (Singh and Kumar, 2019).

Microbial proteases have become more and more attractive to the food industry because of their specific properties, high-production yields, specificity for a certain substrates, high activity as well as being environmentally friendly (Razzaq et al., 2019). In addition to the well-known proteases, novel enzymes, and production techniques are being studied due to increased demands for industrial applications (Samanthan et al., 2006; De Souza et al., 2015; Deniz, 2019).

Prospecting for novel enzymes, Nair and Jayachandran (2019) reported a small group of aspartic proteases, including cathepsin, rennin, and pepsin, which play a major role in the degradation of proteinaceous materials. These enzymes are highly specific for a dipeptide with hydrophobic residues and a β-methylene group. Their applications are well-established in the processing and manufacturing of both traditional and novel food products, where they are extensively used in beverage clarification, cheese manufacturing, and wine preservation.

One of the approaches in the biotechnological applications of proteases is protein mapping after a specific protease analysis; however, concepts like degradomics are also taking on more relevant importance in the use and study of proteases (Tavano, 2013; Tavano et al., 2018).

Lukin (2019) reported the presence of protepsin, an enzyme that belongs to the class of acid proteinases containing carboxyl groups at the active center. It cleaves the peptide bonds formed by aromatic amino acids, such as tyrosine and phenylalanine, but not the esters and amides.

A wide range of perspective benefits can be derived from the use of specific proteases in food processing, including changing the nutritional, bioactive, and functional properties of food proteins that improves digestibility, and sensory quality (such as texture or taste), as well as increasing antioxidant capability and reducing allergenic compounds (Mamo and Assefa, 2018).

9.2.2.4 Selected food applications

9.2.2.4.1 Dairy products

The use of rennet in cheese production is an example of the hydrolytic action of proteases to break a specific peptide bond between phenylalanine and methionine residues in the casein protein present in milk. Hence, microbial proteases contribute significantly to the ripening of cheese. This contribution has a direct effect in reducing processing time and costs while increasing production capacity.

Nevertheless, it was important to solve a major problem, which was temperature liability since chymosin was a relatively unstable enzyme. Hence the development of microbial enzymes from *Mucor miehei* was carried out to overcome this problem. Indeed, proteases from *M. miehei* retain their activity during the maturation stages of cheese-making.

The ripening of cheese is a complex process involving biochemical and certain chemical events that produce the characteristics of a given cheese product, including texture, taste, and aroma. The role of proteases in cheesemaking has been the subject of active research for a long time (Sousa-Gallagher et al., 2001).

The undesirable development of bitterness during cheese ripening, however, is due to the release of bitter non-terminal hydrophobic amino acid residues of oligopeptides. However, the presence of proteases during the ripening of cheeses is not totally undesirable as a protease from *Bacillus amyloliquefaciens* is used to promote flavor formation in cheddar. Lipases from *Mucor miehei* or *Aspergillus niger* are also sometimes used to produce stronger flavors in Italian cheeses before the addition of rennet. This is attributed to the release of free butyric acid.

Bitterness in dairy products is due to the accumulation of hydrophobic peptides as a result of the proteolysis. They are derived from both α_{s1}- and β-caseins, so that the total proteolytic activity of the starter strains can have a significant effect on the development of bitterness (Zhang et al., 2019a). The bitter peptides formed in aged cheddar cheese have been characterized (Karamatsi et al., 2014). Saha and Hayashi (2001) reported early on the debittering of protein hydrolyzates. Zhang et al. (2019b) recently investigated alleviating the problem of bitterness and the instability of UHT milk during shelf-life.

While the development of certain desirable flavors in cheese is a slow process, it results from synergistic enzymatic activities of proteases and lipases (Dwivedi et al., 1973). These enzymes are largely found in the cheese milk as well in the added coagulate, such as the rennet and added exogenous enzymes. The exogenous enzymes play a crucial role in reducing the maturation period as well as in the development of desirable flavors (Kailasapathy and Lam, 2005).

9.2.2.4.2 Meat tenderization

The action of endogenous proteases in meat hung after slaughter not only facilitates its tenderization but also improves nutritional quality and the development of desirable flavors (Ketnawa and Rawdkuen, 2012; Moon, 2019). However, the injection of inactive protease extracts from plants, such as papain, bromelain, and ficin, into the jugular vein of the live animal prior to slaughter would result in the tenderization of its meat (Singh et al., 2018). If the active enzyme was injected, however, it would rapidly kill the animal in an unacceptably inhumane manner, so that an inactive oxidized disulfide form of the enzyme is used (Bekhit et al., 2014). Although this process is effective, it has recently been reported to destroy the animal's heart, liver and kidneys and is therefore no longer practiced. To successfully use exogenous proteases for meat tenderization, it is crucial to determine the enzymatic kinetics and condition of meat, such as temperature, pH, temperature, storage time, and other environmental factors (Bhat et al., 2018).

9.2.2.4.3 Soybean products

The use of inexpensive materials, such as soya protein hydrolysates in such products as soy sauce, has considerable enhanced its value as an additive in food industry (Xu-Jur et al., 2015). The degree of hydrolysis provides a range of applications such as whipping expansion, flavor sources, enhanced mouth feeling and even its use in soft drinks and cured meat. Zhao et al. (2018) indicated that *Aspergillus oryzae* is an excellent strain for industrial application, where its proteases play a key role in the development of flavors during soy sauce fermentation. Indeed, many proteases, including dipeptidase, dipeptidyl aminopeptidase, leucine aminopeptidase, aspartic protease, and extracellular metalloproteinase played important roles during the early stage of soybean fermentation. In addition, leucine aminopeptidase A and extracellular metalloproteinase are the dominant proteolytic enzymes during the later period of fermentation. The authors showed that protease interacts synergistically with amylolytic enzymes resulting in the formation of flavors, such as alcohols, acids, esters, aldehydes, furan, and pyrazines.

9.2.2.4.4 Bakery products

There is an increase in the use of proteases in baking industry due to problems associated with gluten (Melim Miguel et al., 2013). The presence of gluten in the diet for certain people could cause inflammation of the small intestine (Heredia-Sandoval et al., 2016). It is important, therefore, to eliminate gluten totally or subject it to partial hydrolysis. Although there is an increase in the market for gluten-free products, the removal of gluten or its substitution with a product of the same characteristics by the baking industry remains a major challenge. The difficulty in finding a suitable alternative ingredient necessitated the use of proteases to partially hydrolyse gluten to try to solve this problem. This led to the initial use of heat-labile proteases that could be readily inactivated early in the subsequent baking process. While proteases have been shown to reduce gluten immunogenicity to an acceptable level, some quality parameters of the products are significantly affected (Melim Miguel et al., 2013). The weak-gluten flour is suitable for baking biscuits as the dough can be spread thinly and retain decorative impressions. Indeed, the gluten in the flour used in biscuit manufacturing must be extensively broken down to prevent the shrinkage of commercial pie pastry.

9.2.2.4.5 Brewing

Enzymes are used extensively in the brewing industry. They could be endogenous enzymes present in the kernel of the malt or exogenous during the fermentation processes. Hence the yield production of beer and its quality depend largely on the types and activities of enzymes (Gomaa, 2018). There are many types of enzymes intrinsic to the grains used in brewing, but brewers can manipulate two major classes of beer enzymes: proteases and diastases. In order to meet the increasing market demands for better quality, the brewing industry continues to employ additive enzymes, such as exogenous enzymes, that contribute to the changes in the mash profile (dos Santos Mathisa et al., 2017).

9.2.2.4.6 Animal and marine wastes

Proteases are used to recover protein from animals and marine organisms as part of reclaiming waste materials by providing its added value (Guomundsdottir and Bjarnason, 2007). There is an increasing trend in food processing, not only for its economic benefits, but also to protect the environment by using selective enzymes (Fernandes, 2014). The biological diversity of marine species offers a wide range of enzymes with unique properties, including the susceptibility to heat denaturation, high-molecular activity at low temperatures, and their ability to catalyze the hydrolysis of native proteins (Haard, 1992). Hence, there is a great potential for the recovery and use of digestive proteases from marine organism waste (Haard and Simpson, 1994). The possible advantage of using an enzyme from a marine organism with specific applications rather than from other living organisms is very attractive. The digestive proteolytic enzymes, including pepsin, trypsin, chymotrypsin, gastricsin, and elastase, from the waste of marine organisms, are of particular interest to the food industry (Haard, 1992). Proteases from cold adapted ectothermic organisms have interesting applications in the food industry since their temperature and other characteristics differ from those from warm-blooded animals (De Vecchi and Coppes, 1996).

9.2.2.4.7 Cocoa fermentation

The fermentation of cocoa seeds by microbial processes involves the action of enzymes, particularly the proteases. Sousa et al. (2016) showed that protease activity during fermentation, involved the presence of aminopeptidase and carboxypeptidase. The fermentation process is also associated with the formation of appreciable flavors (Carlos-Alayao et al., 2019).

9.2.2.4.8 Blood products

Blood contains haemoglobin, an iron-containing protein composed of four polypepides chains. Each chain is attached to a heme group composed of a porphyrin ring with a central iron. Although blood is highly nutritious and readily available from slaughter-house, few products are available other than black puddings. Utilization of blood by the food industry, however, is very limited because of its color. Proteolytic enzymes, such as subtilisin, are used to hydrolyze blood to separate the non-protein haem molecules. The enriched protein hydrolysates are then used in cured meats, sausages, and luncheon meats. Takenaka et al. (2016) demonstrated the ability of aspartic protease from *Aspergillus repens* to successfully remove the red-pigmented proteins during the fermentation of dried bonito.

9.2.2.4.9 Dried cured meat

The processing of dry-cured meat is complex and involves numerous biochemical reactions, in which muscle proteins undergo an intense proteolysis that produces a large number of small peptides together with high amounts of free amino acids. Hence, controlling the enzyme systems, mainly proteases and lipases, is essential for the processing and/or enhancement of flavor quality of products such as dry-cured ham (Toldra and Flores, 1998). The large amounts of peptides and free amino acids determine the final quality of dry-cured hams (Gallego et al., 2016) with identification of an antioxidant peptide and its quantitation successfully reported (Gallego et al., 2018a). Recently, the use of proteomics in food science has improved the characterization of food products and the control of their final quality despite the fact that food systems are under continuous changes during processing and consumption (Gallego et al., 2018b).

9.2.3 Lipases

9.2.3.1 Generality

Lipases (triacylglycerol acylhydrolases (Ec. 3.1.1.3) are hydrolases that catalyze the hydrolysis of triacylglycerols to glycerol and free fatty acids. They are a versatile group of enzymes which catalyze the hydrolysis or synthesis of a broad range of water-insoluble esters. Lipases carry out an essential role in the digestion, transport, and processing of dietary lipids. They belong to the class of α/β-hydrolases which also contains esterases, acetylcholinesterases, cutinases, carboxylesterases and epoxide hydrolases. In fact, lipases are a subclass of esterases, hydrolase enzymes that, in the presence of water, splits esters into an acid and an alcohol by hydrolysis (Panda and Gowrishankar, 2005; Chahinian and Sadra, 2009). A wide range of different esterases exist that differ in their substrate specificity, protein structure, and biological function.

Most lipases act at a specific position on the glycerol backbone of a lipid substrate, such as triacylglycerols (Fig. 9.13), converting it into a mixture of mono- and diacylglycerols as well as free fatty acids. Nevertheless, the ultimate degree of the hydrolysis of a triacylglycerol, in the presence of lipase and water, is the liberation of three free fatty acids and a single molecule of glycerol. Certain specific lipases are also found in nature, such as phospholipases and sphingomyelinases, which are usually treated separately from "conventional" lipases.

Because the lipid substrate is fat-soluble and the enzyme is water-soluble, this reaction takes place at the oil-water interface (Jain et al., 1979).

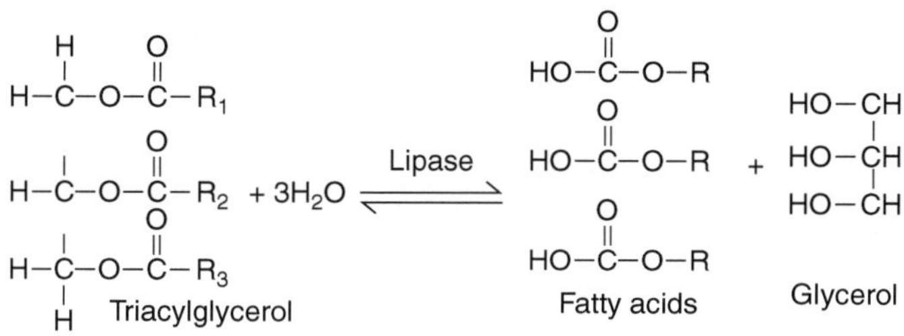

Figure 9.13 Lipase-catalyzed reaction.

Microbial lipases have gained particular prominence by their ability to tolerate extreme temperatures, pH, and organic solvents. Of particular note is their chemo-, regio-, and enantio- selectivity (Verma et al., 2012).

Lipases were originally classified as serine hydrolases as serine was the only common feature in the active site of all lipases enclosed in a highly conserved domain (Antonian, 1988; Widmann et al., 2010). These enzymes are ubiquitous in nature as they are produced by plants, animals, and microorganisms. However, industrial sources of lipases are primarily microbial because of their stability, selectivity, and broad substrate specificity (Dutra et al., 2008; Griebeler et al., 2011). Of the many microbial extracellular lipases, fungal sources are obtained by cultivating in solid-state fermentation, while bacterial and yeast sources require submerged fermentation (Dutra et al., 2008). Continuous research work is aimed at developing strategies for the characterization of fungal lipases that are capable of degrading fatty substances in their applications (Gopinath et al., 2013). The major microbial sources of industrial lipases are *Candida sp.*, *Aspergillus sp.*, *Rhizomucucor sp.*, *Rhizopus sp.*, *Humicola sp.*, *Yarrowia lipolytica,* and *Pseudomas sp.* (Guerrand, 2017).

9.2.3.2 Applications of lipases in food industry

Although lipases represent less than 10% of the global enzyme market, estimated as multibillion dollars, they are used extensively in a wide range of biotechnological applications in food industry, such as fats and oils, dairy and bakery industries, as well as in the pharmaceutical industry (Seitz, 1974; Guerrand, 2017). The ability of lipases to enhance the flavors of different cheese is used extensively by the dairy industry. Such practices include acceleration of cheese ripening as well as lipolysis of butter, fat and cream (Schmidt-Dannert et al., 1994; Sharma et al., 2001). The latter produces short chain fatty acids that enhance the flavor of these products (Vaklhu and Kour, 2006).

Lipases are the most flexible biocatalysts and can be used in a wide range of bioconversion reactions and biotechnological applications such as hydrolysis, interesterification, esterification, alcoholysis and acidolysis. They can also act on a variety of substrates including natural edible oils, synthetic triacylglycerols and esters of fatty acids.

In addition to the food and nutraceutical industries, lipases are used extensively in the detergent industry, the pulp and paper industry, the leather industry, the textile industry, in organic synthesis, as well as in the production of cosmetics and biodiesel (Andualema and Gessesse, 2012). Other estabished novel biotechnological applications of lipases include biopolymer synthesis, treatment of fat-containing waste effluents, as well as in the enantiopure synthesis of pharmaceuticals. Novel lipase genes have been established that will facilitate the production of engineered lipases capable of many new future applications (Aravindan et al., 2007).

Microbial lipases are the only ones commercially available and used in the industry (Horchani et al., 2012). Because of the wide of range of lipase substrates and specificities as well as their high activity, stability and their regio- and/or enantioselectivity, continuous research efforts are aimed at their production and recovery for industrial applications (Vanleeuw et al., 2019). These efforts targeted enzyme kinetics, thermostability and bioactivity (Sharma et al., 2001). The high-level of biomass production of microbial lipases requires not only the efficient overexpression of the corresponding genes, but also a detailed understanding of the molecular mechanisms of gene expression that govern their folding and secretion process (Albayati et al., 2020). The optimization of industrially relevant lipase properties can be achieved by directed evolution (Jaeger and Eggert, 2002).

The use of lipases in industrial applications, however, is still limited by their cost as well as difficulties in their adaptation for selective applications. Research on the production and applications of lipases is expanding, not only in their diverse industrial applications, but also in their structure, classification and sources. Particular emphasis has been made on process optimization, modeling and design that are very critical for further scale-up and industrial implementations of lipases (Samah et al., 2018). The development of low-cost, highly active, stable and specific lipases suitable for a wide range of industrial applications can be achieved by employing protein engineering or directed evolution and rational design as well as the use of structural data for rational modification of lipases (Anobom et al., 2014). However, the screening and the selection of novel lipases with desired and specific properties is time consuming, and costly; hence, efforts are being directed to improve existing lipases by changing the amino acid sequences of the binding lid region of the enzyme that should lead to remarkable enzyme substrate specificity and enantioselectivity improvements (Albayati et al. 2020).

9.2.3.3 Selected bioconversion and food applications

Modification of lipids by lipases has evolved from using the available commercial enzymes to utilizing protein engineering tools (Bernscheeuer, 2013). Indeed, the use of lipases evolved from a simple hydrolysis mechanism of lipids in an aqueous medium to a synthetic one in a hydrophobic medium. The biocatalysis of lipases in a hydrophobic medium is used extensively by industry for the development of structured lipids and nutraceuticals.

9.2.3.3.1 Acidolysis

Acidolysis involves the breakage of a chemical bond by an acid with the hydrogen and anion of the acid become independently attached to the two separated parts. This occurs when triacylglycerols are hydrolyzed in the presence of water, into mono- and diacylglycerols as well as free fatty acids. Lipases are used in acidolysis process (Karam et al., 2009) for the recovery of free fatty acids (Aziz et al., 2015) and the intermediate triacylglycerol hydrolysis, mono- and diacylglycerols (Safari et al., 2009) as well as the esterification of fatty acid alcohol ester with selected nutritional value compounds such as phenolic acids (Lue et al., 2005; Sabally et al., 2005).

9.2.3.3.2 Structured lipids

Transesterification is an excellent example of the industrial application of lipases, each with a different fatty acid specificity, for the safe and effective production of a variety of structured oils (Utama et al., 2019). Utsushi et al. (2009) reported that of 12 lipases examined, 5 had 1,3-regio specificity while 7 lipases had non-regio specificity. The structured lipids were prepared efficiently by transacylation but required the appropriate selection of lipases with higher transacylation of hydrolysis activities of specific fatty acids. Canet et al. (2016) pointed out that the nature of fatty acids and their distribution in the triglycerides greatly influence the transesterification process.

Using tricaprylin and trilinolenin as *substrates.* Bai et al. (2012) reported the synthesis by lipase of medium-long-medium type structured lipids. The lipase-catalyzed synthesis and characterization of structured flaxseed oil was also reported (Khodadadi et al., 2013). Using the response surface methodology, interesterification of flaxseed oil with tricaprylin by lipase was optimized for the production of structured lipids (Khodadadi and Kermasha, 2014a) followed by the modeling of this lipase-catalyzed reaction (Khodadadi and Kermasha, 2014b).

9.2.3.3.3 Nutraceuticals

The potential of lipases as biocatalysts in the production of fatty acid derivatives for the food and nutraceutical industries, such as flavoring esters, fatty acid esters of antioxidants, and structured lipids, is enormous. This is mainly due to their high regio- and stereo-selectivities (Ferreira-Dias et al., 2013). In addition, the enzymatic synthesis of structured phenolic lipids by acidolysis of flaxseed oil with selected phenolic acids has also been reported (Karboune et al., 2008). Using enzymatic acidolysis, Kavadio et al (2018) reported the synthesis of designer triglycerides by incorporating caprylic acid into a fully hydrogenated soybean oil.

Indeed, there has been a marked increase in the use of nutraceuticals and food supplements by consumers. In addition, the increased trend by the food industry in using natural additives and ingredients, such as flavors and fragrances as well food colorants, is a direct response to the consumers' concerns for natural ingredients. Lipases play a major role in the production of these types of products. The development of novel biomolecules with beneficial nutraceutical and physical properties is an example of the current trend of creativity. Safari et al. (2006) reported the enzymatic synthesis by lipase of structured phenolic lipids using selective phenolic acids and triolein as substrate models. Research was also carried out that developed phenolic lipids also using phenolic acids, but with fish liver oil (Sabally et al., 2007), flaxseed oil (Sorour et al., 2012) and krill oil (Aziz et al., 2012). In addition, the successful encapsulation of the developed novel nutraceuticals was also reported (Kermasha et al., 2017).

9.2.3.3.4 Flavors and fragrances

The use of lipases in industrial applications has also attracted special attention for the synthesis of value-added flavors and fragrances, including their role in the liberation of free fatty acid for their bioconversion by other enzymes, such as lipoxygenase and hydroperoxide lyase, into carbonyl compounds (Aziz et al., 2015) as well their interaction with alcohol to form fatty acid acyl alcohol esters (Dhake et al., 2013).

The production of short-chain fatty acids can enhance the flavor of dairy products (Vaklhu and Kour, 2006). This is evident by the biogeneration of short-chain fatty acid alcohol esters with fruity flavors by selected microorganisms (Kermasha et al., 1999; Alvarez-Macarie and Barrati, 2000; Nardi et al., 2002). De Lucia et al. (2019) evaluated the effect of esterase from *Alyciclobacillus acidocaldarius* on lipolysis and volatile compounds profiles of semi-hard cheese. The addition of the esterase increased the degree of lipolysis by 30% and enhanced the relative proportion of short- and medium chain fatty acids. The overall finding suggested that the mechanism of esterase is different from that of lipase with a preference to release of short chain fatty acids.

9.2.3.3.5 Dairy products

Cheese ripening or cheese maturation is a process in the cheese-making. This process is associated the activities many microbial proteases and lipases changing the morphology and physical texture of the cheese as well as developing the taste and flavor of the product. The ability of lipases to enhance the flavors of different cheeses is used extensively by the dairy industry. Such practices include acceleration of cheese ripening as well as lipolysis of butter, fat and cream (Schmidt-Dannert et al., 1994; Sharma et al., 2001).

Yilmaz et al. (2005) reported that the addition of commercial lipase (Piccantase A) to milk prior to the addition of rennet in the processing of Tulum cheese accelerated flavor development over a short ripening period. Rani and Jagtap (2019) investigated the effect of exogenous lipase enzyme on the ripening of Swiss cheese in terms of physico-chemical, textural and sensory characteristics as well as its microstructure. The authors indicated that the addition of microbial lipase may significantly reduce the cost of the cheese production by shortening the ripening period while maintaining of the quality of the final product.

9.2.3.3.6 Egg processing and enzymatic degumming

Phospholipases play an important role in egg processing by improving the emulsifying power of egg lipids (Guerrand, 2017). The enzymatic conversion of egg yolk phospholipids to lyso-phospholipids enhances its emulsification properties which reduces the amount of egg yolk needed in dressings and mayonnaise-like products. The application of phospholipases in the edible oil industry eliminates impurities in the crude oil that could impact the sensory and visual appearance of the oil as well as its stability. Phospholipids, referred to as gums, were traditionally removed during refining by thermochemical processes (Unger, 2011). Enzymatic hydrolysis of the phospholipids at the water-oil interface, however, prevents the formation of an emulsion which facilitates the release of the oil. This process, known as enzymatic degumming, is widely practiced in the edible oil industry as it not only reduces the amount of energy and water used but also produces higher oil yields.

9.2.3.3.7 Dry cured meat

The processing of dry-cured meat is complex and involves numerous biochemical reactions, where muscle proteins undergo an intense proteolysis that produces a great number of small peptides as well as free amino acids. Hence, controlling the enzyme systems, mainly proteases and lipases, is essential for the processing and/or enhancement of flavor quality of the product (Toldra and Flores, 1998). To maintain the quality of such products and to inhibit enzymes involved in the oxidation process, it is essential that in addition to the maintenance of glutathione peroxidase activity, is also to inhibit phospholipase as well as neutral and acid lipase, and promote the superoxide dismutase activity (Sheng et al., 2019).

9.2.3.3.8 Cold-active enzymes

An interesting example of the efforts made for new applications of lipases is the development of cold-active enzymes that have potential applications in reactions requiring lower temperatures. Čanak et al. (2015) investigated the lipases obtained from the cultures of *Candida, Yarrowia, Rhodotorula,* and *Pichia* genera, recovered from refrigerated poultry meat; the experimental data showed that lipase recovered from the culture of *Candida zeylanoides* may have the potential to become an important industrial enzyme.

9.2.3.3.9 Food and toxic wstes

One of the major challenges is the ability to distinguish within the carboxylesterases, or carboxyl ester hydrolases, two groups of enzymes, namely nonspecific esterases (EC 3.1.1.1) and lipases (EC 3.1.1.3). They differ from each other by their substrate specificity. Esterases hydrolyze solutions of water-soluble short acyl chain esters while lipases hydrolyze the water-insoluble long chain triacylglycerols (Chahinian and Sadra, 2009). There are significant structural differences, as unlike esterases, lipases have hydrophobic amino acid residues in the vicinity of their active sites. The difference between the two groups is important in term of their applications. In addition to their major role in the degradation of natural materials and industrial pollutants, such as cereal wastes, plastics, and other toxic chemicals, esterases have useful applications in agriculture, food, and pharmaceutical industries (Panda and Gowrishankar, 2005).

9.3 Oxidoreductases

Oxidoreductases are a group of enzymes that catalyze oxidation-reduction reactions by incorporating such cofactors as heme, flavin, and metal ions. The redox reaction they catalyze involves transferring electrons from a large number of electron-donors (reductants) to a variety electron acceptor (oxidant) molecules (Gygli and van Berkel, 2015). To carry out these reactions, oxidoreductases use a variety of redox centers (Munro et al., 2000). Compared to hydrolases, the applications of oxidoreductases are fewer with the main one being glucose oxidase (EC 1.1.3.4) while catalase (EC 1.11.1.6) has a more limited use. Other important oxidoreductases used industrially include lipoxygenase (EC 1.13.11.12) and laccase (EC 1.13.11.12).

9.3.1 Glucose oxidase

Glucose oxidase (GOD: β-D-glucose:oxygen 1-oxidoreductase; glucose aerodehydro-genase; E.C. 1.1.3.4) is a very important oxidoreductase enzyme. It first discovered by Muller (1928) and is one of the most widely studied enzymes because of its extensive industrial applications (Wong et al., 2008; Dubey et al.; 2017; Shaikh Sumaiya and Trivedi, 2015). In the presence of molecular oxygen, it catalyzes the oxidation of β-D-glucose to gluconic acid. The primary source of this enzyme is fungal, *Aspergillus niger*, in which glucose oxidase was shown to be a glycoprotein composed of two identical 80 kDa subunits bound with two FAD coenzymes. The high specificity of glucose oxidase for β-D-glucose results in its oxidation to D-gluconolactone and hydrogen peroxide, both of which breakdown spontaneously and catalytically to gluconic acid and water, respectively. Commercial glucose oxidase contains catalase as it is found together

$$\beta\text{-D-glucose} + O_2 \rightarrow \text{D - gluconolactone} + H_2O_2$$
$$2H_2O_2 \rightarrow 2H_2O + O_2$$
$$\delta\text{ - gluconolactone} + H_2O \rightarrow \text{Gluconic acid}$$

with glucose oxidase in the mycelium cell wall of *Aspergillus niger*. Catalase catalytically breaks down hydrogen peroxide preventing the accumulation of hydrogen peroxide which can inactivate glucose oxidase (Bao et al., 2003; Kleppe, 1966). In addition, the increase in acidity from D-gluconic acid can also inhibit glucose oxidase (Morin et al., 2004).

The GRAS (Generally Regarded as Safe) status of glucose oxidase under FDA classification (FDA/CFSAN, 2002a,b) allows its extensive use in the food industry. One of the major uses of glucose oxidase is in the removal of glucose during the production of dried egg powder by adding it to the liquid egg prior to spray drying. The formation of gluconic acid blocks the reducing end of glucose rendering it unable to participate in the Maillard reaction with the amino groups of amino acids and protein and form undesirable brown products or melanoidins (Sisak et al., 2006). An additional benefit is that hydrogen peroxide produced in this reaction destroys or inhibits any microorganisms naturally present in the liquid egg (Dobbonie et al., 1955). This can be done in a batch process, although a continuous process is possible by immobilizing the glucose oxidase preparation.

Glucose oxidase is also an effective antioxidant as it removes oxygen in converting glucose to gluconic acid. This can prevent oxidative rancidity in such high-fat food products such as mayonnaise and salad dressings (Isakien and Adler-Nissen, 1997). It is also effective in removing oxygen in wines and beers to protect and maintain taste and flavor (Labuza and Breene, 1989). By removing oxygen in the headspace of canned and bottled foods, glucose oxidase maintains an anaerobic environment that prevents microbial growth (Kirk et al., 2002).

Another application of glucose oxidase is replacing potassium bromate ($KBrO_3$), an effective maturing oxidizing agent used in breadmaking to strengthen gluten to improve the bread texture. Potassium bromate, a proven carcinogen, is no longer permitted in foods and has now been replaced by glucose oxidase in many countries (Moore and Chen, 2006). As an effective antioxidant it also produces hydrogen peroxide which enhances the elasticity and viscosity of the dough improving both bread texture and volume (Rasiah et al., 2005).

9.3.2 Catalase

Catalase (EC 1.11.1.6), another oxidoreductase, quenches reactive oxygen species such as hydrogen peroxide, a by-product of aerobic oxidation. It is composed of four polypeptide each containing over 600 amino acids with four porphyrin heme groups containing a central iron atom (Kaushal et al., 2018). In the production of Swiss cheese, hydrogen peroxide is used to destroy pathogenic bacteria while preserving the natural milk enzymes important to flavor. This process, known as cold pasteurization, however, requires the removal of hydrogen peroxide prior to cheesemaking. This is carried out by catalase which efficiently removes hydrogen peroxide from the milk samples (Geciova et al., 2002).

A particular useful application of glucose oxidase/catalase is in active packaging designed to prevent the oxidation of food products (Vermeiren et al., 1999). Together in food wrappers glucose oxidase scavenges oxygen while catalase is needed to remove hydrogen peroxide, considered a dangerous by-product (Grigoras, 2017; Nikola and Marco, 2015).

9.3.3 Lipoxygenase

Lipoxygenase (linoleate: oxygen oxidoreductase, EC 1.13.11.12) (LOX), includes a group of non-heme metal-containing dioxygenases (Fig. 9.14). While the enzyme was originally observed in plants, it is widely distributed in both fungi and animals (Axelrod, 1974; Eskin et al., 1977; Hayward et al., 2017; Heshof et al. 2016). In the presence of oxygen, lipoxygenase catalyzes the regio- and stereo-specific peroxidation of polyunsaturated fatty acids with a cis, cis-1,4-pentadiene containing structure (Vora et al., 2018). The enzyme was first reported in soybean flour by Bohn and Hass (1928) as the factor responsible for the bleaching of wheat flour pigments. Because the pigment involved was carotene, it was originally referred to as carotene oxidase. Subsequent research showed that the enzyme responsible was a fat oxidase, lipoxygenase, that catalyzed the peroxidation of polyunsaturated fatty acids (Axelrod, 1974). These names were consolidated to

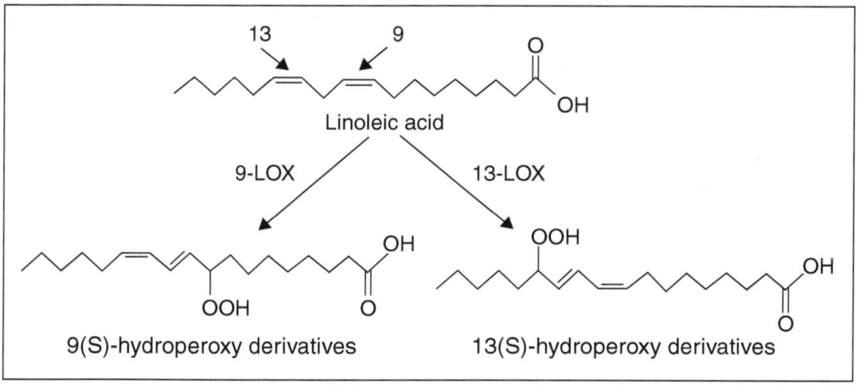

Figure 9.14 Lipoxygenase conversion of linoleic acid into 9- and 13 hydroperoxide isomers (Wikipedia).

lipoxygenase which covered the combined activities of this enzyme (Hayward et al., 2017).

In addition to its role in bleaching wheat flour pigments, lipoxygenase also acts as an oxidative improving agent in the development of the dough texture. In particular, oxidation of the flour fatty acids by lipoxygenases oxidizes the thiol groups in glutenins. This facilitates the formation of disulfide bonds between the glutenin proteins essential for the formation of gluten (Poppa et al., 2006; Permyakova and Trifamov, 2011). Consequently, the enzyme-active soybean flour as a dough improver is a standard ingredient for the production of white bread (Faubion and Hoseney, 1981; Luning et al. 1991). The synthesis of polyunsaturated fatty acid hydroperoixdes by lipoxygensae (Bisakowski et al., 1997), is considered as essential for the formation flavor precursors (Hall et al., 2004 and 2005; Aziz et al., 2014).

9.3.4 Laccase

Laccase (*p*-diphenol dioxygen oxidoreductase, EC 1.10.3.2), a polyphenol oxidase, is a member of the multicopper protein family. In the presence of molecular oxygen, it degrades wood components such as lignin (Claus, 2003; Dittmer et al., 2004; Durao et al., 2006). Laccase was originally discovered in the exudates of the Japanese lacquer tree in 1883 but was subsequently identified as a fungal enzyme (Yoshida, 1883; Levine, 1965). Fungal sources are very rich in laccase, particularly *Ascomyces*, *Basidiomyces,* and *Deuteromyces* (Yang et al., 2017) It is used to stabilize and improve the color of fruit juices by preventing the occurrence of enzymatic browning from the polyphenols present (Ribiero et al., 2010; Verma et al., 2018). The formation of haze or sediment involves the interaction between proteins and polyphenols which can be readily prevented by a combination of laccase and filtration (Moore and Chen, 2006; Nerifar et al., 2009). Recent studies showed that lactase immobilized on inexpensive carriers proved very effective in fruit juice clarification (De Souza Bezerra et al., 2015).

Laccase, as an alternative to physical adsorbents, is used extensively for wine stabilization (Minussi et al., 2007). The selective removal of polyphenols from the wine is crucial to prevent any undesirable changes in organoleptic (color and taste) properties (Osma et al., 2010). The use of laccase in wine cork stoppers oxidatively reduces cork taint and/or astringency generally associated with aged bottled wine (Conrad et al., 2000). Laccase is also used to enhance the storage life of beer by preventing haze formation as well as reducing the oxygen content, particularly at the end of the beer production (Mathiasen, 1991; Conrad et al., 2000).

9.4 Isomerases

9.4.1 Glucose isomerase

This enzyme is responsible for the large-scale production of sweet syrups by catalyzing the reversible isomerisation of glucose to fructose (Eskin, 1990). It converts glucose, released from corn starch by a combination of heat, chemicals, α-amylase, and glucoamylase, to fructose for the production of high fructose corn syrup (HFCS). The sweetness, solubility, acidity and inexpensiveness of HFCS has made it a very popular ingredient in a wide variety of processed food products. However, coincident with its increased consumption are many health concerns, such as a dramatic rise in obesity, diabetes, cardiovascular diseases, and metabolic syndrome (Parker et al., 2010).

Initially, the commercial manufacture of glucose isomerase was far too expensive and uncompetitive with the production of invert sugar from sucrose. This changed with the discovery by Japanese researchers of a strain of *Streptomyces* sp. in which glucose isomerise could be induced by xylan or xylose (Takasaki, 1972, 1974; Takasaki and Tanabe, 1971). A large number of *Streptomyces* species as well as other bacterial sources have since been identified (Bhosale et al., 1996; Chanitnun and Pinphanichakarn, 2012). As a result, this enzyme is now designated glucose isomerase (D-xylose-aldose-ketose-isomerase EC. 5.3.1.5) as it catalyzes the reversible isomerisation of D-glucose and D-xylose to the corresponding keto sugars, D-fructose and D-xylulose. Glucose isomerase produced from a nongenetically modified *Streptromyces murinus* strain NZYM-GA by Novozymes A/S was recently approved for food use by the European Safety Authority in 2018.

9.5 Lyases

Lyases catalyze addition and elimination reactions that break bonds between a carbon atom and other atoms such carbon, oxygen, or sulfur. α-Acetolactone decarboxylase, hydroperoxide lyase and pectate lyase, are three enzymes in this category with industrial applications.

9.5.1 α-Acetolactone decarboxylase (ALDC)(EC 4.1.1.5)

The volatile flavor compound diacetyl (2,3-butanediol), is found in a wide variety of foods (Krogerus and Gibson, 2013). Its presence in beer, however,

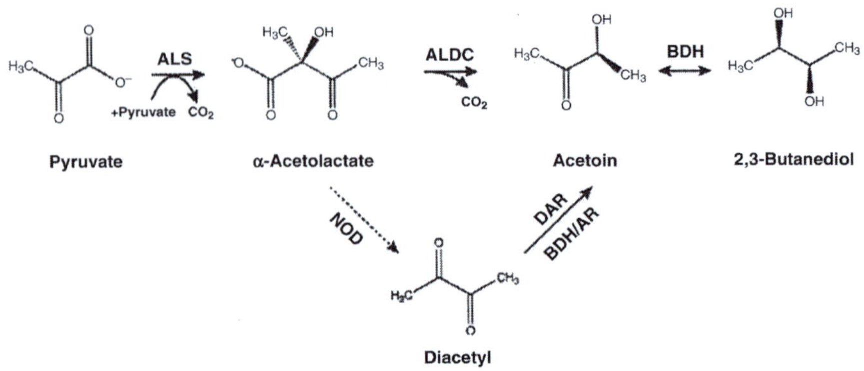

Figure 9.15 Scheme showing the α-acetolactate pathway from pyruvate. ALS,α-acetolactate synthase; ALDC, α-acetolactate decarboxylase; BDH, 2,2-butanediol dehydrogenase; AR, acetoin reductase; DAR, dioacyl rductase; NON, nonoxidizing nonenzymatic decarboxylation (Ji et al., 2019).

is a particular problem due to its strong buttery flavor (Garcia et al., 1996). Diacetyl, a vicinal diketone, is formed from acetolactate, an intermediate in the biosynthesis of valine by yeast. α-Acetolactone decarboxylase (ALDC) (4.1.1.5), an enzyme produced by a modified strain of *Bacillus subtilis*, directly converts α-acetolactone to acetoin, thereby reducing the production of diacetyl. (Fig. 9.15) Acetoin is then converted to 2,3-butanediol (BD) by butanediol dehydrogenase (BDH) or acetoin reductase.

α-Acetolactate formed from pyruvate can undergo nonoxidize non-enzymatic oxidative decarboxylase (NOD) to diacetyl which can then be converted to acetoin by diacetyl reductase (DAR) (Fig. 9.6). Consequently, ALDC plays an important role in the accelerating the maturation of beer by reducing diacetyl formation.

ALDC was first isolated from an *Enterobacter aerogenes* strain 1033 and found to effectively remove α-acetolactone and α-aceto-α-hydroxybutyratelactone from freshly fermented beer reducing both diacetyl and 2,3-pentanedione levels below the taste threshold, respectively (Godfredsen and Ottensen, 1982; Godfredsen et al., 1984). An encapsulated ALDC by Dulieu et al. (2000) improved its performance in controlling beer fermentation. A more recent study by Choi et al. (2015) reported a 25% and 15% reduction in the diacetyl content in beer brewed in South Korea from two-row barley (Jinyang) and six-row barley (Dahyang) with added ALDC. The decreased levels in diacetyl proved very acceptable by the microbrewery. A more detailed description of the structure and mechanism of ALDC was published by Marlow et al. (2013). In addition to its importance in brewing, ALDC is used for the synthesis of enantiometrically pure diols (Yan et al., 2009).

9.5.2 Hydroperoxide lyase (4.2.1.92)

Hydroperoxide lyases {12,13-hydro-lyase [(9Z)-(13S)-12,13-epoxyocta-deca-9,11-dienoate-forming] EC. 4.2.1.92} are enzymes that catalyze the cleavage of C-C bonds in the hydroperoxides of fatty acids (Fig. 9.16). They belong to the cytochrome P450 enzyme family (CYP74C and CYP74B) (Grechkin et al., 2006). Polyunsaturated fatty acids such as linoleic and linolenic acids are susceptible to the formation of hydroperoxides on contact with oxygen in air. Hydroperoxides are highly reactive functional groups since they contain an oxidant (O-O bond) adjacent to a reductant (C-H bonds). When flanked by olefins, the hydroperoxides can be induced to rearrange to form a hemiacetal. It is this reaction that is catalyzed by hydroperoxide lyases. The resulting aldehydes are notable as fragrances, green leaf volatiles, and antifeedants (Matsui, 2006). Using circular dichroism (CD), the secondary structure of purified greem bell hydroperoxide lyase cloned in yeast *Yarrowia lipolytica,* was confirmed by Santiago-Gómez et al. (2008) in selected organic media.). Another study by Tran Thanh et al. (2008)

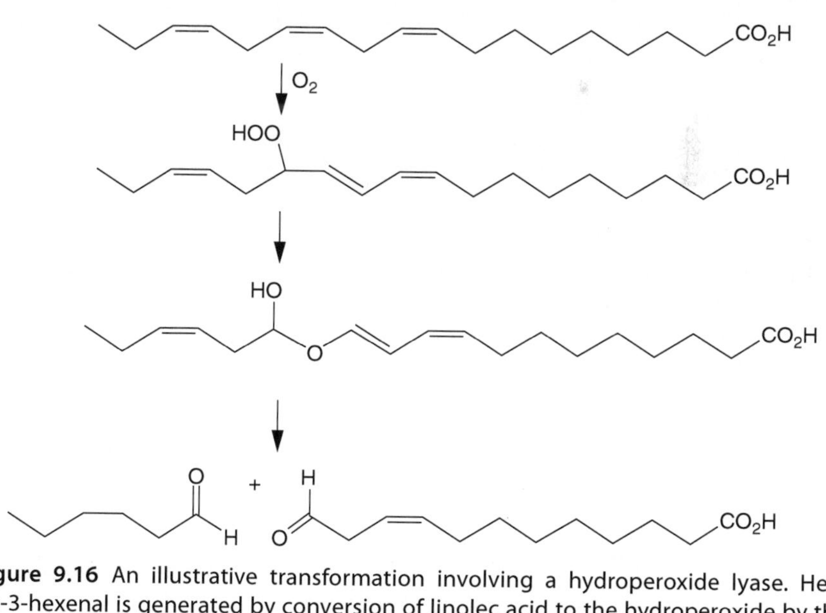

Figure 9.16 An illustrative transformation involving a hydroperoxide lyase. Here cis-3-hexenal is generated by conversion of linolec acid to the hydroperoxide by the action of a lipoxygenase followed by the lyase-induced formation of the hemiacetal (Wikipedia).

investigated the mechanism for the production of aroma compounds by a recombinant hydroperoxide lyase in the yeast *Yarrowia lipolytica*.

Hydroperoxide lyase, in combination with lipoxygenase, catalyze the production of selected flavor compounds (Hall et al., 2008; Patui et al., 2010). One such volatile compound, hexanal, is an invaluable natural food aroma compound produced by the synergistic action of both of these enzymes (Husson et al., 2002; Marczy et al., 2002). Modeling hexanal production in oxido-reducing conditions by yeast *Yarrowia lipolytica* expressing hydroperoxide lyase from green bell pepper was reported by Santiago-Gómez et al. (2009).

De novo synthesis of natural flavors and aroma (Vincenti et al., 2019) is thus carried out by lipoxygenase system in association with other enzymatic systems particularly hydroperoxide lyase (Kermasha et al., 2002; Aziz et al., 2014). Progress in genetic engineering is expected to identify metabolic bottlenecks by creating novel high-yielding microbial strains (Berger, 2009).

9.5.3 Pectate lyase (EC 4.2.2.10)

Pectate lyases cleave by β-elimination (Fig. 9.7C), the α-1-4 galacturonate units at the non-reducing of the pectin chain, forming 4,5-unsaturated oligogalacturonates (Pedrolli et al., 2009). These enzymes, obtained from fungi, were shown to have optimum activity in acid and neutral media while bacterial sources were more active in alkaline conditions. Zhou et al. (2015) successfully engineered an alkaline pectate lyase from *Bacillus* sp. Strain M16-56 to improve its thermostability. Using the directed evolution method, they produced a mutant (EA) with a 24-fold increase in t_{50} value at 50°C compared to the wild type enzyme as well as an increase in activity of 23.3%. Pectate lyases are used extensively for clarifying fruit juices and are preferred by industry as they do not produce toxic methanol.

9.6 Conclusion

The overall information gathered in this chapter provides a comprehensive review of the main groups of enzymes, carbohydrases, proteases, lipases and fatty acid esterases, oxidases as well as lyases. The novel approaches and strategies discussed in this book will hopefully enhance the biotechnological use of enzymes to ensure a safe and nutritious food supply for an increasing world population. Green processing with enzymes is not only energy saving but will also help to maintain a sustainable environment. The next decade will see the development of a whole new generation of artificial enzymes that will not only change our understanding of catalysis but also have a major impact on food processing, medicine and pharmacology.

References

Aiyer, P.V., 2005. Amylases and their applications. African J. Biotechnol. 4, 1525–1529.

Alaiyad, S., Conant, K., 2019. A role of matrix metalloproteasees in antidepressant efficacy. Front. Mol. Neurosci., (Doi: 10.3389/fnmol.2019.00117).

Albayati, S.H., Masomian, M., Ishak, S.N.H., Mohamad Ali, M.S., Thean, A.L., Sheriff, F.B.M., Noor, N.B.M., Abd Rahman, R.N.Z., 2020. Main structural targets for engineering lipase substrate specifity. Catal., (Doi: 10.3390/catal.10770747).

Alvarez-Macarie, E., Barrati, J., 2000. Short chain flavour ester synthesis by a new esterase from *Bacillus licheniformis*. J. Mol. Catal. B: Enzym. 10, 377–383.

Andualema, B., Gessesse, A., 2012. Microbial lipases and their industrial applications: Review. Biotechnology 11, 100–118.

Antonian, E., 1988. Recent advances in the purification characterization and structure determination of lipases. Lipids. 23, 1101–1106.

Aravindan, R., Anbumathi, P., Viruthagiri, T., 2007. Lipase applications in food industry. Indian J. Biotechnol. 6, 141–158.

Anobom, C.D., Pinheiro, A.S., De-Andrade, R.A., Aguieiras, E.C.G., Andrade, G.C., Moura, M.V., Almeida, R.V., Freire, D.M., 2014. From structure to catalysis: Recent developments in the biotechnological applications of lipases, In: Hmidet, N. (Ed.) Recent development, in: Production and Biotechnological Application of Microbial Enzymes, Hindawi Publisher, Open Access, Vol. 2014 (Doi: org/10.1155/2014/684506).

Axelrod, B., 1974. Lipoxygenases. In: Whitake, J.R. (Ed.), Food Related Enzymes. Advances in Chemistry, 136, American Chemical Society, Washington, D.C, pp. 324–328.

Aziz, S., Dutilleul, Kermasha, S., 2012. Lipase-catalyzed transesterification of krill oil and 3,4-dihydroxyphenyl acetic acid in solvent-free medium using response surface methodology. J. Mol. Catal. B: Enzym. 84, 189–197.

Aziz, M., Ben Akacha, N., Husson, F., Kermasha, S., 2014. Synthesis of linoleic acid hydroperoxides as flavor precursors, using selected substrate sources. J. Amer. Oil Chem. Soc. 91, 1867–1876.

Aziz, M., Husson, F., Kermasha, S., 2015. Optimization of the hydrolysis of safflower oil for the production of linoleic acid, used as flavor precursor. Int. J. Food Sci., http://dx.doi.org/10.1155/2015/594238.

Bai, S., Aziz, S., Khodadadi, Bou-Metri, C., St-Louis, R., Kermasha, S., 2012. Lipase-catalyzed synthesis of medium-long-medium type structured lipids using tricaprylin and trilinolenin as substrate models. J. Amer. Oil Chem. Soc. 90, 377–389.

Bajpai, P., 1999. Application of enzymes in the pulp and paper industry. Biotechnol. Prog. 15, 147–157.

Bao, J., Furumoto, K., Yoshimoto, M., Fukunaga, K., Nakao, K., 2003. Competitive inhibition by hydrogen peroxide produced in glucose oxidation catalyzed by glucose oxidation. Biochem. Engineer. 13, 69–72.

Barbosa, A.M., Dekker, R.F.H., Giece, E.C., 2010. Bioactive Oligosaccharides: Production, Biological Functions and Potential Commercial Applications. Nova Science Publishers, New York, N.Y.

Beg, Q.K., Gupta, P., 2003. Purification and characterization of an oxidation stable, thiol-dependent, serine alkaline protease from *Bacillus mojarensis*. Biochemistry 39, 203–209.

Beg, Q.K., Kapoor, M., Mahajan, l., Hoomdal, G.S., 2001. Microbial xylanases and their industrial applications: A review. Appl. Microbiol. Biotechnol. 56, 326–338.

Bekhit, A.A., Hopkins, D.L., Geesink, G., Behkit, A.A., Franks, P., 2014. Exogenous proteases for meat tenderizaion. Crit. Rev. Food Sci. Nutr. 54, 1012–1031.

Ben Messaoud, E., Ben Ammar, Y., Mellouli, L., Bejar, S., 2002. Thermostable pullulanase type 1 from new isolated *Bacillus thermoleavorans* U105: Cloning, sequencing and expression of the gene in *E. coli*. Enzym. Microb. Technol. 31, 827–832.

Bender, H., Wallenfels, K., 1961. Untersuchungen an pullulan II specifischer abbau durch ein bakterielles enzyme. Biochem. Z. 34, 79–95.

Berger, R.G., 2009. Biotechnology of flavours: The next generation. Biotechnol. Lett. 31, 1651, (Doi: 10.1007/s10529-009-0083-5).

Bernscheeuer, U.T., 2013. Enzymes in lipid modification: From classical biocatalysis with commercial enzymes to advanced protein engineering tools. Oilseed Fat Crops Lipids (OCL) 20, 45–49.

Bertoft, E., 2017. Understanding starch structure: Recent progress. Agronomy 7, 56.

Bezerra, C.S., Lemos, C.M.G.D.F., Souza, M.D., Gonclaves, L.R.B., 2016. Enzyme immobilization onto renewable polymeric matrixes. Past, present, and future trends. J. Appl. Polym. Sci. 132, 42125.

Bhat, Z.F., Morton, J.D., Mason, S.L., Bekhit, A.E.-D., 2018. Applied and emerging methods for meat tenderization: A comprehensive perspective. Compr. Rev. Food Sci. Food Safety 17, 941–959.

Bhosale, S.H., Rao, M.B., Deshpande, V.V., 1996. Molecular and industrial aspects of glucose isomerase. Microbiol. Rev. 60, 280–300.

Bisakowski, B., Perraud, X., Kermasha, S., 1997. Characterization of hydroperoxides and carbonyl compounds obtained by lipoxygenase extracts of selected microorganisms. Biosci. Biochem. Biotech. 61, 1262–1269.

Bogyo, H., 2006. Metalloproteases see the light. Nature Chem. Biol. 2, 229–230.

Bohn, R.M., Hass, L.W., 1928. In: Sumner, J.B., Somers, G.F. (Eds.), Chemistry and Methods of Enzymes. Academic Press, New York, N.Y.

Bommarhus, A.S., Paye, M.F., 2013. Stabilizing biocatalysts. Chem. Soc. Rev. 42, 6534–6565.

Čanak, I., Berkics, A., Bajcsi, N., Kovacs, M., Belak, A., Tepari, c´.R., Maraz, A., Mrša, V., 2015. Purification and characterization of a NoveCold-active lipase from the yeast *Candida zeylanoides*. J. Mol. Microbiol. Biotechnol. 25, 403–411.

Canet, A., Bonet-Ragel, K., Benaiger, M.D., Valero, F., 2016. Lipase-catalysed transesterification: Viewpoint of the mechanism and influence of free fatty acids. Biomass Bioener. 85, 94–99.

Cao, S., Xu, P., Ma, Y., Yao, X., Zong, M., Li, X., Lou, W., 2016. Recent advances in immobilized enzymes on carriers. Chin. J. Ta. J. Catal. 37, 1814–1823.

Carlos López-Otín, C., Bond, J.S., 2008. Proteases: Multifunctional enzymes in life and disease. J. Biol. Chem. 283, 30433–30437.

Carlos-Alayao, E.M., Idrago-Vasquez, G., Siche, R., Cardenas-Toro, F.P., 2019. Formation of aromatic compounds precursors during fermentation of Criollo and Forastero cocoa. Heliyou 5, (Doi: 10.10161/j.heliyou.2019.e01157).

Cerda-Costa, N., Gomis-Ruth, F.X., 2014. Archiyecture and function of metallopeptidase catalytic domains. Protein Sci. 23, 123–144.

Chahinian, H., Sadra, L., 2009. Distinction between esterases and lipases: Comparative biochemical properties of sequence-related carboxylesterases. Protein Peptide Lett. 16, 1149–1161.

Chanitnun, K., Pinphanichakarn, P., 2012. Glucose(xylose) isomerise production by *Streptomyces* sp. CH7 grown on agricultural residues. Braz. J. Microbiol. 43, 1084–1093.

Chaudhri, A., Suneetha, V., 2012. Microbially derived pectinases: Review. J. Pharm. Biol. Sci. 2, 01–05.

Chakraborti, S., Mandal, M., Das, S., Mandal, A., Chakraborti, T., 2003. Regulation of matrix metalloproteinases: An overview. Mol. Cell Biochem. 253, 269–285.

Choi, E.J., Ahn, H.W., Kim, W.J., 2015. Effect of α-acetolactate decrboxylase on diacetyl content of beer. Food Sci. Technol. 24, 1373–1380.

Claus, H., 2003. Laccases and their occurrence in prokaryotes. Arch. Microbiol. 179, 145–150.

Conrad, L.S., Sponholz, W.R., Berker, O., 2000. Treatment of cork with a phenol oxidizing enzyme. US Patent, 6152966.

Contesini, F.J., de Alencar Figuera, J., Kawaguti, H.Y., de Barros Fernandes, P.C., de Oliviera Carvalho, P., devGraca Nascimento, M., Sato, H.H., 2013. Potential applications of carbohydrases immobilization in the food industry. Int. J. Mol. Sci. 14, 1335–1369.

Dahlqvist, A., Asp, N.-G., Burvall, A., Rausing, H., 1977. Hydrolysis of lactose in milk and whey with minute amounts of lactase. J. Dairy Res. 44, 541–548.

Dalby, P.A., 2003. Optimising enzyme function by directed evolution. Curr. Opin. Struct. Biol. 13, 500–505.

De Lucia, V., Perotti, M.C., Wolf, I.V., Minardi, C.A., Mandrich, L., 2019. The addition of the thermophilic esterase EST2 influences the fatty acids and volatile compound profiles of semi hard cheeses. Food Sci. Technol. 39, (Doi: 10.1590/fst.06018).

De Souza Bezerra, T.M., Bassan, J.C., de Oliviera Santos, V.T., Ferraz, A., Monti, R., 2015. Covalent immobilization of laccase in green coconut fiber and use in clarification of apple juice. Proc. Biochem. 50, 417–423.

De Souza, P.M., De Assis, B., Bittercourtm, M.L., Caporara, C.C., de Freitas, M., de Almeidair, P.C., Silviera, D., Fonseca, Y.M., Fereira, X.M., Filho, Jr., A.P., Megalhaes, P.O., 2015. A biotechnology perspective of fungal proteases. Brazilian J. Microbiol. 46, 337–346.

De Vecchi, S., Coppes, Z., 1996. Marine fish digestive proteases: Relevance to food industry and the south west atlantic region: A review. J. Food Biochem. 20, 193–214.

Deng, Y., Misselwitz, B., Dai, N., Fax, M., 2015. Lactose intolerance in adults: Biological mechanisms and dietary management. Nutrients 7, 8020–8035.

Deniz, I., 2019. Production of microbial proteases for food industry. In: Parameswaran, B., Varijani, S., Raveendran, S. (Eds.), Green Bio-Process: Enzyme, Environment, and Sustainability. Springer Singapore, pp. 9–14.

Dhake, K.P., Thakare, D.D., Bhanage, B.M., 2013. Lipase: A potential biocatalyst for the synthesis of valuable flavour and fragrance ester compounds. Flavour Fragrance J. 28, 71–83.

Dhimann, S.S., Garg, G., Mahajan, R., Garg, N., Sharma, J., 2009. Single layout and mixed layout enzymatic processes for biobleaching of kraft pulp. Bioresour. Technol. 100, 4736–4741.

DiCosimo, R., McAuliffe, J., Populouse, A.J., Bohlmann, G., 2013. Industrial use of immobilized enzymes. Chem. Soc. Rev. 42, 6437–6474.

Dittmer, N.T., Suderman, R.J., Jiang, H., Zhu, Y.-C., Gorman, M.J., Kramer, K.J., Kanost, M.R., 2004. Characterization of cDNAs encoding putative laccase-like multicopper oxidases and developmental expression in the tobacco hornworm, *Manduca sexta*, and the malaria mosquito, *Anopheles gambia*. Insect Biochem. Mol. Biol. 34, 29–41.

Dobbonie, D., Uyttendaele, M., Debevere, J., 1955. Antibacterial activity of the glucose oxidase/glucose system in liquid whole egg. J. Food Proteins 58, 273–279.

dos Santos Mathisa, T.R., Diaz Lopez, M.C.R., Amorim-Oliveiro, C., Correa de Carvelho, R., de Carvelho Marquees, F.F., Camporese Servulo, E.F., 2017. Influence of mashing profile curve and addition of proteases on the composition of the wort and beer. Food Process. Technol. 5, 282–286.

Dubey, M.K., Zehra, A., Aamir, M., Meena, M., Ahriwal, L., Singh, S., Shukla, S., Upadhyay, R., Bueno-Mari, R., Bajpai, V.K., 2017. Improvement strategies, cost effective production, and potential applications of fungal glucose oxidase (GOD): Current updates. Front. Microbiol. 8, 1032.

Dulieu, C., Moll, M., Boudrant, J., Poncelet, D., 2000. Improved performances and control of bear fermentation using encapsulated α-acetolactate decarboxylase and modeling. Biotechnol. Progress. 16, 958–966.

Durao, P., Bento, I., Fernandes, A.T., Melo, E.P., Lindley, P.E., Martins, L.O., 2006. Pertubations of the T1 copper site in the CotA laccase from *Bacillus subtilis*: structural, biochemical, enzymatic and stability studies. J. Biol. Inorg. Chem. 11, 514–526.

Dutra, J.C.V., de, C., Terzi, S.C., Bevilaqua, J.V., Damasco, M.C., Couri, S., Langone, M.A., Senna, L.P., 2008. Lipase production in solid-state fermentation monitoring biomass growth of *Aspergillus niger* using digital image processing. Appl. Biochem. Biotechnol. 147, 63–75.

Dwivedi, B.K., Shahani, K.M., Arnold, R.G., 1973. The role of enzymes in food flavor. Part I. Crit. Rev. Food Technol. 3, 457–478.

Eskin, N.A.M., 1990. Biochemistry of Foods, second ed. Academic Press, New York.

Eskin, N.A.M., Grossman, S., Pinsky, A., 1977. Biochemistry of lipoxygenase in relation to food quality. CRC Crit. Rev. Food Sci. Nutr. 9, 1–40.

Esmaeili, S., Noorolahi, Z., 2017. Application of starch processing enzymes in food technology: A review. Carpathian J. Food Sci. Technol. 9, 114–127.

Faubion, J.M., Hoseney, R.C., 1981. Lipoxygenase: its biochemistry and role in breadmaking. Cereal Chem. 58, 175–180.

FDA/CFSAN, 2002a. Agency Response Letter: GRAS Notice No. GRN 000106.

FDA/CFSAN, 2002b. Agency Response Letter: GRAS Notice No. GRN 000089.

Fernandes, P., 2014. Marine enzymes and food industry: Insight on existing and potential interactions. Front. Marine Sci., (Doi: 3389/fmars.2014.00046).

Ferreira-Dias, S., Sandoval, G., Plou, F., Valero, F., 2013. The potential use of lipases in the production of fatty acid derivatives for the food and nutraceutical industries. Electronic J. Biotechnol. 16, (Doi: org/10.2225/vol16-issue3-fulltext-5).

Flores-Gallegos, A.C., Delgado-Garcia, M. Ascacio-Valdes, J.A., Villareal-Morales, S., Michel-Michel, M.R. Aguilar-Gonzalez-, C.N., Rodriguez-Harrera, R., 2019 Hydrolases of halophilic origin with importance for the food industry, In: *Enzymes in Food Biotechnology, Production, Applications, and Future Prospects*, Academic Press, New York., N.Y., pp. 197–219.

Fukiu, K., Fukiu, Y., Moriyama, T., 1974. Purification and properties of dextran sucrase and invertase from *Streptococcus mutans*. J. Bacteriol. 118, 796–804.

Gallego, M., Mora, L., Toldra, F., 2016. Peptidomics as a tool for quality control in dry-cured ham processing. J. Proteomics 147, 98–107.

Gallego, M., Mora, L., Toldra, F., 2018a. Characterisation of the antioxidant peptide AEEEYPDL and its quantification in Spanish dry-cured ham. Food Chem. 2258, 8–15.

Gallego, M., Mora, L., Toldra, F. 2018b. Perspectives in the use of peptidomics in ham. Proteomics, 18 (Doi: 10.1002/pmic.201700422).

Garcia, A.I., Garcia, L.A., Diaz, M., 1996. Modelling of diacetyl products during beer fermentation. J. Inst. Brew. 100, 179–183.

Garg, G., Singh, A., Kaur, A., Mahajan, R., 2016. Microbial pectinases: An ecofriendly tool for industries. Biotechnol. 6, 47–59.

Geciova, J., Bury, D., Jelen, P., 2002. Methods of disruption of microbial cells for potential use in the dairy industry-a review. Int. Dairy J. 12, 541–553.

Giese, E.C., Decker, R.F.H., Barbosa, A.M., de Lourdes Corradi da Silva, M., da Silva, R., 2011a. Production of β-(1,3)-glucanases by *Trichoderma harzianum* Rifai: Optimization and application to produce gluco-oligosaccharides from paramylon and pustulan. Ferment. Technol 1, 1-5.

Giese, E.C., Decker, R.F.H., Barbosa, A.M., Scariminio, I.S., Barbosa, A.M., da Silva, R., 2011b. Comparison of β-1,3-glucanase production by *Botryosphaeria rhodina* MAMB-05 and *Trichoderma harzianum* Rifai and its optimization using a statistical mixture-design. Biochem. Engineer. J. 53, 239–243.

Godfredsen, S.E., Ottesen, M., 1982. Maturation on beer with α-acetolactate decarboxylase. Carlsberg Res. Commun. 47, 93–102.

Godfredsen, S.E., Rasmussen, A.M., Ottesen, M., Rafn, P., Peitersen, N., 1984. Occurrence of α-acetolactate decarboxylases among lactic acid bacteria and their utilization for maturation for beer. Appl. Microb. Biotechnol. 20, 23–28.

Gomaa, A.H., 2018. Application enzymes in brewing. J. Nutr. Food Sci. Forecast 1, 1002.
Gopinath, S.C.B., Anbu, P., Lakshmipriya, T., Hilda, A., 2013. Strategies to characterize fungal lipases for applications in medicine and dairy industry. Biomed. Res. Int., (Doi: org/10.1155/2013/154549).
Grechkin, A.N., Brühlmann, F., Mukhtarova, L.S., Gogolev, Y.V., Hamberg, M., 2006. Hydroperoxide lyases (CYP74C and CYP74B) catalyze the homolytic isomerization of fatty acid hydroperoxides into hemiacetals. Biochim. Biophys. Acta 1761, 1419–1428.
Griebeler, N., Polloni, A.E., Remonatto, D., Arbter, F., Vardanega, R., Di Luccio, M., de Oliveira, D., Treichel, H., Casian, R.L., Regio, E., Ninow, J.L., 2011. Isolation and screening of lipase-producing fungi with hydrolytic activity. Food Bioprocess Technol. 4, 578–586.
Grigoras, A.G., 2017. Catalase immobilization: A review. Biochem. Engineer. J. 117, 1–20.
Grosova, Z., Rosenberg, M., Rebros, M., 2007. Perspectives and applications of immobilised β-galactosidase in food industry: A review. Czech. J. Food Sci. 26, 1–14.
Guerrand, D., 2017. Lipases industrial applications: Focus on food and agroindustries. OCL. 24(4), D403 (Doi: org/10.1051/ocl/2017031).
Guomundsdottir, A., Bjarnason, J., 2007. Applications of cold-adapted proteases in the food industry. In: Rastall, R. (Ed.), Novel Enzyme Technology for Food Applications. Woodhead Publishing Series in Food Science, Technology and Nutrition, Elsevier, Amsterdam, The Netherland, pp. 205–214.
Gupta, R.G., Beg, Q.K., Lorenz, P., 2002. Bacterial alkaline proteases: Molecular approaches and industrial applications. Appl. Microbiol. Biotechnol. 59, 15–32.
Gygli, G., van Berkel, W.J.H., 2015. Oxizymes for biotechnology. Curr. Biotechnol. 4, 100–110.
Haard, N.F., 1992. A review of proteolytic enzymes from marine organisms and their application in the food industry. J. Aquatic Food Product Technol. 1, 17–35.
Haard, N.F., Simpson, B.K., 1994. Proteases from aquatic organisms and their uses in the food industry. In: Martin, A.M. (Ed.), Fisheries Processing: Biotechnological Applications. Springer-Verlag, New York, N.Y, pp. 132–154.
Hall, C.E., Husson, F., Kermasha, S., 2004. Characterization of an enriched lipoxygenase extract from *Aspergillus niger* in terms of specificity and nature of flavor precursors production. J. Mol. Catal. B: Enz. 29, 201–209.
Hall, C.E., Husson, F., Kermasha, S., 2005. Production of flavour precursors by *Penicillium candidum* using selected polyunsaturated fatty acids. Flavour Fragrance J. 20, 690–702.
Hall, C.E., Karboune, S., Husson, F., Kermasha, S., 2008. Stabilization of an enzymatic extract from *Penicillium camemberti* containing lipoxygenase and hyderoperoxide lyase activities. Process Biochem. 43, 258–264.
Harju, M., Kallioinen, H., Tossavainen, O., 2012. Lactose hydrolysis and other conversions in dairy products: Technical aspects. Int. Dairy J. 22, 104–109.
Harris, A.D., Ramalingam, C., 2019. Xylanases and its application in food industry: A review. J. Exp. Sci. 1, 1–11.
Hayward, S., Cilliersm, T., Swart, P., 2017. Lipoxygenases: From isolation to application. Comp. Rev. Food Sci. Food Safety 16, 194–211.
Hedstrom, E., 2002. Serine protease mechanism and specificity. Chem. Rev. 102, 4501–5624.
Heredia-Sandoval, N.G., Valencia-Tapia, M.Y., Calderon de la Barca, A.M., Isla-Rubio, A.R., 2016. Microbial proteases in baked goods: Modification of gluten and effects on immunogenicity and product quality. Foods 5, (Doi: 10.3390/foods5030059).
Heshof, R., de Oraaff, L.H., Villavardo, J.J., Silvestre, A.J.D., Haamman, T., Dalsgaard, T.K., Buchert, J., 2016. Industrial potential of lipoxygenase. CRC Crit. Rev. Biotechnol. 35, 665–674.
Hizikuri, S., Takeda, Y., Yasuda, M., 1981. Multi-branched nature of amylose and the action of debranching enzymes. Carbohydr. Res. 94, 204–213.

Hodges, J.K., Cao, S., Cladis, D.P., Weaver, C.M., 2019. Lactose intolerance and bone health: The challenge of ensuring adequate calcium intake. Nutrients 11, 718.

Horchani, M., Aissa, I., Ouertani, S., Zarai, Z., Gargouri, Y., Sayari, A., 2012. *Staphylococcal* lipases: Biotechnological applications. J. Mol. Catal. B: Enzym. 76, 125–132.

Hussain, Q., 2010. β-Galactosidases and their potential applications: A review. Crit. Rev. Biotechnol. 30, 41–62.

Husson, F., Thomas, M., Kermasha, S., Belin, J.M., 2002. Effect of linoleic acid induction on the production of 1-octen-3-ol by the lipoxygenase and hydroperoxide lyase activities of *Penicillium camemberti*. J. Mol. Catal. B: Enzym. 19-20, 363–369.

Isakien, A., Adler-Nissen, I., 1997. Antioxidative effect of glucose oxidase and catalase in mayonnaises of different oxidative susceptibility. 1. Product trials. Lebensm. Wiss. Technol. 30, 811–816.

Jablonska-Typuc, A.J., Matejczyk, M., Rosochacki, S., 2016. Matrix metalloproteases (MMPs), the main extracellular matrix (ECM) enzymes in collagen degradation as target for anticancer drugs. J. Enzym. Inhib. Med. Chem. 31, 177–183.

Jaeger, K.-E., Eggert, T., 2002. Lipase for biotechnology. Curr. Opin. Biotechnol. 13, 390–397.

Jain, J.L., Jain, S., Jain, N., 1979. Fundamental of Biochemistry, seventh ed. New Delhi, India: S. Chand & Company Ltd., pp. 197–224.

Jayani, R.S., Saxena, S., Gupta, R., 2005. Micrbrobial pectolytic enzymes: A review. Process Biochem. 40, 2911–2944.

Ji, F., Feng, Y., Li, M., Lomg, F., Yang, Y., Wang, T., Wang, J., Bao, Y., Xue, S., 2019. Structure and catalytic mechanistic insight into *Enterobacter aerogenes* acetolactate decarboxylase. Enzym. Microb. Technol. 126, 9–17.

John, J., 2017. Amylases: Bioprocess and potential applications: A review. International J. Bioinform. Biol. Sci. 5, 41–50.

Kadi, N., Crouzet, J., 2008. Transglycosylation reaction of endoxylanases from *Thrichoderma Longibrachiatum*. Food Chem. 105, 466–474.

Kailasapathy, K., Lam, S.H., 2005. Applications of encapsulated enzymes to accelerate cheese ripening. International Dairy J. 15, 929–939.

Kapoor, M., Beg, Q.K., Bhushan, B., Singh, K., Dadhich, K.S., Hoodal, G.S., 2001. Application of an alkaline and thermostable polygalcturonase from Bacius sp. NG0cp-2 in degumming of ramie (Boehmeria nivea) and sun hemp (Crotalaria juncea) bast fibers. Process Biochem. 36, 863–867.

Karam, R., Karboune, S., St-Louis, R., Kermasha, S., 2009. Lipase-catalyzed acidolysis of fish liver oil with dihydroxyphenylacetic acid in organic solvent media. Process Biochem. 44, 1193–1199.

Karamatsi, K., Kokkinidou, S., Ronningen, I., Peterson, D.G., 2014. Identification of bitter peptides in aged cheddar cheese. J. Agric. Food Chem. 62, 8034–8041.

Karboune, S., St-Louis, R., Kermasha, S., 2008. Enzymatic synthesis of structured phenolic lipids by acidolysis of flaxseed oil with selected phenolic acids. J. Mol. Catal. B: Enz. 52-53, 96–105.

Kareen, S.O., Adebowale, A.A., 2007. Clarification of orange juice by crude fungal pectinease from citrus peel. Niger. Food J. 25, 130–136.

Kashyap, D.R., Vohra, P.K., Chopra, S., Tewari, R., 2001. Applications pf pectinae in commercial sector. A review. Bioresour. Technol. 77, 215–227.

Kaushal, J., Mejhandia, S., Sigh, G., Daina, A., Arya, K., 2018. Catalase enzyme: Application in bioremediation and food industry. Biocatal. Agric. Biotechnol. 16, 192–199.

Kavadio, M.R., Yadav, M.G., Odaneth, A.A., Lali, A.M., 2018. Synthesis of designer triglycerides by enzymatic acidolysis. Biotechnol. Report, e00246 (Doi: 10.1016/j.btre.2018.e00246).

Kermasha, S., Bisakowski, B., Morin, A., Ismail, S., 1999. Biogeneration of short-chain fatty acid esters by *Pseudomonas fragi* CRDA 037. Biocatal. Biotransform. 17, 269–282.

Kermasha, S., Perraud, X., Bisakowski, B., Husson, F., 2002. Production of flavor compounds by hydroperoxide lyase from enzymatic extract of *Penicillium* sp. J. Mol. Catal. B: Enzym. 19-20, 479–487.

Kermasha, S., Aziz, S., Gill, J., Neufeld, R., 2017. Microencapsulation of esterified krill oil, using complex coacervation. J. Microencapsul. 35, 36–48.

Kertesz, Z.I., 1951. *The Pectic Substances*, Interscience Publishers, Inc., New York, N.Y.

Ketnawa, S., Rawdkuen, S., 2012. Application of bromelain extract for muscle foods tenderization. Food Nutr. Sci. 2, (Doi: 10.4236/fns.2011.25055).

Khan, M., Nakkeeran, E., Kumar, S.U., 2013. Potential application of pectinaes in developing functional foods. Ann. Rev. Food Sci. Technol. 4, 21–34.

Khodadadi, M., Kermasha, S., 2014a. Optimization of lipase-catalyzed interesterification of flaxseed oil and tricaprylin using response surface methodology. J. Am. Oil Chem. Soc. 91, 395–403.

Khodadadi, M., Aziz, S., St-Louis, R., Kermasha, S., 2013. Lipase-catalyzed synthesis and characterization of flaxseed oil-based structured lipids. J. Funct. Foods 5, 424–433.

Khodadadi, M., Kermasha, S., 2014b. Modeling lipase-catalyzed interesterification of flaxseed oil and tricaprylin for the synthesis of structured lipids. J. Mol. Catal. B: Enz. 102, 33–40.

Kibra, A., 1982. Enzymes and their uses in processed apple industry. A review. Process. Biochem. 23, 25–41.

Kirk, O., Borchert, T.V., Fuglsang, C.C., 2002. Industrial enzyme applications. Curr. Opin. Biotechnol. 13, 345–351.

Kleppe, K., 1966. The effect of hydrogen peroxide on glucose oxidase from *Aspergillus niger*. Biochem. 5, 139–143.

Kohli, P., Kalia, M., Gupta, R., 2015. Pectin methylesterases: A review. J. Bioprocess. Biotechniques 5, (DOI: 10.4172/2155-9821.10000227).

Kotwal, S.M., Shankar, V., 2009. Immobilized invertase. Biotechnol. Adv. 27, 311–322.

Krigshauser, G., Liebl, W., 2000. Pullulanase from hyperthermophilic bacterium, *Thermotoga maritime*. Purification by cyclodextrin affinity chromatography. J. Chromatogt. 737, 245–251.

Krogerus, K., Gibson, B.R., 2013. Anniversary review. Diacetyl and its control during brewing fermentation. J. Inst. Brew. 119, 86–97.

Kuhad, R.C., Singh, A., 1993. Lignocellulose biotechnology: Current and future prospects. Crit. Rev. Biotechnol. 13, 151–172.

Kulshrestha, s., Tyagi, P., Sindhi, V., Yadavilli, K.S., 2013. Invertase and its applications. J. Pharm. Res. 7, 792–797.

Kumar, P., Satyanatayana, T., 2009. Microbial glucoamylases: Characteristics and applications. Crit. Rev. Biotechnol. 29, 225–255.

Labuza, T.P., Breene, W.M., 1989. Applications of "active packaging" for improvement of shelf life and nutritional quality of fresh and extended shelf-life foods. J. Food Process. Preserv. 13, 1–69.

Levine, W.G., 1965. Laccase: A review. Biochemistry of Copper. Academic Press, New York. N.Y, pp. 371–385.

Ling, H.S., Chuan, L.T., Mohammed, R., Ariff, A.B., 2009. Characterization of pullulanase type II from *Bacillus cereus* H1.5. Amer. J. Biochem. Biotechnol. 5, 170–179.

Lopez-Otin, C., Bond, J.S., 2008. Proteases: Multifunctional enzymes in life and diseases. J. Biol. Chem. 283, 30433–30437.

Lue, B.-M., Karboune, S., Yeboah, F.K., Kermasha, S., 2005. Lipase-catalyzed esterification of cinnamic acid and oleyl alcohol in organic solvent media. J. Chem. Technol. Biotechnol. 80, 462–468.

Lukin, A., 2019. Application and comparison of proteolytic enzyme preparations in technology of protein hydrolyzates. Food Sci. Technol. 40, (Doi: 0000.0003-4753-3210).

Luning, P.A., Roozen, J.R., Most, R.A.F.J., Posthumus, M.A., 1991. Volatile composition of white bread using enzyme active soya flour as improver. Food Chem. 41, 81–91.
Madhaven, A., Sindhu, R., Binod, P., Sukumaran, R.K., Pandey, A., 2017. Strategies for design of improved biocatalysts for industrial applications. Bioresource Technol. 245 (Part B), 1304–1313.
Makky, E.A., Yusoff, M.M., 2015. Bioeconomy: pectinases purification and application of fermented waste from *Thermomyces lanuginosus*. J. Med. Bioengineer. 4, 76–80.
Malakar, R., Tiwari, A., Malviya, S.N., 2010. Pullulanase: A potential enzyme for industrial application. Int. J. Biomed. Res. 1, 10–20.
Mamo, J., Assefa, F., 2018. The role of microbial aspartic protease enzyme in food and beverage industries. Open Access, Hindawi Publisher, vol. 2018. doi: 10.1155/2018/7957269.
Marczy, J.S., Nemeth, A.S., Samu, Z., Hager-Veress, A., Szajani, B., 2002. Production of hexanal from hydrolyzed sunflower oil by lipoxygenase and hydroperoxide lyase enzymes. Biotechnol. Lett. 24, 1673–1675.
Marlow, V.A., Rea, D., Najmudin, S., Wills, M., Fulop, V., 2013. Structure and mechanism of acetolactate decarboxylase. ACS Chem. Biol. 8, 2339–2344.
Marnett, A.B., Craik, C.S., 2005. Papa's got a brand-new tag: Advances in identification of proteases and their substrates. Trends Biotechnol. 23, 59–64.
Martin, K., McDougall, B.M., Mclloroy, S., Chen, J., Seviour, R.J., 2007. Biochemical and molecular biology of extracellular fungal beta-(1,3)- and beta-(1,6)-glucanases. FEMS Microbiol. Rev. 31, 168–192.
Mathiasen, T.E., 1991. Laccase and beer storage. PCT Interational Application W09521240, A2.
Matsui, K., 2006. Green leaf volatiles: Hydroperoxide lyase pathway of oxylipin metabolism. Curr. Opin. Plant Biol. 9, 274–280.
McCormick, R.D., 1976. A nutritious alternative to the lactose-intolerant consumer. Food Product Develop. 10, 17.
Melim Miguel, A.S., Martins-Meyer, T.S., da Costa-Figueiredo, E.V., Lobo, W.P., Dellamora-Ortiz, G.M., 2013. Enzymatic bakery: Current and future trends. Open Access, (Doi: 10.5772/53168).
Micheli, F., 2001. Pecin methylesterases: Cell wall enzymes with important roles in plant physiology. Trends Plant Sci. 6, 414–419.
Minussi, P.C., Rossi, M., Bologna, L., Rotilio, F., Pastore, G.M., Duran, N., 2007. Phenols removal in musts: strategy for wine stabilization by laccase. J. Mol. Catal. B. Enzym. 45, 102–107.
Moller, M.S., Svensson, B., 2016. Structural biology of starch-dergrading enzymes and their regulation. Curr. Opin. Struct. Biol. 40, 33–42.
Moon, S.S., 2019. Effect of proteolytic enzymes and ginger extract on tenderization of M. *pectoralis profundus* from holstein steer. Korean J. Food Sci. Anim. Resour. 38, 143–151.
Moore, M.M., Chen, T., 2006. Mutagenicity of bromated. Implications for cancer risk assessment. Toxicol. 221, 190–196.
Morin, J., Ganzalez, M.P., Vazquez, J.A., Pastrana, L., Murado, M.A., 2004. A mathematical model for glucose oxidase kinetics, including inhibitory, deactivant and diffusional effects, and their interactions. Enzym. Microb. Technol. 34, 513–522.
Mohammadi Panah, F., Hamedi, J., Dehhaghi, M., 2015. Halophilic bacteria: Potentials and applications in biotechnology. In: Maheshwari, D.K., Saraf, M. (Eds.). Beneficial Usages of Halophilic Microorganisms, pp. 277-321. Doi: 101007/978-3-319-14595-2-11.
Muller, D., 1928. Oxidation von glukose mit extraklen aus *Aspergillus niger*. Biochem. Z. 199, 136–170.
Munro, A.W., Taylor, P., Tipton, P.A., 2000. Curr. Opin. Biotechnol. 11, 369–376.
Murthy, P.S., Nardo, M.M., 2011. Improvement of robusta coffee fermentation with microbial enzymes. Eur. J. Appl. Sci. 3, 130–139.

Myrback, K., 1960. Invertase. In: Boyer, P.D., Lardy, H.A., Myrback, K. (Eds.), Enzymes. Vol. 4, Academic Press, New York. N.Y., p. 379.
Nadeem, H., Rashid, M.H., Siddique, M.H., Azeem, F., Muzammid, S., Javed, M.R., Ali, M.A., Rasul, I., Riaz, M., 2015. Microbial invertases: A review on kinetics, thermodynamics, physiochemical properties. Process Biochem. 50, 1202–1210.
Nair, I.C., Jayachandran, K., 2019. Aspartic proteases in food industry. In: Parameswaran, B., Varijani, S., Raveendran, S. (Eds.), Green Bio-Process: Enzyme, Environment, and Sustainability. Springer Singapore, pp. 15–30.
Nardi, M., Fiez-Vandal, C., Tailliez, P., Monnet, V., 2002. The EstA esterase is responsible for the main capacity of *Lactococcus lactis* to synthesize short chain fatty acid esters *in vitro*. J. Appl. Microbiol. 93, 994–1002.
Nerifar, M., Ghorbel, P., Kamoun, A., Baklouti, S., Mokni, A., Jaouani, A., Ellouz, S.C., 2009. Effective clarification of pomegranate juice using laccase treatmemt optimized by response surface methodology followed by filtration. J. Food Proc. Engineer. 34, 1190–1219.
Nigam, P., Singh, D., 1995. Enzyme and microbial systems involved in starch processing. Enzym. Microb. Technol. 17, 770–778.
Nikola, L., Marco, W.E., 2015. Catalases as biocatalysts in technical application: Current state and perspectives. Appl. Microbiol. Biotechnol. 59, 3351–3357.
Onwulata, C.I., Rao, D.R., Vankineni, P., 1989. Relative efficiency of yogurt sweet acidophilus milk, hydrolysed-lactose milk, and a commercial lactase tablet in alleviating lactose maldigestion. Am. J. Clin. Nutr. 49, 1232–1237.
Osma, J.F., Toca-Herrera, T.L., Rodriguez-Cuoto, S., 2010. Uses of laccases in the food industry. Enzyme Res. Hundawi 10, 981761. doi: 10.4061/2010/91876.
Otten, L.G., Quax, W.J., 2005. Directed evolution: Selecting today's biocatalysts. Biomol. Engineer. 22, 1–9.
Panda, T., Gowrishankar, B.S., 2005. Production and applications of esterases. Appl. Microbial. Biotechnol. 67, 160–169.
Pang, Z., Olaka, K., Maola, T., Hidaka, K., Isijima, S., Oda, M., Ohnishi, M., 2005. Structure of β-glucan oligome from laminarim and its effect on human monocytes to inhibit the proliferation of U937 cells. Biosci. Biotechnol. Biochem. 69, 553–558.
Park, S.H., Na, Y., Kim, J., Dal Kang, S., Park, K.H., 2018. Properties and applications of starch modifying enzymes for use in the baking industry. Food Sci. Biotechnol. 27, 299–312.
Parker, K., Colas, M., Nwosu, V.C., 2010. High fructose corn syrup: Production, uses and health concerns. Biotechnol. Mol. Biol. Rev. 5, 71–78.
Patui, S., Braidot, E., Peresson, C., Tubaro, F., Mizzau, M., Rabiei, Z., 2010. Lipoxygenase and hydroperoxide lyase activities in two olive varieties from Northern Italy. Eur. J. Lipid Sci. 112, 780–790.
Pedrolli, D., Monteiro, A.P.C., Carmona, E.C., 2009. Pectin and pectinases: Production, characterization and industrial production of microbial pectinolytic enzymes. Open Biotechnol. J. 3, 9–18.
Permyakova, M.D., Trifamov, V.A., 2011. Effect of soybean lipoxygenase on baking propertiesd of wheat flour. Appl. Biochem. Microbiol. 47, 315–320.
Poppa, L., Schafer, W., Freund, W., 2006. Oxidation and flour maturation. In: Poppa, L., Schafer, W., Freund, W. (Eds.), Future of Flour: A Compendum of Flour Improvements. Bergen/Dumme, Germany, Agrimedia, Germany, pp. 7–21.
Praveen, K.G., Suneetha, V.A., 2014. A cocktail enzyme-pectinase from fruit industrial dump sites: a review. Res. J. Pharm. Biol. Chem. Sci. 5, 1252–1258.
Rani, S., Jagtap, S., 2019. Acceleration of Swiss cheese ripening by microbial lipase without affecting its quality characteristics. J. Sci. Food Technol. 56, 497–506.

Rasiah, J.A., Sutton, K.H., Low, F.L., Lin, H.M., Gerar, J.A., 2005. Crosslinking of wheat dough by glucose oxidase and the resulting effects on bread and croissants. Food Chem. 89, 325–332.

Razzaq, A., Shamsi, S., Ali, A., Ali, Q., Sajjad, M., Malik, A., Ashraf, M., 2019. Microbial proteases applications. Front. Bioengineer. Biotechnol. 7, 110, (Doi: 10.3389/fbioe.2019.00110).

Revilla, I., Ganzalez-San Jose, M.L., 2003. Addition of pectolytic enzymes: an enological practice that improves the chromaticity and stability of red wines. Int. J. Food Sci. Technol. 38, 29–36.

Ribiero, D.S., Henrique, S.M.B., Oliviera, L.S., Macedo, G.A., Fleuri, L.F., 2010. Enzymes in juice processing: A review. Int. J. Food Sci. Technol. 25, 635–661.

Richmond, M.L., Gray, J.I., Stine, C.M., 1981. Beta-galactosidase: Review of recent research related to technological applications, nutritional concerns, and immobilization. J. Dairy Sci. 64, 1759–1771.

Riviera-Sagredo, F., Canada, F.J., Jimene-Barbero, J., Martin-Lomas, M., 1992. Substrate specificity of small-intestinal lactase. Assessment of the role of the substrate hydroxyl groups. Eur. J. Biochem. 209, 415–422.

Rombouts, F.M., Pilnik, W., 1980. Pectic enzymes. In: Rose, A.H. (Ed.), Microbial Enzymes and Bioconversions. Academic Press, New York, N.Y, pp. 227–272.

Roy, I., Prasad, S., 2018. Converting enzymes into tools of industrial importance. Recent Pat. Biotechnol. 112, 33–56.

Rubio, D., Alejo, A., Rodriguez, I., Salas, M.L., 2003. Polyprotein processing protease of African swine fever virus: Purification and biochemical characterization. J. Virol. 77 (7), 4444–4448.

Sabally, K., Karboune, S., Yeboah, F.K., Kermasha, S., 2005. Enzymatic esterification of dihydrocaffeic acid with linoleyl alcohol in organic solvent media. Biocatal. Biotransform. 23, 37–44.

Sabally, K., Karboune, S., St-Louis, R., Kermasha, S., 2007. Lipase-catalyzed synthesis of phenolic lipids from fish liver oil and dihydrocaffeic acid. Biocatal. Biotransform. 25, 211.

Safari, M., Karboune, S., St-Louis, R., Kermasha, S., 2006. Enzymatic synthesis of structured phenolic lipids by incorporation selected phenolic acids into triolein. Biocatal. Biotransform. 24, 272–279.

Safari, M., Karboune, S., Kermasha, S., 2009. Intermediate production of mono- and diolein by immobilized lipase from *Candida antarctica*. Iranian J. Biotechnol. 7, 67–74.

Saha, B., Hayashi, K., 2001. Debittering of protein hydrolyzates. Biotechnol. Adv. 19, 355–370.

Saini, R., Singh, H., Dahiya, A., 2017. Amylases: Characteristics and industrial applications. J. Pharmacog. Phytochem. 6, 1865–1871.

Samah, N., Revathi, D., Sheelu, G., Rani, K.Y., Sridhar, S., Mehtab, V., Sumana, C., 2018. Recent advances on sources and industrial applications of lipases. Biotechnol. Progr. 34, 5–28.

Samanthan, A., Larroche, C., Pandey, A., 2006. Microbiology and industrial biotechnology of food grade proteases: A perspective. Food Technol. Biotechnol. 44, 211–220.

Santiago-Gómez, M.P., Kermasha, S., Nicaud, J.-M., Belin, J.-M., Husson, F., 2008. Secondary structure conformation of hydroperoxide lyase from green bell pepper, cloned in *Yarrowia lipolytica*, and its activity in selected media. J. Mol. Catal. B: Enzym. 52-53, 128–132.

Santiago-Gómez, M.P., Tran Thanh, H., De Coninck, J., Cachon, R., Kermasha, S., Belin, J.-M., Gervais, P., Husson, F., 2009. Modeling hexanal production in oxido-reducing conditions by the yeast *Yarrowia lipolytica*. Process Biochem. 44, 1013–1018.

Schilling, O., Overall, C.M., 2008. Proteome-derived, database-searchable peptide libraries for identifying protease cleavage sites. Nat. Biotechnol. 26, 685–694.

Schmidt-Dannert, C., Sztajer, H., Stocklein, W., Menge, W., Schimid, R.D., 1994. Screening purification and properties of thermophilic lipase from *Bacillus thermocatenulatus*. Biochem, Biophys. Acta 1214, 43–53.

Scrimshaw, N.S., Murray, E.B., 1988. The acceptability of milk and milk products in populations with a high prevalence of lactose intolerance. Amer. J. Clin. Nutr. 48, 1079–1159.

Seitz, W., 1974. Industrial applications of microbial lipases: A review. J. Am. Oil Chem. Soc. 51, 12–16.

Shahenvaz, A., Pranaw, K., Tiwari, R., Khare, S.K., 2019. Recent development in the uses of asparaginase in food enzymes. In: Parameswaran, B., Varijani, S., Raveendran, S. (Eds.), Green Bio-Process: Enzyme, Environment, and Sustainability. Springer Singapore, pp. 55–81.

Shaikh Sumaiya, A., Trivedi, R., 2015. A review on glucose oxidase. Inter. J. Curr. Microbia. Appl. Sci. 4, 636–642.

Sharma, R., Christi, Y., Banerjee, U.C., 2001. Production, purification, characterization and applications of lipases. Biotechnol. Adv. 19, 627–662.

Shen, H.-B., Chou, K.-C., 2009. Identification of proteases and their types. Anal. Biochem. 385, 153–160.

Sheng, X., Qao, J., Chen, Z., 2019. Changes in lipase and antioxidant enzyme activities during processing of Cantonese susage with D-sodium erythrobate. J. Food Qual., (Doi: org/10.1155/2019/1671603).

Simpson, B.K., Rui, X., Klomkao, S., 2012. Enzymes in food processing. In: Simpson, B.K. (Ed.), Food Biochemistry and Food Processing. second ed. Wiley-Blackwell, Oxford, U.K., pp. 181–206.

Singh, P., Kumar, S., 2019. Microbial enzymes. In: Kuddos, M. (Ed.), Enzymes in Food Biotechnology: Production, Applications, and Future Prospects. Academic Press, New York, N.Y., pp. 5–20.

Singh, T.A., Sarangi, P.K., Singh, N.J., 2018. Tenderisation of meat by bromelain enzyme extracted from pineapple wastes. Int. J. Curr. Microbiol. App. Sci. 7, 3256–3264.

Sisak, C., Csanádi, Z. E., Ronay, E., Szaján, B., 2006. Elimination of glucose in egg white while immobilized glucose oxidase. Enzym. Microb. Technol. 39, 1002–1007.

Sorour, N., Karboune, S., Saint-Louis, R., Kermasha, S., 2012. Lipase-catalyzed synthesis of structured phenolic lipids in solvent-free system using flaxseed oil and selected phenolic acids as substrates. J. Biotechnol. 158, 128–136.

Sousa, L.S., Rocha, F., de, S., Silveira, P.T., de, S., Bispo, E., de, S., Soares, S.E., 2016. Enzymatic activity of proteases and its isoenzymes in fermentation process in cultivars of cocoa (*Theobroma cacao* L.) produced in south Bahia, Brazil. Food Sci. Technol. 36, (Doi: 10.1590/1678-457x.10916).

Sousa-Gallagher, M.J., Ardo, Y., McSweeney, P.L.H., 2001. Advances in the study of proteolysis during cheese ripening. Int. Dairy J. 11, 327–345.

Stryer, L., Berg, J.M., Tymoczke, J. Gatto, G., 2019. Catalytic strategies. In: *Biochemistry*, Palgrave Macmillan, London, U.K., pp. 262–271.

Suri, S., Kumar, V., Prasad, R., Tanwar, B., Goyal, A., Kaur, S., Gat, Y., Kumar, A.m., Kaur, J., Singh, D., 2019. Considerations for development of lactose-free food. J. Nutr. Intermed. Metab. 15, 27–34.

Takasaki, Y., 1972. On the separation of sugars. Agric. Biol. Chem. 36, 2575–2577.

Takasaki, Y., 1974. Formation of glucose isomerase by *Streptomyces* sp. Agric. Biol. Chem. 38, 667–668.

Takasaki, Y., Tanabe, O., 1971. Enzymatic method for converting glucose in glucose syrups, US Patent 3,616,221.

Takenaka, S., Umeda, M., Senba, H., Koyama, D., Tanaka, K., Yoshida, K.-I., Doi, M., 2016. Heterologous expression and characterization of the *Aspergillus* aspartic protease involved in the hydrolysis and decolorization of red-pigmented proteins. J. Sci. Food Agric. 97, 95–101.

Tallant, C., Marrero, A., Gomis-Ruth, F.X., 2010. Matrix metalloproteases: Fold and function of their catalytic domains. Biochim. Biophys. Acta 1803, 20–28.

Tapre, A.R., Jain, R.K., 2014. Pectinases: Enzymes for fruit processing industry. Int. Food Res. J. 21, 447–453.

Tavano, O.L., 2013. Protein hydrolysis using proteases: An important tool for food biotechnology. J. Mol. Catal. B. Enzym. 90, 1–11.

Tavano, O.L., Berneguer-Murcia, A., Secundo, F., Fernandez-Lafuenta, R., 2018. Biotechnological applications of proteases in food technology. Compr. Rev. Food Sci. Food Safety 17, 412–436.

Tester, R.F., Karkalas, J., Qi, X., 2004. Starch-composition, fine structure and architecture. J. Cereal Sci. 39, 151–165.

Tian, G.W., Chen, M.H., Zaltsman, A., Citovsky, V., 2006. Pollen-specific pectin methylesterases involved in pollen tube growth. Dev. Biol. 294, 83–91.

Toldra, F., Flores, M., 1998. The role of muscle proteases and lipases in flavor development during the processing of dry-cured ham. Crit. Rev. Food Sci. Nutr. 38, 331–352.

Tran Thanh, H., Vergoignan, C., Cachon, R., Kermasha, S., Gervais, P., Nguyen, T.X.S., Belin, J.-M., Husson, F., 2008. Recombinant hydroperoxide lyase for the production of aroma compound: Effect of substrate on the yeast *Yarrowia lipolytica*. J. Mol. Catal. B: Enzym. 52-53, 146–152.

Tufvesson, P., Leima-Ramos, J., Nordblad, M., Woodley, J.M., 2010. Guidelines and cost analysis for catalyst production in biocatalytic processes. Org. Process Res. Develop. 15, 266–274.

Unger, E., 2011. Processing. In: Daun, J.K., Eskin, N.A.M., Hickling, D. (Eds.), Canola: Chemistry, Production, Processing, and Utilization. AOCS Press, Urbana, IL, pp. 178–179.

Utama, Q.D., Sitanggang, A.B., Adawiyah, D.R., Hariyadi, P., 2019. Lipase-catalyzed interesterification for the synthesis of medium-long-medium (MLM) structured lipids: A Review. Food Technol. Biotechnol. 57, 305–318.

Utsushi, A., Kanda, A., Hara, S., 2009. Lipase specificity in the transacylation of triacylglycerin. J. Oleo Sci. 58, 123–132.

Vaklhu, J., Kour, A., 2006. Yeast lipase: Enzyme purification, biochemical properties and gene cloning. Electronic J. Biotechnol. 9, 69–85.

Van Der Maarel, M.J.E.C., Van Der Veen, B., Uitdehaag, J.C.M., Leemhuis, H., Dijkhuizen, L., 2002. Properties and applications of starch-converting enzymes of the α-amylase family. J. Biotechnol. 94, 137–155.

Vanleeuw, E., Winderickx, S., Thevisseu, K., Lagrain, B., Dusselier, M., Cammue, B.P.A., Sels, B.F., 2019. Substrate specificity of *Candida rugosa* lipase and its industrial application. Sustain. Chem. Engineer. 7, 15828–15844.

Verma, A., Shirkot, P., Dhiman, K., Rana, N., Joshi, V.K., 2018. Bacterial laccase mediated inhibition of enzymatic browning in apple juice and its sensory evaluation. Int. J. Curr. Microbiol. Appl. Sci. 7, 3371–3381.

Verma, N., Thakur, S., Bhatt, A.K., 2012. Microbial lipases: Industrial applications and properties: A review. Int. Res. J. Biol. 1, 88–92.

Verma, S., Dixit, R., Pandey, K.C., 2016. Cysteine proteases: Modes of activation and prospects as pharmacological targets. Front. Pharmacol., (Doi: 10.3389/fphar.2016.00107).

Vermeiren, L., Devlieghere, F., van Beest, M., De Kenijf, N., Debevere, J., 1999. Developments in the active packaging of foods. Trends Food Sci. Technol. 10, 77–86.

Vincenti, S., Mariani, M., Alberti, J.-C., Jacopini, S., de Caraffa, V.B.-B., Borti, L., Maury, J., 2019. Biocatalytic synthesis of natural green leaf volatiles using the lipoxygenase metabolic pathway. Catalysts 9, 873, (Doi: 10.3390/catal9100873).

Virupakshi, K., Kyu, K.L., Tanticharoen, M., 2005. Purifuication and properties of a xylan-binding endoxylanase from alkaophilic *Bacillus* sp. strain K-1. Appl. Environ. Microbiol. 65, 694–697.
Vora, D., Hasabnis, S., Gutka, D.J., 2018. Biochemical role of lipoxygenase in development of off-flavors, bleaching of pigments and dough formation. J. Biotechnol. Biochem. 4, 52–67.
Waschkowitz, T., Rockstroh, S., Rolf, D., 2009. Isolation and characterization of metalloproteases with a novel domain structure by construction and screening of metqgenomic libraries. Appl. Environ. Mictobiol., (Doi: 10.1128/AEM.02136-08).
Widmann, M., Juhl, P.M., Jurgen, P., 2010. Structural classification by the lipase engineering database: A case study of *Candida antarctica* lipase A. BMC Genomics 11, (Doi: org/10.1186/1471-2164-11-123).
Wong, C.M., Wong, K.H., Chen, X.D., 2008. Glucose oxidase: Natural occurrence, function, properties and industrial applications. Appl. Microbiol. Biotechnol. 78, 927–938.
Xu-Jur, L., Zheng-Yuu, W., Chao-Yi, C., Jun, Y., Wen-Xue, Z., 2015. Increasing protease activities and antioxidant properties of Koji for soy sauces brewing by adding medical herb Rhodiola rosea. Int. J. Food Engineer., (Doi: 10.155/jife-2015-0187).
Yan, Y., Lee, C.C., Liao, J.C., 2009. Enantioselective synthesis of pure (*R*,*R*)-2,3-butanediol in *Eschrichia coli* with stereospecific secondary alcohol dehydrogenase. Org. Biomol. Chem. 7, 3914–3917.
Yang, J., Li, W., Ng, T.B., Deng, X., Lin, J., Ye, X., 2017. Laccases: Production, expression regulation, and applications in pharmaceutical biodegradation. Front. Microbiol. 8, 832.
Yilmaz, G., Ayar, A., Akin, N., 2005. The effect of microbial lipase on the lipolysis during the ripening of Tulum cheese. J. Food Engineer. 68, 269–274.
Yoshida, H., 1883. LXIII. Chemistry of lacquer (Urushi). Part 1. Communication from the chemical sciety of Tokio. J. Chem. Soc. (Tokyo) 43, 472–477.
Zambare, V., 2010. Solid state fermentation of *Aspergillus flavus* for glucoamylase production on agro residues. Int. J. Life Sci. 4, 16–25.
Zhang, C., Biji, E., Svensson, B., Herttingen, K., 2019b. Extracellular protease AprX from *Pseudomonas* and its spoilage potential UHT milk: A review. Compr. Rev. Food Sci. Food Safety, 4, 834-852.
Zhang, X., Zeng, Z., Zhang, J., Sarwar, A., Aziz, T., Yang, Z., 2019a. Change of proteolysis and sensory profile during ripening of Cheddar-style cheese as influenced by a microbial rennet from rice wine. Food Sci. Nutr. 7, 1540–1550.
Zhang, Y., Zhong, Q., 2018. Freeze-dried capsules prepared from emulsions with encapsulated lactase as a potential delivery system to control lactose hydrolysis in milk. Food Chem. 241, 397–402.
Zarghi, H., 2018. Application of xylanase and β-glucanase to improve nutrient utilization in poultry fed cereal base diets: Used of enzymes in poultry diet. Insights Enzym. Res. 2, 11–18.
Zhao, G., Ding, L.-L., Yao, Y., Cao, Y., Pan, Z.-H., Kong, D.-H., 2018. Extracellular proteome analysis and flavor formation during soy sauce fermentation. Front Microbiol. 9, (Doi: 10.3383/fmicb.2018.01872).
Zhou, C., Ye, T., Yue, Y., Ma, Y., 2015. Directed evolution and structural analyses of alkaline pectate lyasefrom the alkaphilic bacterium *Bacillus* sp. Strain N16T to improve its thermostability for efficient ramie degumming. Appl. Environ. Microbiol. 81, 5714–5723.
Zubeyde, B., Uyarm, F., Dogru, M., Alkan, H., 2008. Production of extra-cellular alkaline alpha amylase by solid state fermentation with newly isolated Bacillus sp. Prep. Biochem. Biotechnol. 38, 184–190.

Index

Note: Page numbers followed by "f" indicate figures, "t" indicate tables.

F

G

H

I

S

T

U

V

W

X

Y

Z

CPI Antony Rowe
Eastbourne, UK
July 08, 2022